Alexander Apelblat

Bessel and Related Functions

Also of interest

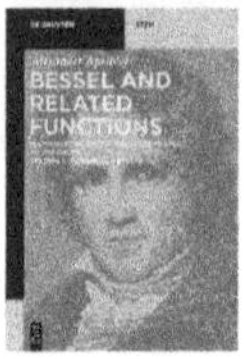

Bessel and Related Functions.
Mathematical Operations with Respect to the Order
Volume 2: Numerical Results
Alexander Apelblat, 2020
ISBN 978-3-11-068163-5, e-ISBN (PDF) 978-3-11-068247-2,
e-ISBN (EPUB) 978-3-11-068347-9, Set-ISBN 978-3-11-068323-3

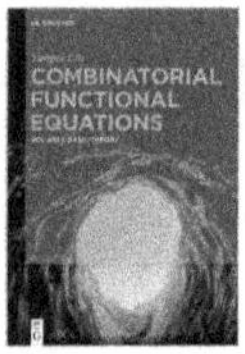

Combinatorial Functional Equations.
Basic Theory
Yanpei Liu, 2019
ISBN 978-3-11-062391-8, e-ISBN (PDF) 978-3-11-062583-7,
e-ISBN (EPUB) 978-3-11-062422-9

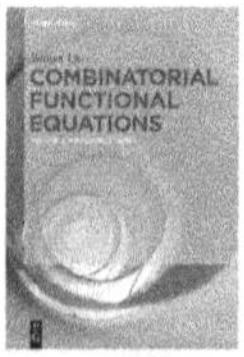

Combinatorial Functional Equations.
Advanced Theory
Yanpei Liu, 2020
ISBN 978-3-11-062435-9, e-ISBN (PDF) 978-3-11-062733-6,
e-ISBN (EPUB) 978-3-11-062480-9

Tensor Numerical Methods in Scientific Computing
Series: Radon Series on Computational and Applied Mathematics 19
Boris N. Khoromskij, 2018
ISBN 978-3-11-037013-3, e-ISBN (PDF) 978-3-11-036591-7,
e-ISBN (EPUB) 978-3-11-039139-8

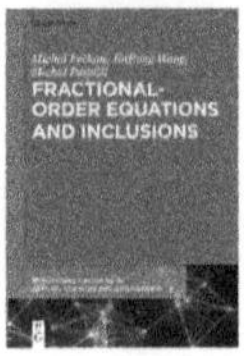

Fractional-Order Equations and Inclusions
Series: Radon Series on Computational and Applied Mathematics 19
Michal Fečkan, JinRong Wang, Michal Pospíšil, 2017
ISBN 978-3-11-052138-2, e-ISBN (PDF) 978-3-11-052207-5,
e-ISBN (EPUB) 978-3-11-052155-9

Alexander Apelblat

Bessel and Related Functions

Mathematical Operations with Respect to the Order
Volume 1: Theoretical Aspects

DE GRUYTER

Author
Prof. Dr. Alexander Apelblat
Ben-Gurion University of the Negev
Department of Chemical Engineering
Beer Sheva, Israel
apelblat@bgu.ac.il

ISBN 978-3-11-068157-4
e-ISBN (PDF) 978-3-11-068164-2
e-ISBN (EPUB) 978-3-11-068181-9

Library of Congress Control Number: 2020933594

Bibliographic information published by the Deutsche Nationalbibliothek
The Deutsche Nationalbibliothek lists this publication in the Deutsche Nationalbibliografie; detailed bibliographic data are available on the Internet at http://dnb.dnb.de.

Typesetting: Integra Software Services Pvt. Ltd.
Printing and binding: CPI books GmbH, Leck

www.degruyter.com

To Ira and Yoram

Preface

I encountered infinite series of the Bessel functions of integer orders rather early, when in the seventies of the previous century I did my PhD thesis. I solved the Navier–Stokes and the Darcy partial differential equations in cylindrical geometry. These equations expressed the physical situation which included the blood flow in fine capillaries with permeable walls and the plasma movement into surrounding tissue space. This was a mathematical modeling of the capillary-tissue fluid exchange in a human body (microcirculation). For the first time, a detailed distribution of velocities and pressures in tissue regions, induced by the blood flow in capillaries, was determined. In the solution of this mathematical problem, an important part plied series of the Bessel functions. The sums of infinite series of the Bessel functions of integer orders were calculated by using a central computer, with a help of punched cards, and by employing old versions of Fortran programs [1].

In the next step in my engineering career, this time in yearly eighties, I met with the Bessel and the Airy functions in the context of the heat and mass transfer coupled with homogeneous and heterogeneous chemical reactions of the first order. Analytical solutions were proposed for various types of flows and, in different geometric configurations. Analogous problems of the heat and mass transfer, but taking also into account the axial diffusions were also considered. These physical problems were solved by applying the Laplace transform methods, and they included infinite integrals with integrands containing the Bessel and the Airy functions [2, 3].

However, my most important involvement with the Bessel functions took place in 1985 when together with late Naftali Krivitsky we investigated the integral representations of integrals and derivatives with respect to the order of the Bessel functions, the integral Bessel function and the Anger function [4]. In a similar way, I later examined the mathematical properties of the Struve functions and the Kelvin functions [5, 6].

There is a huge literature devoted to various properties of the Bessel or related functions and the subject continue to be active nowadays. However, one topic, is much less in the center of interest of mathematicians or theoretical physicists, the mathematical operations with respect to the order of the Bessel functions. This is rather strange situation, because the first investigations on this subject are already mentioned in the fundamental Watson treatise on the Bessel functions [7], which was published at beginning of previous century.

Since these few results and not many derived later are dispersed in mathematical literature, it seems desirable to present them in a more systematic, much visible form. Personally, I was intrigued by the mathematical operations with respect to the order or with respect to the parameters of other special functions, and for many years I had an intention to write a small book on the subject which would be devoted to the Bessel and related functions. Over many years I collected material for

https://doi.org/10.1515/9783110681642-202

such project, but the present two volume book is a result of my intense efforts during the last three years.

During the last two decades appeared some new results dealing with various properties of the Bessel functions, they are not included in standard monographs or books on the subject, and therefore I decided to place them also in this book. Some less known in the literature topics when the order plays role of independent variable are also incorporated. To make reading of presented material easy, in consistent form, one chapter is devoted entirely to the basic properties of the Bessel and related functions.

This book includes primarily material connected with mathematical operations with the respect to the order of the Bessel and related functions, but it also contains many formulas, equations and mathematical techniques which will be of interest in different scientific disciplines, and in solving various technical problems. Evidently, this book is directed first of all to mathematicians who have interest in properties of special functions, but it is also relevant for researchers and graduate students in mathematics, mathematical physics, physical chemistry and some engineering disciplines. It is also expected that it will stimulate new research and more interest on the subject which is considerably less known in mathematical literature or applied in mathematical practice.

This book is divided into two parts (Part 1 – Theoretical Aspects and Part 2 – Numerical Results), and it provides the up-to-date survey of mathematical operations with regard to the order of the Bessel and related functions. Part 1 contains four Chapters, References and tree Appendices. Introduction which is considered as Chapter 1, gives a short historical look at the available on the subject literature. In Chapter 2, are presented briefly the basic properties of the Bessel and related functions. They are given for convenience of readers, but also because these formulas are intensively used throughout this book. The presentation of material is arranged in such way, to make a location of desired formulas easy. In Chapter 2, large section is devoted to solutions of the second or higher order differential equations which can be expressed in terms of the Bessel functions. Some of these solutions were derived by Hayek in terms of the Bessel-Clifford functions [8], but here they are expressed in terms of the Bessel functions. Chapter 3 includes a survey of investigations devoted to mathematical operations with the order of Bessel and related functions. Besides, a new material is also added, which includes integrals and higher order derivatives with respect to the order and other related topics. All presented functional expressions in Chapter 3 are coming from the literature, and therefore they are given without detailed derivations or proofs. Most of material in Chapter 3 was compiled from handbooks dealing with special functions [9–21], monographs dedicated to the Bessel functions [7, 22–26], tabulations of series, infinite integrals and integral transforms [15, 27–41], and few important contributions published in physical and mathematical journals [42–57]. Chapter 4 is entirely devoted to mathematical operations dealing with the order of Bessel and related functions. The order as an independent variable participates in differentiation and integration operations, in the location and evaluation of order zeros and interrelations between arguments and orders. Different approaches and procedures are illustrated, and advantages associated with operational techniques are demonstrated. Two sections in Chapter 4 are devoted to zeros of the Bessel functions which are considered as a function of the order, and to zeros of the Bessel and related functions with respect to the argument when the order is changed. The last subject has been popular during many decades, as the special case when the Bessel functions have the same argument and order. The benefit to use the shifted Dirac delta function which is defined as a limit, when the order of the Bessel function is variable, is also thoroughly illustrated in Chapter 4. Appendix A contains miscellaneous integrals of the Bessel and related functions compiled from the recent literature, but not included in the main tabulations of integrals of special functions. The integral representations of special functions as the limits of infinite integrals are given in Appendix B. Notation and definitions of special functions is presented in Appendix C.

Part 2 contains four Chapters where are reported numerical results which include the first, second and third derivatives with respect to the order at fixed values of arguments, and in some cases the involved functions. They are presented in tables and in a graphical form. These derivatives were calculated for the first time for the Bessel and related functions.

I appreciate very much Professor Marija Běster-Rogaě from Department of Physical Chemistry, Ljubljana University, Ljubljana, Slovenia and Professor Francesco Mainardi, Department of Physics and Astronomy, University of Bologna, Bologna, Italy who helped me when I had difficulty to obtain publications associated with discussed here topics. I am also greatly indebted to Professor Larry Glasser from Department of Physics and Mathematics, Clarkson University, Potsdam, N.Y. who derived a general formula that permitted to express the Bessel – Clifford functions in terms of the Bessel functions. All numerical calculations in this book were performed by using MATHEMATICA program. I am thankful to Dr. Juan Luis Gonzáles-Santander, Valencia, Spain who introduced me to this program.

Time spent in the preparation of this book was evidently taken from my family, and therefore, I am truly grateful to my wife Ira and my son Yoram for their continuous encouragement, profound understanding that I am not always available at home, and for so many years of constant assistance and patience.

Alexander Apelblat
January 2017 – August 2019.

Contents

1 Introduction

Starting from the end of the seventeenth century, in the context of solution of the Riccati-type differential equations, first elements of the present-day Bessel functions appeared. They are associated with different mathematical problems solved by the Bernoulli brothers, Leibniz, Euler, Lagrange, Laplace, Fourier, Poisson and others. These functions are named after the German astronomer Friedrich Wilhelm Bessel (1784–1846) who during the period 1822–1824 had shown that the radius vector in planetary motion can be expressed by the expansion similar to that established for the Bessel functions, he also obtained many other results associated with them. During the second part of the nineteenth century, a very important progress had been achieved in the theory of the Bessel functions and more generally of the cylindrical functions. This is a result of investigations performed by famous mathematicians Jacobi, Schlömilch, Lommel, Neumann, Sonine, Hankel, Struve and Weber. Significant contributions can also be attributed to Rayleigh, Lamb, Duhamel, Stokes and Thomson (Lord Kelvin), who solved physical and engineering problems, usually in the field of hydrodynamics, heat transfer, strength of materials and electricity.

In these investigations, basic properties of the Bessel functions, such as representations of these functions by series expansions, recurrence formulas, asymptotic expansions and zeroes of the Bessel functions have been established. They also included definite and indefinite integrals with integrands having the Bessel functions, the integral representations of the Bessel functions, differential equations solved in terms of the Bessel functions, tabulations of the Bessel and related auxiliary functions and many other results. There is a little doubt that from special functions, the Bessel functions are the most frequently investigated and applied in mathematical, physical and engineering studies.

All knowledge about the Bessel functions until the year 1922 is compiled and discussed by G.N. Watson in his monumental monograph *A Treatise on the Theory of Bessel Functions* [7]. Evidently, this heavy book of 804 pages became some kind of "bible" to anyone who was or is involved with the study of the Bessel functions. Similarly, as with other special functions, with an increasing modernization of society in the twentieth century, expressed by needs to solve many problems in mathematical physics and in various branches of engineering, the investigations dealing with the Bessel functions flourished. The Bessel functions are essential tools in electrodynamics, field theory, quantum and molecular physics, laser optics, pattern recognition and many other scientific and industrial disciplines.

The most important results of performing investigations are summarized in a number of books that are entirely devoted to the Bessel functions and their associated integrals. Besides, in mathematical handbooks that are dedicated to properties of special functions, [7, 9, 10, 12, 17, 18, 20–28, 33–35, 39–41, 58, 59], special chapters are devoted to the Bessel and related functions. In addition, a continuous

https://doi.org/10.1515/9783110681642-001

study of the Bessel functions' properties produced a huge number of papers that are dispersed in mathematical, physical and engineering journals.

From an enormous spectrum of discussed subjects that are associated with the Bessel functions, one topic found only limited interest in mathematical literature. The Bessel functions are defined as analytic functions of two variables, their argument and order. Almost all efforts have been directed to properties of the Bessel functions connected with arguments at fixed orders. On the contrary, from a concise historic survey given here, it is clear that a number of investigations directed towards operations such as differentiation or integration with respect to the order of the Bessel functions is very small, and they received little attention in the mathematical literature. The Bessel functions of real or complex order are mentioned many times in the Watson monograph, but usually the order is not considered as an independent variable. However, there are few exceptions, the first case is when the asymptotic expansion of the Bessel functions of large order is discussed; and in the second case, the Bessel functions of equal order and argument are examined [7]. There are also few occasions when the first derivatives with respect to the order are used to obtain the second solution of the Bessel differential equation and this is also extended to various products of the Bessel functions. These results are derived by direct differentiations with regard to the order and by using basic properties of the Bessel functions. There is an example of considerable beauty: it came from the investigation of Ramanujan, his very famous integral published in 1920, where the integration variable v is the order of Bessel functions [7]:

$$\int_{-\infty}^{\infty} J_{\mu+v}(a)\, J_{\lambda-v}(a)\, dv = J_{\lambda+\mu}(2a) \qquad a>0 \quad ; \quad \mathrm{Re}(\lambda+\mu)>1 \tag{1.1.1}$$

Considering the operations with regard to the order, an allied subject – the Kontorovich–Lebedev integral transform – should also be noted [18]. This transformation has a large number of applications in mathematical physics, but it is far from the material discussed here; hence, it is beyond the scope of this book. In view of its applicable character, it should be treated separately.

The history of investigations directed towards operations with regard to the order of the Bessel functions is rather short, and therefore probably all of them are mentioned here. They include four distinct subjects, the first rather large topic is dedicated to the representation of derivatives of Bessel functions with respect to the order, usually in terms of series or other functions [9, 18]. The second group of investigations is dealing with numerical calculations, tabulations of the Bessel functions as a function of the order, and determination of the first and second derivatives for particular values of the order [42, 43, 49, 55]. Few studies of the third subject are devoted to locations of the order of zeros at fixed arguments of the Bessel functions [60–72]. And finally, by applying the Laplace transform techniques, the fourth topic deals with an integration when the order serves as an integration variable [4]. It is worthwhile to note that Müller [73], G. Petiau [23] and others introduced a special, simple notation for differentiation with regard to the order of Bessel functions as $J_\nu^*(z)$. It replaces the explicit form of differentiation $\partial J_\nu(z)/\partial \nu$, but both notations are simultaneously used in the literature.

As mentioned earlier, only few results dealing with derivatives of the Bessel functions with respect to the order can be found in the Watson treatise [7]. The first derivatives appeared in the context of definitions of the Bessel functions with an integer order and when orders ν and arguments z are equal or nearly equal. Watson also reports values of functions and their first derivatives for orders and arguments being integers. From many available tables in his book, it is possible to obtain the change caused when the order is altered at fixed value of an argument (e.g. see Mitra [74, 75]). In particular cases of the order, $\nu = \pm 1/2$ and $\nu = \pm 3/2$, Airey in 1928 [42] tabulated values of $\partial J_\nu(x)/\partial \nu$. In 1935, he performed [43] a series of more extensive calculations, which included the second derivatives $\partial^2 J_\nu(x)/\partial \nu^2$ and the functional dependence of the Bessel functions at fixed values of the order $\nu = 0$, $\nu = \pm 1/2$ and $\nu = 1$. Based on the recurrening relations of the Bessel functions, Müller [73], in 1940 reported general formulas for the first derivatives of the Bessel functions with respect to the order. In 1958, Oberhettinger [56] established expressions for the first derivatives at $\nu = \pm (1/2 + n)$, $n = 0, 1, 2, 3, \ldots$ in terms of the exponential sine and cosine integrals. In 1960, Lee and Radosevich [55], in the $1 \le z \le 15$ range, tabulated values of $\partial J_\nu/\partial \nu$ for $\nu = n + \alpha$; $n = 1, 2, 3$ and 4 and $\alpha = 1/4, 1/3, 2/3$ and $3/4$. In 1963, Erber and Gordon [49] calculated the first derivatives with respect to the order of the modified Bessel function $\partial I_\nu(x)/\partial \nu$ for $\nu = \pm 1/3$ in the $0 \le z \le 2.5$ range and derived a number of asymptotic expressions for various ranges of argument z. Order derivatives for integer values of ν at $\nu = 1/2$ were also expressed in terms of the sums of Bessel functions [18]. In 1977, Wienke [57] gave, explicitly, the first and second order derivatives in terms of the Neumann function expansion. In 2005, after a long interruption, Brychkov and Geddes [44] derived in the closed form, the formulas for the first derivatives of $J_\nu(z)$, $Y_\nu(z)$, $I_\nu(z)$ and $K_\nu(z)$ for $\nu = \pm 1/2 \pm n$ and $n = 0, 1, 2, 3, \ldots$. They extended their results to the Struve functions $H_\nu(z)$ and $L_\nu(z)$ for $\nu = \pm 1/2 \pm n$ and to the integral Bessel functions $Ji_\nu(z)$, $Yi_\nu(z)$ and $Ki_\nu(z)$ for $\nu = \pm n$. Some results are expressed in the form of integral representations using the Meijer G-function and the generalized hypergeometric functions. In 2007, Brychkov [45] derived the first derivatives with respect to the order of the Weber $\mathrm{E}(z)$ and the Anger $\mathbf{J}(z)$ functions for $\nu = \pm n$. This work was continued in 2016 by considering the Kelvin functions $ber_\nu(z)$, $bei_\nu(z)$, $ker_\nu(z)$ and $kei_\nu(z)$, and the second and third derivatives of the Bessel functions $J_\nu(z)$, $Y_\nu(z)$, $I_\nu(z)$ and $K_\nu(z)$. During the same period, few new results associated with the order derivatives of the Bessel functions were presented by Dunster [48]. In 1954, Cooke, [46] in a small note, gave three new integrals of the Ramanujan type and his work continued during the period 1993–1997 by Fényes [50–52, 76]. Watson, [57, 77] in 1916, started to investigate behaviour of the Bessel functions of equal order ν and argument z. The same topic was considered later in 1992 by Lorch [78].

In the context of the theory of scattering and transmission of electromagnetic waves, Magnus and Kotin [68] in 1960, Keller et al [66] in 1963, Cohen [64] in 1964, Streifer [70–72] and Cochran [65] in 1965 determined ν – values that satisfy the equations of the type $J_\nu(\nu) = 0$ and $\partial J_\nu(\nu)/\partial\nu = 0$. Similar problems were discussed by Franz and his co-workers [79, 80] during the period 1954–1957. They presented solutions in the form of Green functions applicable to cylinder and sphere geometries. However, the first investigation on a similar mathematical problem was performed much earlier, during the period 1918–1919, by Watson [57]. He considered the transmission of electrical waves around the Earth. The Coulomb work [47] from 1936 also belongs to this group of investigations.

An application of the Laplace transformation techniques was initiated by van der Pol [81, 82] in 1929 by using the convolution theorem to evaluate integrals with respect to the order. A more extensive use of the Laplace transform is demonstrated in 1985 by the present author and Kravitsky [4]. They established integral representations of derivatives and integrals with respect to the order of the Bessel, Anger and Bessel integral functions. The Laplace transformation technique was also applied in the case of the Struve functions in 1989 [5] and the Kelvin functions in 1991 [6]. During the 1998–2004 period, the present author showed that the Lamborn [83] expression for the shifted Dirac delta function

$$\delta(z-1) = \lim_{\nu\to\infty} [J_\nu(\nu z)] \tag{1.1.2}$$

permits to evaluate integral representations of elementary and special functions, asymptotic relations, various integrals and limits of series [29, 84, 85].

2 Properties of the Bessel and Related Functions

2.1 Definitions and Notations of the Bessel and Related Functions and Associated Differential Equations

The basic properties of Bessel and related functions are compiled here from books devoted to special mathematical functions and particularly from these dedicated solely to the Bessel functions

$J_\nu(z)$ – ***Bessel function of the first kind of order*** ν

$Y_\nu(z)$ – ***Bessel function of the second kind of order*** ν **(also called as the Neumann or the Weber functions)**

The order ν and argument z can be arbitrary complex or a real number, but

$$Y_\nu(z) = \frac{J_\nu(z)\cos(\pi\nu) - J_{-\nu}(z)}{\sin(\pi\nu)} \quad ; \quad \nu \neq 0, 1, 2, 3, \ldots \tag{2.1.1}$$

$$Y_\nu(z) = \lim_{\nu \to n}[J_\nu(z)] = \left[\frac{\partial J_\nu(z)}{\partial\nu} - (-1)^n \frac{\partial J_{-\nu}(z)}{\partial\nu}\right]_{\nu=n} \quad n = 0, 1, 2, 3, \ldots \tag{2.1.2}$$

Bessel functions are solutions of the Bessel differential equation

$$z^2 \frac{d^2w(z)}{dz^2} + z\frac{dw(z)}{dz} + (z^2 - \nu^2)w(z) = 0 \tag{2.1.3}$$

$$[z\,w(z)']' + \left(z - \frac{\nu^2}{z}\right)w(z) = 0 \tag{2.1.4}$$

where

$$w(z) = A\,J_\nu(z) + B\,Y_\nu(z) \equiv C_\nu(z) \tag{2.1.5}$$

and A and B are integration constants.

$I_\nu(z)$ – **modified *Bessel function of the first kind of order*** ν

$K_\nu(z)$ – **modified *Bessel function of the second kind of order*** ν **(McDonald functions)**

The modified Bessel functions are solutions of the following Bessel differential equation:

$$z^2 \frac{d^2w(z)}{dz^2} + z\frac{dw(z)}{dz} - (z^2 + \nu^2)w(z) = 0 \tag{2.1.6}$$

where

$$w(z) = A\,I_\nu(z) + B\,K_\nu(z) \equiv Z_\nu(z) \tag{2.1.7}$$

https://doi.org/10.1515/9783110681642-002

$H_\nu^{(1)}(z)$ – ***Hankel function of the first kind of order v***
$H_\nu^{(2)}(z)$– ***Hankel function of the second kind of order v***

$$H_\nu^{(1)}(z) = J_\nu(z) + i\,Y_\nu(z) \tag{2.1.8}$$

$$H_\nu^{(2)}(z) = J_\nu(z) - i\,Y_\nu(z) \tag{2.1.9}$$

$\mathbf{H}_\nu(z)$ – ***Struve functions of order v***
$\mathbf{L}_\nu(z)$ – ***modified Struve functions of order v***

The Struve functions of order ν are particular solutions of the following differential equation

$$z^2\frac{d^2w(z)}{dz^2} + z\frac{dw(z)}{dz} + (z^2 - \nu^2)w(z) = \frac{4\left(\frac{z}{2}\right)^{\nu+1}}{\sqrt{\pi}\,\Gamma\left(\nu+\frac{1}{2}\right)} \tag{2.1.10}$$

$s_{\mu,\nu}(z)$ – ***Lommel functions of the order μ and v*** provided that

$\mu \pm \nu \neq -1, -2, -3, \ldots$

$S_{\mu,\nu}(z)$ – ***Lommel functions of the order μ and v*** with arbitrary values *μ and ν*

These functions are particular solutions of the following differential equation

$$z^2\frac{d^2w(z)}{dz^2} + z\frac{dw(z)}{dz} + (z^2 - \nu^2)w(z) = z^\mu \tag{2.1.11}$$

$\mathbf{J}_\nu(z)$ – ***Anger functions of the order v***

The Anger function satisfies the following differential equation:

$$z^2\frac{d^2w(z)}{dz^2} + z\frac{dw(z)}{dz} + (z^2 - \nu^2)w(z) = \left(\frac{z-\nu}{\pi}\right)\sin(\pi\nu) \tag{2.1.12}$$

and is also defined by the finite integral

$$\mathbf{J}_\nu(z) = \frac{1}{\pi}\int_0^\pi \cos(\nu t - z\sin t)\,dt \tag{2.1.13}$$

$\mathbf{E}_\nu(z)$ – ***Weber functions of the order v***

The Weber function satisfies the following differential equation

$$z^2\frac{d^2w(z)}{dz^2}+z\frac{dw(z)}{dz}+(z^2-\nu^2)w(z)=-\frac{[z+\nu+(z-\nu)\cos(\pi\nu)]}{\pi} \tag{2.1.14}$$

and is also defined by the integral

$$E_\nu(z)=\frac{1}{\pi}\int_0^\pi \sin(\nu t-z\sin t)\,dt \tag{2.1.15}$$

$\mathrm{ber}_\nu(z)$ – **Kelvin function, the real part of the Bessel function $J_\nu(z)$ of order ν and of the argument $i^{3/2}z$**

$\mathrm{bei}_\nu(z)$ – **Kelvin function, the imaginary part of the Bessel function $J_\nu(z)$ of order ν and of the argument $i^{-3/2}z$**

The order ν is a real, non-negative number and

$$\mathrm{ber}_0(z)=\mathrm{ber}(z)\quad;\quad \mathrm{bei}_0(z)=\mathrm{bei}(z) \tag{2.1.16}$$

The Kelvin $\mathrm{ber}_\nu(z)$ and $\mathrm{bei}_\nu(z)$ functions satisfy the following differential equation:

$$z^2\frac{d^2w(z)}{dz^2}+z\frac{dw(z)}{dz}-(iz^2+\nu^2)w(z)=0 \tag{2.1.17}$$

where

$$w(z)=A\,J_\nu(i^{1/2}z)+B\,K_\nu(i^{1/2}z) \tag{2.1.18}$$

They can be defined in terms of the Bessel functions as

$$J_\nu(zi^{3/2})=J_\nu(ze^{3/4\pi i})=\mathrm{ber}_\nu(z)+i\,\mathrm{bei}_\nu(z) \tag{2.1.19}$$

$$J_\nu(zi^{-3/2})=J_\nu(ze^{-3/4\pi i})=\mathrm{ber}_\nu(z)-i\,\mathrm{bei}_\nu(z) \tag{2.1.20}$$

and

$$\mathrm{ber}_\nu(z)=\frac{1}{2}\left[J_\nu\left(-\frac{1+i}{\sqrt{2}}z\right)+J_\nu\left(\frac{-1+i}{\sqrt{2}}z\right)\right]=\mathrm{Re}\left\{J_\nu\left(\frac{-1+i}{\sqrt{2}}z\right)\right\} \tag{2.1.21}$$

$$\mathrm{bei}_\nu(z)=\frac{1}{2}\left[J_\nu\left(-\frac{1+i}{\sqrt{2}}z\right)-i\,J_\nu\left(\frac{-1+i}{\sqrt{2}}z\right)\right]=\mathrm{Im}\left\{J_\nu\left(\frac{-1+i}{\sqrt{2}}z\right)\right\} \tag{2.1.22}$$

where Re{ . . . } denotes the real part and Im{ . . . } the imaginary part of the Bessel functions.

$\mathrm{ker}_\nu(z)$ – **Kelvin modified function, the real part of the Bessel function $K_\nu(z)$ of order ν and of the argument $i^{1/2}z$**

$\mathrm{kei}_\nu(z)$ – **Kelvin modified function, the imaginary part of the Bessel function $K_\nu(z)$ of order ν and of the argument $i^{-1/2}z$**

$$\mathrm{ker}_0(z) = \mathrm{ker}(z) \quad ; \quad \mathrm{kei}_0(z) = \mathrm{kei}(z) \tag{2.1.23}$$

$$i^{-\nu} K_\nu(z i^{1/2}) = e^{-\pi \nu i/2} K_\nu(z e^{\pi i/4}) = \mathrm{ker}_\nu(z) + i\,\mathrm{kei}_\nu(z) \tag{2.1.24}$$

$$i^{\nu} K_\nu(z i^{-1/2}) = e^{\pi \nu i/2} K_\nu(z e^{-\pi i/4}) = \mathrm{ker}_\nu(z) - i\,\mathrm{kei}_\nu(z) \tag{2.1.25}$$

and

$$\mathrm{ker}_\nu(z) = \mathrm{Re}\left\{ \left[\cos\left(\frac{\pi\nu}{2}\right) - i \sin\left(\frac{\pi\nu}{2}\right)\right] K_\nu\left(\frac{1+i}{\sqrt{2}} z\right)\right\} \tag{2.1.26}$$

$$\mathrm{kei}_\nu(z) = \mathrm{Im}\left\{ \left[\cos\left(\frac{\pi\nu}{2}\right) - i \sin\left(\frac{\pi\nu}{2}\right)\right] K_\nu\left(\frac{1+i}{\sqrt{2}} z\right)\right\} \tag{2.1.27}$$

The Kelvin functions $\mathrm{her}_\nu(z)$ and $\mathrm{hei}_\nu(z)$ are $\mathrm{ker}_\nu(z)$ and $\mathrm{kei}_\nu(z)$ functions multiplied by factor $\pm 2/\pi$

$$\mathrm{her}_0(z) = \mathrm{her}(z) \quad ; \quad \mathrm{hei}_0(z) = \mathrm{hei}(z) \tag{2.1.28}$$

$$\mathrm{her}_\nu(z) = \frac{2}{\pi} \mathrm{kei}_\nu(z) \tag{2.1.29}$$

$$\mathrm{hei}_\nu(z) = -\frac{2}{\pi} \mathrm{ker}_\nu(z) \tag{2.1.30}$$

Ai(z) – **Airy function is expressed in terms of the modified Bessel function $K_\nu(z)$ of order $\nu = 1/3$ and the argument $2/3z^{3/2}$.**
Bi(z) – **Airy function is expressed in terms of the modified Bessel functions $I_\nu(z)$ of order $\nu = \pm 1/3$ and the argument $2/3z^{3/2}$.**

The Airy functions are particular solutions of the following differential equation:

$$\frac{d^2 w(z)}{dz^2} - z w(z) = 0 \tag{2.1.31}$$

and

$$w(z) = A\,\mathrm{Ai}(z) + B\,\mathrm{Bi}(z) \tag{2.1.32}$$

$$\mathrm{Ai}(0) = \frac{1}{\sqrt{3}} \mathrm{Bi}(0) = \frac{1}{3^{2/3}\Gamma\left(\frac{2}{3}\right)} \tag{2.1.33}$$

$$\mathrm{Ai}'(0) = -\frac{1}{\sqrt{3}} \mathrm{Bi}'(0) = -\frac{1}{3^{1/3}\Gamma\left(\frac{1}{3}\right)} \tag{2.1.34}$$

The Airy functions Ai(z), Ai($-z$), Bi(z) and Bi($-z$), with $z > 0$, are usually presented in the form

$$\mathrm{Ai}(z) = \frac{\sqrt{z}}{3}\left[I_{-1/3}(\zeta) - I_{1/3}(\zeta)\right] = \sqrt{\frac{z}{3\pi^2}}\, K_{1/3}(\zeta) \quad ; \quad \zeta = \frac{2}{3} z^{3/2} \tag{2.1.35}$$

$$\mathrm{Ai}'(z) = -\frac{z}{3}\left[I_{-2/3}(\zeta) - I_{2/3}(\zeta)\right] = -\frac{z}{\sqrt{3\pi^2}}\, K_{1/3}(\zeta) \tag{2.1.36}$$

$$\mathrm{Ai}(-z) = -\frac{\sqrt{z}}{3}\left[J_{-1/3}(\zeta) + J_{1/3}(\zeta)\right] \tag{2.1.37}$$

$$\mathrm{Ai}'(-z) = \frac{z}{3}\left[J_{2/3}(\zeta) - J_{-2/3}(\zeta)\right] \tag{2.1.38}$$

$$\mathrm{Bi}(z) = \sqrt{\frac{z}{3}}\left[I_{-1/3}(\zeta) + I_{1/3}(\zeta)\right] \tag{2.1.39}$$

$$\mathrm{Bi}(-z) = \sqrt{\frac{z}{3}}\left[J_{-1/3}(\zeta) - J_{1/3}(\zeta)\right] \tag{2.1.40}$$

$$\mathrm{Bi}'(-z) = \frac{z}{\sqrt{3}}\left[J_{-2/3}(\zeta) + J_{2/3}(\zeta)\right] \tag{2.1.41}$$

or in terms of infinite integrals

$$\mathrm{Ai}(z) = \frac{1}{\pi}\int_0^\infty \cos\left(\frac{t^3}{3} + zt\right) dt \tag{2.1.42}$$

$$\mathrm{Ai}(-z) = \frac{1}{\pi}\int_0^\infty \cos\left(\frac{t^3}{3} - zt\right) dt \tag{2.1.43}$$

$$\mathrm{Bi}(z) = \frac{1}{\pi}\int_0^\infty \left[e^{-t^3/3 + zt} + \sin\left(\frac{t^3}{3} + zt\right)\right] dt \tag{2.1.44}$$

$$\mathrm{Bi}(-z) = \frac{1}{\pi}\int_0^\infty \left[e^{-t^3/3 - zt} + \sin\left(\frac{t^3}{3} - zt\right)\right] dt \tag{2.1.45}$$

Gi(z) – **Scorer function is expressed in terms of the Airy functions.**

The Scorer functions Gi(z) are particular solutions of the following differential equation

$$\frac{d^2 w(z)}{dz^2} - z w(z) = -\frac{1}{\pi} \tag{2.1.46}$$

and

$$w(z) = A\,\mathrm{Ai}(z) + B\,\mathrm{Bi}(z) + \mathrm{Gi}(z) \tag{2.1.47}$$

$$\mathrm{Gi}(0) = \frac{1}{3}\mathrm{Bi}(0) = \frac{1}{\sqrt{3}}\mathrm{Ai}(0) = \frac{1}{3^{7/6}\Gamma(\frac{2}{3})} \tag{2.1.48}$$

$$\mathrm{Gi}'(0) = -\frac{1}{\sqrt{3}}\mathrm{Ai}'(0) = \frac{1}{3^{5/6}\Gamma(\frac{1}{3})} \tag{2.1.49}$$

Hi(z) – **Scorer function is expressed in terms of Airy functions.**

The Scorer functions Hi(z) are particular solutions of the following differential equation

$$\frac{d^2w(z)}{dz^2} - zw(z) = \frac{1}{\pi} \tag{2.1.50}$$

and

$$w(z) = A\,\mathrm{Ai}(z) + B\,\mathrm{Bi}(z) + \mathrm{Hi}(z) \tag{2.1.51}$$

$$\mathrm{Hi}(0) = \frac{2}{3}\mathrm{Bi}(0) = -\frac{2}{\sqrt{3}}\mathrm{Ai}(0) = \frac{2}{3^{7/6}\Gamma(\frac{2}{3})} \tag{2.1.52}$$

$$\mathrm{Hi}'(0) = \frac{2}{3}\mathrm{Bi}'(0) = -\frac{2}{3}\mathrm{Ai}'(0) = \frac{2}{3^{5/6}\Gamma(\frac{1}{3})} \tag{2.1.53}$$

The Scorer functions are expressed in terms of infinite integrals in the following way

$$\mathrm{Gi}(z) = \frac{1}{\pi}\int_0^{\infty} \sin\left(\frac{t^3}{3} + zt\right) dt \tag{2.1.54}$$

$$\mathrm{Gi}(z) = 3^{-3/2}\left(\frac{2z}{\pi}\right)^2 \int_0^{\infty} \frac{K_{1/3}(t)}{\zeta^2 - t^2} dt \quad ; \quad \zeta = \frac{2}{3}z^{3/2} \tag{2.1.55}$$

$$\mathrm{Gi}(-z) = \mathrm{Bi}(-z) - 3^{-3/2}\left(\frac{2z}{\pi}\right)^2 \int_0^{\infty} \frac{K_{1/3}(t)}{\zeta^2 + t^2} dt \tag{2.1.56}$$

$$\mathrm{Gi}(-z) = \mathrm{Bi}(-z) - \frac{2\sqrt{z}}{3\pi} S_{0,1/3}\left(\frac{2}{3}z^{3/2}\right) \tag{2.1.57}$$

$$\mathrm{Hi}(z) = \mathrm{Bi}(z) - \mathrm{Gi}(z) \tag{2.1.58}$$

$$\mathrm{Hi}(z) = \frac{1}{\pi}\int_0^{\infty} e^{-t^3/3 + zt} dt \tag{2.1.59}$$

$$\mathrm{Hi}(-z) = 3^{-3/2}\left(\frac{2z}{\pi}\right)^2 \int_0^{\infty} \frac{K_{1/3}(t)}{\zeta^2 + t^2} dt \tag{2.1.60}$$

$$\mathrm{Hi}'(z) = \frac{1}{\pi}\int_0^\infty t e^{-t^3/3 + zt}\, dt \tag{2.1.61}$$

$F_\nu(z)$ and $G_\nu(z)$ are **the Bessel–Clifford functions of order** $\boldsymbol{\nu}$ defined by

$$F_\nu(z) = z^{-\nu/2} C_\nu(2\sqrt{z}) \quad ; \quad C_\nu(z) = J_\nu(z), Y_\nu(z) \tag{2.1.62}$$

$$G_\nu(z) = z^{-\nu/2} Z_\nu(2\sqrt{z}) \quad ; \quad Z_\nu(z) = I_\nu(z), K_\nu(z) \tag{2.1.63}$$

These functions satisfy the following differential equations

$$z^2 \frac{d^2 w(z)}{dz^2} + (\nu + 1) z \frac{dw(z)}{dz} \pm z w(z) = 0 \tag{2.1.64}$$

Solutions of differential equations when expressed in terms of the Bessel–Clifford functions can be interrelated by

$$\begin{aligned} w(z) &= z^\alpha F_\nu(Az^\beta) = \frac{z^\alpha}{(Az^\beta)^{\nu/2}} C_\nu(2\sqrt{Az^\beta}) \\ w(z) &= h(z)\, F_\nu[g(z)] = h(z)\, g(z)^{-\nu/2} F_\nu[2\sqrt{g(z)}] \end{aligned} \tag{2.1.65}$$

$\mathrm{Ji}_\nu(z)$, $\mathrm{Yi}_\nu(z)$, $\mathrm{Ii}_\nu(z)$ and $\mathrm{Ki}_\nu(z)$ are **the integral Bessel functions of order** $\boldsymbol{\nu}$ defined by

$$\mathrm{Ji}_\nu(z) = -\int_z^\infty \frac{J_\nu(t)}{t}\, dt \tag{2.1.62}$$

$$\mathrm{Yi}_\nu(z) = -\int_z^\infty \frac{Y_\nu(t)}{t}\, dt \tag{2.1.63}$$

$$\mathrm{Ii}_\nu(z) = -\int_0^z \frac{I_\nu(t)}{t}\, dt \tag{2.1.64}$$

$$\mathrm{Ki}_\nu(z) = -\int_z^\infty \frac{K_\nu(t)}{t}\, dt \tag{2.1.65}$$

Frequently in the literature these functions are defined when minus sign is omitted.

The integral defining $\mathrm{Ii}_\nu(t)$ converges only for restricted range of orders. This function is rarely used, an example worth mentioning is when the integrand is replaced by $(I_0(t) - 1)/t$ [9].

$\mathrm{Ji}_{n,k}(z)$and $\mathrm{Ji}_n^{(k)}(z)$ are **the** $\boldsymbol{k}$**–repeated integrals of the Bessel functions** $J_n(z)$ and $J_n(z)/z$, and $\mathrm{Ki}_{n,k}(z)$and $\mathrm{Ki}_n^{(k)}(z)$ are **the** $\boldsymbol{k}$**–repeated integrals of the modified Bessel functions** $K_n(z)$ and $K_n(z)/z$, and are defined as follows:

$$\mathrm{Ji}_{n,k}(z) = \int_0^z dt \ldots \int_0^z J_n(t)dt \quad ; \quad n \geq 0 \tag{2.1.66}$$

$$\mathrm{Ji}_n^{(k)}(z) = \int_0^z dt \ldots \int_0^z \frac{J_n(t)}{t} dt \quad ; \quad n > 0 \tag{2.1.67}$$

$$\mathrm{Ki}_{n,k}(z) = \int_0^z dt \ldots \int_0^z K_n(t)dt \quad ; \quad n \geq 0 \tag{2.1.68}$$

$$\mathrm{Ki}_n^{(k)}(z) = \int_0^z dt \ldots \int_0^z \frac{K_n(t)}{t} dt \quad ; \quad n > 0 \tag{2.1.69}$$

$\mathrm{Ki}_k(z)$ are **the k–repeated integrals** (also called **the Bickley functions**) defined by

$$\mathrm{Ki}_0(z) = K_0(z) \tag{2.1.70}$$

$$\mathrm{Ki}_k(z) = \int_z^\infty \mathrm{Ki}_{k-1}(t)\,dt \quad ; \quad z > 0 \quad ; \quad k \geq 1 \tag{2.1.71}$$

2.2 Interrelations, Recurrence Relations, Differential and Integral Formulas for the Bessel $J_\nu(z)$ and $Y_\nu(z)$ and the Hankel $H_\nu^{(1)}(z)$ and $H_\nu^{(2)}(z)$ functions

$$J_\nu(iz) = i^\nu I_\nu(z) = \left[\cos\left(\frac{\pi\nu}{2}\right) + i\sin\left(\frac{\pi\nu}{2}\right)\right] I_\nu(z) \tag{2.2.1}$$

$$J_\nu(-z) = (-1)^\nu J_\nu(z) = [\cos(\pi\nu) + i\sin(\pi\nu)] J_\nu(z) \tag{2.2.2}$$

$$J_{-\nu}(z) = J_\nu(z)\cos(\pi\nu) - Y_\nu(z)\sin(\pi\nu) \tag{2.2.3}$$

$$Y_n(z) = \lim_{\nu \to n} [\cot(\pi\nu) J_\nu(z) - \csc(\pi\nu) J_{-\nu}(z)] \quad ; \quad n = 0, \pm 1, \pm 2, \pm 3 \ldots \tag{2.2.4}$$

$$Y_{-\nu}(z) = J_\nu(z)\sin(\pi\nu) + Y_\nu(z)\cos(\pi\nu) \tag{2.2.5}$$

$$Y_\nu(z) + Y_{-\nu}(z) = \cot\left(\frac{\pi\nu}{2}\right)[J_\nu(z) - J_{-\nu}(z)] \tag{2.2.6}$$

$$Y_\nu(z) - Y_{-\nu}(z) = -\tan\left(\frac{\pi\nu}{2}\right)[J_\nu(z) + J_{-\nu}(z)] \tag{2.2.7}$$

$$H_{-\nu}^{(1)}(z) = e^{\pi\nu i} H_\nu^{(1)}(z) \tag{2.2.8}$$

$$H_{-\nu}^{(2)}(z) = e^{-\pi\nu i} H_\nu^{(2)}(z) \tag{2.2.9}$$

$$C_\nu(z) = J_\nu(z), Y_\nu(z), H_\nu^{(1)}(z), H_\nu^{(2)}(z)$$

$$C_{-n}(z) = (-1)^n C_n(z) \quad ; \quad n = 1, 2, 3, \ldots \tag{2.2.10}$$

$$C_n(z) = (-1)^n C_n(-z) \tag{2.2.11}$$

$$C_{-n}(z) = C_n(-z) \tag{2.2.12}$$

$$C_{\nu-1}(z) + C_{\nu+1}(z) = \frac{2\nu}{z} C_\nu(z) \tag{2.2.13}$$

$$\frac{d}{dz}[C_0(z)] = -C_1(z) \tag{2.2.14}$$

$$\frac{d}{dz}[z\, C_1(z)] = z\, C_0(z) \tag{2.2.15}$$

$$\frac{d\, C_\nu(z)}{dz} = \frac{C_{\nu-1}(z) - C_{\nu+1}(z)}{2} \tag{2.2.16}$$

$$\frac{d\, C_\nu(z)}{dz} = \frac{\nu}{z} C_\nu(z) - C_{\nu+1}(z) \tag{2.2.17}$$

$$\frac{d\, C_\nu(z)}{dz} = C_{\nu-1}(z) - \frac{\nu}{z} C_\nu(z) \tag{2.2.18}$$

$$\frac{d}{dz}[z^\nu C_\nu(z)] = z^\nu C_{\nu-1}(z) \quad ; \quad \nu \geq 0 \tag{2.2.19}$$

$$\frac{d}{dz}[z^\nu C_\nu(az)] = a\, z^\nu C_{\nu-1}(az) \tag{2.2.20}$$

$$\frac{d}{dz}[z^{-\nu} C_\nu(z)] = -z^{-\nu} C_{\nu+1}(z) \quad ; \quad \nu \geq 0 \tag{2.2.21}$$

$$\frac{d}{dz}[z^{-\nu} C_\nu(az)] = -a\, z^{-\nu} C_{\nu+1}(az) \tag{2.2.22}$$

$$\left(\frac{d}{z\, dz}\right)^k [z^\nu C_\nu(z)] = z^{\nu-k} C_{\nu-k}(z) \quad ; \quad k = 1, 2, 3, \ldots \tag{2.2.23}$$

$$\left(\frac{d}{z\, dz}\right)^k [z^{-\nu} C_\nu(z)] = (-1)^k z^{-\nu-k} C_{\nu+k}(z) \quad ; \quad k = 1, 2, 3, \ldots \tag{2.2.24}$$

$$\int C_1(z)\, dz = -C_0(z) \tag{2.2.25}$$

$$\int z\, C_0(z)\, dz = z\, C_1(z) \tag{2.2.26}$$

$$\int (1 + z^2)\, C_0(z)\, dz = z\, C_0(z) + z^2\, C_1(z) \tag{2.2.27}$$

$$\int z^{\nu} C_{\nu-1}(z)dz = z^{\nu} C_{\nu}(z) \tag{2.2.28}$$

$$\int z^{\nu} C_{\nu}(z)dz = 2^{\nu-1}\sqrt{\pi}\,\Gamma\left(\nu+\frac{1}{2}\right) z\,[C_{\nu}(z)H'_{\nu}(z) - C'_{\nu}(z)H_{\nu}(z)] \tag{2.2.29}$$

$$\nu \neq -\frac{1}{2}$$

$$\int z^{-\nu} C_{\nu+1}(z)dz = -z^{-\nu} C_{\nu}(z) \tag{2.2.30}$$

$$\int z \ln z\, C_0(z)dz = C_0(z) + z \ln z\, C_1(z) \tag{2.2.31}$$

$$\int [2nz^{n-1} + z^{n+1} \ln z\, C_n(z)]\, dz = z^n C_n(z) + z^{n+1} \ln z\, C_{n+1}(z) \tag{2.2.32}$$

$$\int z^{\nu} C_{\nu}(z)dz = 2^{\nu-1}\sqrt{\pi}\,\Gamma\left(\nu+\frac{1}{2}\right) z\,[C_{\nu}(z)H_{\nu-1}(z) - C_{\nu-1}(z)H_{\nu}(z)] \tag{2.2.33}$$

$$\int z^{\lambda} C_{\nu}(z)dz = (\lambda+\nu-1)\, z\, C_{\nu}(z)\, S_{\lambda-1,\nu-1}(z) - z\, C_{\nu-1}(z)\, S_{\lambda,\nu}(z) \tag{2.2.34}$$

$$\int_0^z t^{-\nu} J_{\nu+1}(t)dt = \frac{1}{2^{\nu}\Gamma(\nu+1)} - z^{-\nu} J_{\nu}(z) \tag{2.2.35}$$

$$\int_0^z t^{\nu} Y_{\nu-1}(t)dt = \frac{2^{\nu}\Gamma(\nu)}{\pi} + z^{\nu} Y_{\nu}(z) \quad ; \quad \operatorname{Re}\nu > 0 \tag{2.2.36}$$

$$[J_{\nu}(z)]^2 - J_{\nu-1}(z)J_{\nu+1}(z) > \frac{1}{\nu+1}[J_{\nu}(z)]^2 \quad ; \quad \nu > 0 \quad ; \quad z - \text{real} \tag{2.2.37}$$

$$\int_0^z \ln t\, J_1(t)\, dt = \ln 2 - \gamma - \ln z\, J_0(z) + \mathrm{Ji}_0(z) \tag{2.2.38}$$

2.3 Interrelations, Recurrence Relations, Differential and Integral Formulas for the Modified Bessel $I_\nu(z)$ and $K_\nu(z)$ Functions

$$I_n(-z) = (-1)^n I_n(z) \quad ; \quad n = 0, \pm 1, \pm 2, \pm 3 \ldots \tag{2.3.1}$$

$$I_{-n}(z) = I_n(z) \quad ; \quad n = 1, 2, 3, \ldots \tag{2.3.2}$$

$$I_{\nu}(-z) = [\cos(\pi\nu) + i \sin(\pi\nu)] I_{\nu}(z) \tag{2.3.3}$$

$$I_{-\nu}(z) = I_{\nu}(z) + \frac{2}{\pi}\sin(\pi\nu)\, K_{\nu}(z) \tag{2.3.4}$$

$$I_\nu(iz) = i^\nu J_\nu(z) = \left[\cos\left(\frac{\pi\nu}{2}\right) + i\sin\left(\frac{\pi\nu}{2}\right)\right] J_\nu(z) \tag{2.3.5}$$

$$I_{\nu-1}(z) - I_{\nu+1}(z) = \frac{2\nu}{z} I_\nu(z) \tag{2.3.6}$$

$$\frac{d}{dz}[I_0(z)] = I_1(z) \tag{2.3.7}$$

$$\frac{dI_\nu(z)}{dz} = \frac{I_{\nu-1}(z) + I_{\nu+1}(z)}{2} \tag{2.3.8}$$

$$\frac{dI_\nu(z)}{dz} = I_{\nu-1}(z) - \frac{\nu}{z} I_\nu(z) \tag{2.3.9}$$

$$\frac{dI_\nu(z)}{dz} = I_{\nu+1}(z) + \frac{\nu}{z} I_\nu(z) \tag{2.3.10}$$

$$\left(\frac{d}{zdz}\right)^k [z^\nu I_\nu(z)] = z^{\nu-k} I_{\nu-k}(z) \quad ; \quad k = 1, 2, 3, \ldots \tag{2.3.11}$$

$$\left(\frac{d}{zdz}\right)^k [z^{-\nu} I_\nu(z)] = z^{-\nu-k} I_{\nu+k}(z) \tag{2.3.12}$$

$$\int z^{\nu+1} I_\nu(z) dz = z^{\nu+1} I_{\nu+1}(z) \tag{2.3.13}$$

$$\int z^{-\nu} I_{\nu+1}(z) dz = z^{-\nu} I_\nu(z) \tag{2.3.14}$$

$$\int_0^z t^{-\nu} I_{\nu+1}(t) dt = z^{-\nu} I_\nu(z) - \frac{1}{2^\nu \Gamma(\nu+1)} \tag{2.3.15}$$

$$\int z^{1-\nu} I_\nu(z) dz = z^{1-\nu} I_{\nu-1}(z) \tag{2.3.16}$$

$$\int z^\nu I_{\nu-1}(z) dz = z^\nu I_\nu(z) \quad ; \quad \operatorname{Re}\nu > 0 \tag{2.3.17}$$

$$\int z^\nu I_\nu(z) dz = 2^{\nu-1}\sqrt{\pi}\,\Gamma\left(\nu + \frac{1}{2}\right) z\,[I_\nu(z) L_{\nu-1}(z) - I_{\nu-1}(z) L_\nu(z)] \tag{2.3.18}$$

$$K_\nu(z) = \frac{\pi}{2}\left[\frac{I_{-\nu}(z) - I_\nu(z)}{\sin(\pi\nu)}\right] \quad ; \quad \nu \neq 0, \pm 1, \pm 2, \pm 3, \ldots \tag{2.3.19}$$

$$K_n(z) = \frac{\pi}{2}\left[\frac{\frac{\partial}{\partial\nu}(I_{-\nu}(z) - I_\nu(z))}{\frac{\partial}{\partial\nu}(\sin(\pi\nu))}\right]_{\nu=n} \quad ; \quad n = 0, 1, 2, 3, \ldots \tag{2.3.20}$$

$$K_\nu(z) = K_{-\nu}(z) \tag{2.3.21}$$

$$K_{\nu-1}(z)-K_{\nu+1}(z) = -\frac{2\nu}{z}K_\nu(z) \tag{2.3.22}$$

$$\frac{d}{dz}[K_0(z)] = -K_1(z) \tag{2.3.23}$$

$$\frac{dK_\nu(z)}{dz} = -\frac{K_{\nu-1}(z)+K_{\nu+1}(z)}{2} \tag{2.3.24}$$

$$z\frac{dK_\nu(z)}{dz} + \nu K_\nu(z) = -z\,K_{\nu-1}(z) \tag{2.3.25}$$

$$z\frac{dK_\nu(z)}{dz} - \nu K_\nu(z) = -z\,K_{\nu+1}(z) \tag{2.3.26}$$

$$\left(\frac{d}{zdz}\right)^k [z^\nu K_\nu(z)] = (-1)^k z^{\nu-k} K_{\nu-k}(z) \quad ; \quad k=1,2,3,\ldots \tag{2.3.27}$$

$$\left(\frac{d}{zdz}\right)^k [z^{-\nu} K_\nu(z)] = (-1)^k z^{-\nu-k} K_{\nu+k}(z) \tag{2.3.28}$$

$$\int z^{\nu+1} K_\nu(z)\,dz = -z^{\nu+1} K_{\nu+1}(z) \tag{2.3.29}$$

$$\int z^{1-\nu} K_\nu(z)dz = -z^{1-\nu} K_{\nu-1}(z) \tag{2.3.30}$$

$$\int_0^z t^\nu K_{\nu-1}(t)dt = 2^{\nu-1}\Gamma(\nu) - z^\nu K_\nu(z) \quad ; \quad \mathrm{Re}\,\nu > 0 \tag{2.3.31}$$

$$\int_z^\infty t^{-\nu} K_{\nu+1}(t)dt = z^{-\nu} K_\nu(z) \tag{2.3.32}$$

$$\int z^\nu K_\nu(z)dz = 2^{\nu-1}\sqrt{\pi}\,\Gamma\left(\nu+\frac{1}{2}\right) z\,[K_\nu(z)L_{\nu-1}(z) + K_{\nu-1}(z)L_\nu(z)] \tag{2.3.33}$$

2.4 Interrelations, Recurrence Relations, Differential and Integral Formulas for the Struve $H_\nu(z)$ and the Modified Struve $L_\nu(z)$ Functions

$$H_n(-z) = (-1)^{n+1} H_n(z) \quad ; \quad n = 0, \pm 1, \pm 2, \pm 3, \ldots \tag{2.4.1}$$

$$H_{-1}(z) = \frac{2}{\pi} - H_1(z) \tag{2.4.2}$$

$$H_{-2}(z)=H_2(z)-\frac{2}{\pi z}-\frac{2z}{3\pi} \tag{2.4.3}$$

$$H_{\nu-1}(z)+H_{\nu+1}(z)=\frac{2\nu}{z}H_\nu(z)+\frac{\left(\frac{z}{2}\right)^\nu}{\sqrt{\pi}\,\Gamma\left(\nu+\frac{3}{2}\right)} \tag{2.4.4}$$

$$H_\nu(z)=\frac{2^{1-\nu}s_{\nu,\nu}(z)}{\sqrt{\pi}\,\Gamma\left(\nu+\frac{1}{2}\right)} \tag{2.4.5}$$

$$H_\nu(z)=Y_\nu(z)+\frac{2^{1-\nu}S_{\nu,\nu}(z)}{\sqrt{\pi}\,\Gamma\left(\nu+\frac{1}{2}\right)} \tag{2.4.6}$$

$$\frac{\mathrm{d}H_0(z)}{\mathrm{d}z}=\frac{2}{\pi}-H_1(z) \tag{2.4.7}$$

$$\frac{\mathrm{d}}{\mathrm{d}z}[zH_1(z)]=zH_0(z) \tag{2.4.8}$$

$$\frac{\mathrm{d}H_\nu(z)}{\mathrm{d}z}=H_{\nu-1}(z)-\frac{\nu}{z}H_\nu(z) \tag{2.4.9}$$

$$\frac{\mathrm{d}H_\nu(z)}{\mathrm{d}z}=\frac{\nu}{z}H_\nu(z)-H_{\nu+1}(z)+\frac{\left(\frac{z}{2}\right)^\nu}{2\sqrt{\pi}\,\Gamma\left(\nu+\frac{3}{2}\right)} \tag{2.4.10}$$

$$\frac{\mathrm{d}H_\nu(z)}{\mathrm{d}z}=\frac{H_{\nu-1}(z)-H_{\nu+1}(z)}{2}+\frac{\left(\frac{z}{2}\right)^\nu}{2\sqrt{\pi}\,\Gamma\left(\nu+\frac{3}{2}\right)} \tag{2.4.11}$$

$$\frac{d}{dz}[z^\nu H_\nu(z)]=z^\nu H_{\nu-1}(z) \tag{2.4.12}$$

$$\frac{d}{dz}[z^{-\nu}H_\nu(z)]=\frac{1}{2^\nu\sqrt{\pi}\,\Gamma\left(\nu+\frac{3}{2}\right)}-z^{-\nu}H_{\nu+1}(z) \tag{2.4.13}$$

$$\int_0^z t^{1+\nu}H_\nu(t)\,dt=z^{1+\nu}H_{\nu+1}(z) \quad ; \quad \nu>-\frac{1}{2} \tag{2.4.14}$$

$$\int_0^z t^{1-\nu}H_\nu(t)\,dt=\frac{z}{2^{\nu-1}\sqrt{\pi}\,\Gamma\left(\nu+\frac{1}{2}\right)}-z^{1-\nu}H_{\nu-1}(z) \tag{2.4.15}$$

$$\int_0^z t^{-\nu}H_\nu(t)\,dt=\frac{z}{2^\nu\sqrt{\pi}\,\Gamma\left(\nu+\frac{3}{2}\right)}-z^{-\nu}H_\nu(z) \tag{2.4.16}$$

$$L_\nu(z)=-ie^{-\pi\nu i}H_\nu(iz) \tag{2.4.17}$$

$$L_1(z)=\frac{2}{\pi}-H_1(z) \tag{2.4.18}$$

$$L_2(z) = \frac{2z}{3\pi} - H_2(z) \tag{2.4.19}$$

$$L_{-1}(z) = \frac{2}{\pi} + L_1(z) \tag{2.4.20}$$

$$L_{-2}(z) = L_2(z) - \frac{2}{\pi z} + \frac{2z}{3\pi} \tag{2.4.21}$$

$$L_{\nu-1}(z) - L_{\nu+1}(z) = \frac{2\nu}{z} L_\nu(z) + \frac{\left(\frac{z}{2}\right)^\nu}{\sqrt{\pi}\,\Gamma\left(\nu+\frac{3}{2}\right)} \tag{2.4.22}$$

$$\frac{dL_0(z)}{dz} = L_1(z) + \frac{2}{\pi} \tag{2.4.23}$$

$$\frac{dL_\nu(z)}{dz} = \frac{L_{\nu-1}(z) + L_{\nu+1}(z)}{2} + \frac{\left(\frac{z}{2}\right)^\nu}{2\sqrt{\pi}\,\Gamma\left(\nu+\frac{3}{2}\right)} \tag{2.4.24}$$

$$\frac{d}{dz}\left[z^\nu L_\nu(z)\right] = z^\nu L_{\nu-1}(z) \tag{2.4.25}$$

$$\frac{d}{dz}\left[z^{-\nu} L_\nu(z)\right] = \frac{1}{2^\nu \sqrt{\pi}\,\Gamma\left(\nu+\frac{3}{2}\right)} + z^{-\nu} L_{\nu+1}(z) \tag{2.4.26}$$

$$\int z \ln z\, L_0(z)\, dz = z \ln z\, L_1(z) - L_0(z) + \frac{2z}{\pi} \tag{2.4.27}$$

$$\int_0^z L_1(z)\, dt = L_0(z) - \frac{2z}{\pi} \tag{2.4.28}$$

2.5 Interrelations, Recurrence Relations and Differential Formulas for the Lommel $s_{\mu,\nu}(z)$ and $S_{\mu,\nu}(z)$ Functions

$$s_{\mu,-\nu}(z) = s_{\mu,\nu}(z) \tag{2.5.1}$$

$$s_{\mu+2,\nu}(z) = z^{\mu+1} - \left[(\mu+1)^2 - \nu^2\right] s_{\mu,\nu}(z) \tag{2.5.2}$$

$$s_{\mu,\nu}(z) = \frac{z}{2\nu}\left[(\mu+\nu-1)\, s_{\mu-1,\nu-1}(z) - (\mu-\nu-1)\, s_{\mu-1,\nu+1}(z)\right] \tag{2.5.3}$$

$$s_{\nu,\nu}(z) = 2^{\nu-1}\sqrt{\pi}\,\Gamma\left(\nu+\frac{1}{2}\right) H_\nu(z) \tag{2.5.4}$$

$$s_{-1,\nu}(z) = -\frac{\pi}{2\nu}\csc(\pi\nu)\left[J_\nu(z) + J_{-\nu}(z)\right] \tag{2.5.5}$$

$$s_{0,\nu}(z) = \frac{\pi}{2}\csc(\pi\nu)\left[J_\nu(z) - J_{-\nu}(z)\right] \tag{2.5.6}$$

$$\frac{ds_{\mu,\nu}(z)}{dz} = (\mu-\nu-1)\, s_{\mu-1,\nu-1}(z) - \left(\frac{\nu}{z}\right) s_{\mu,\nu}(z) \tag{2.5.7}$$

$$\frac{ds_{\mu,\nu}(z)}{dz} = \frac{(\mu+\nu-1)}{2}\, s_{\mu-1,\nu-1}(z) + \frac{(\mu-\nu-1)}{2}\, s_{\mu-1,\nu+1}(z) \tag{2.5.8}$$

$$\frac{d}{dz}\left[z^{\nu} s_{\mu,\nu}(z)\right] = (\mu+\nu-1)\, s_{\mu-1,\nu-1}(z) \tag{2.5.9}$$

$$S_{\lambda,\nu}(z) = s_{\mu,\nu}(z) + 2^{\mu-1}\Gamma\left(\frac{\mu-\nu+1}{2}\right)\Gamma\left(\frac{\mu+\nu+1}{2}\right) \left[\sin\left(\frac{\pi(\mu-\nu)}{2}\right) J_\nu(z) - \cos\left(\frac{\pi(\mu-\nu)}{2}\right) Y_\nu(z)\right] \tag{2.5.10}$$

$$S_{\mu,-\nu}(z) = S_{\mu,\nu}(z) \tag{2.5.11}$$

$$S_{\nu,\nu}(z) = 2^{\nu-1}\sqrt{\pi}\,\Gamma\left(\nu+\frac{1}{2}\right)\left[H_\nu(z) - Y_\nu(z)\right] \tag{2.5.12}$$

$$S_{\nu,\nu}(z) = s_{\nu,\nu}(z) - 2^{\nu-1}\sqrt{\pi}\,\Gamma\left(\nu+\frac{1}{2}\right) Y_\nu(z) \tag{2.5.13}$$

$$S_{0,\nu}(z) = \frac{\pi}{2}\csc(\pi\nu)\left[\boldsymbol{J}_\nu(z) - \boldsymbol{J}_{-\nu}(z) - J_\nu(z) + J_{-\nu}(z)\right] \tag{2.5.14}$$

$$S_{-1,\nu}(z) = \frac{\pi}{2\nu}\csc(\pi\nu)\left[J_\nu(z) + J_{-\nu}(z) - \boldsymbol{J}_\nu(z) - \boldsymbol{J}_{-\nu}(z)\right] \tag{2.5.15}$$

$$S_{1,\nu}(z) = 1 + \nu^2 S_{-1,\nu}(z) \tag{2.5.16}$$

$$S_{\mu+2,\nu}(z) = z^{\mu+1} - \left[(\mu+1)^2 - \nu^2\right] S_{\mu,\nu}(z) \tag{2.5.17}$$

$$S_{\mu,\nu}(z) = \frac{z}{2\nu}\left[(\mu+\nu-1)\, S_{\mu-1,\nu-1}(z) - (\mu-\nu-1)\, S_{\mu-1,\nu+1}(z)\right] \tag{2.5.18}$$

$$S_{0,-1}(z) = \frac{1}{z} \tag{2.5.19}$$

$$\frac{dS_{\mu,\nu}(z)}{dz} = (\mu-\nu-1)\, s_{\mu-1,\nu-1}(z) - \left(\frac{\nu}{z}\right) S_{\mu,\nu}(z) \tag{2.5.20}$$

$$\frac{dS_{\mu,\nu}(z)}{dz} = (\mu-\nu-1)\, S_{\mu-1,\nu+1}(z) + \left(\frac{\nu}{z}\right) S_{\mu,\nu}(z) \tag{2.5.21}$$

$$\frac{dS_{\mu,\nu}(z)}{dz} = \frac{(\mu+\nu-1)}{2}\, S_{\mu-1,\nu-1}(z) + \frac{(\mu-\nu-1)}{2}\, S_{\mu-1,\nu+1}(z) \tag{2.5.22}$$

2.6 Interrelations, Recurrence Relations, Differential and Integral Formulas for the Anger $J_\nu(z)$ and the Weber $E_\nu(z)$ Functions

$$\boldsymbol{J}_n(z) = J_n(z) \quad ; \quad n = 0, \pm 1, \pm 2, \ldots \tag{2.6.1}$$

$$\boldsymbol{J}_{\nu-1}(z) + \boldsymbol{J}_{\nu+1}(z) - \frac{2\nu}{z}\boldsymbol{J}_\nu(z) = -\frac{2\sin(\pi\nu)}{\pi z} \tag{2.6.2}$$

$$\boldsymbol{J}_\nu(z) = \cot(\pi\nu)\,\boldsymbol{E}_\nu(z) - \csc(\pi\nu)\boldsymbol{E}_{-\nu}(z) \tag{2.6.3}$$

$$\boldsymbol{J}_{-\nu}(z) = \cos(\pi\nu)\,\boldsymbol{J}_\nu(z) + \sin(\pi\nu)\boldsymbol{E}_\nu(z) \tag{2.6.4}$$

$$\boldsymbol{J}_\nu(z) = \frac{1}{\pi}\sin(\pi\nu)\,[s_{0,\nu}(z) - \nu s_{-1,\nu}(z)] \tag{2.6.5}$$

$$\frac{d\boldsymbol{J}_\nu(z)}{dz} = \frac{\boldsymbol{J}_{\nu-1}(z) - \boldsymbol{J}_{\nu+1}(z)}{2} \tag{2.6.6}$$

$$\frac{d\boldsymbol{J}_\nu(z)}{dz} = \left(\frac{\nu}{z}\right)\boldsymbol{J}_\nu(z) - \boldsymbol{J}_{\nu+1}(z) - \frac{\sin(\pi\nu)}{\pi z} \tag{2.6.7}$$

$$\int_0^z t^\nu \boldsymbol{J}_\nu(t)\,dt = 2^{\nu-1}\sqrt{\pi}\,\Gamma\left(\nu + \frac{1}{2}\right) z\,[\boldsymbol{J}_\nu(z)\boldsymbol{L}_{\nu-1}(z) - \boldsymbol{J}_{\nu-1}(z)\boldsymbol{L}_\nu(z)] \tag{2.6.8}$$

$$\mathrm{Re}\,\nu > -\frac{1}{2}$$

$$\int_0^z \boldsymbol{J}_0(t)\,dt = z\boldsymbol{J}_0(z) + \frac{\pi z}{2}\,[\boldsymbol{J}_0(z)\boldsymbol{L}_1(z) - \boldsymbol{J}_1(z)\boldsymbol{L}_0(z)] \tag{2.6.9}$$

$$\boldsymbol{E}_{-2n}(z) = \boldsymbol{E}_{2n}(z) \tag{2.6.10}$$

$$\boldsymbol{E}_{2n}(z) = n\sum_{k=0}^{n}(-1)^k \frac{\Gamma(2n-k)}{\Gamma(k+1)\Gamma(n-k+1)}\left(\frac{z}{2}\right)^{k-n} H_{n-k}(z) \tag{2.6.11}$$

$$n = 1, 2, 3, \ldots$$

$$\boldsymbol{E}_{\nu-1}(z) + \boldsymbol{E}_{\nu+1}(z) - \frac{2\nu}{z}\boldsymbol{E}_\nu(z) = \frac{2\,[\cos(\pi\nu) - 1]}{\pi z} \tag{2.6.12}$$

$$\boldsymbol{E}_\nu(z) = \csc(\pi\nu)\,\boldsymbol{J}_{-\nu}(z) - \cot(\pi\nu)\,\boldsymbol{J}_\nu(z) \tag{2.6.13}$$

$$\boldsymbol{E}_{-\nu}(z) = \cos(\pi\nu)\,\boldsymbol{E}_\nu(z) - \sin(\pi\nu)\,\boldsymbol{J}_\nu(z) \tag{2.6.14}$$

$$\boldsymbol{E}_\nu(z) = \boldsymbol{J}_\nu(z)\tan\left(\frac{\pi\nu}{2}\right) - \frac{2}{\pi}s_{0,\nu}(z) \tag{2.6.15}$$

$$\boldsymbol{E}_\nu(z) = -\frac{1+\cos(\pi\nu)}{\pi}s_{0,\nu}(z) - \frac{\nu\,[1-\cos(\pi\nu)]}{\pi}s_{-1,\nu}(z) \tag{2.6.16}$$

$$\frac{d\boldsymbol{E}_\nu(z)}{dz} = \frac{\boldsymbol{E}_{\nu-1}(z) - \boldsymbol{E}_{\nu+1}(z)}{2} \tag{2.6.17}$$

$$\frac{d\boldsymbol{E}_\nu(z)}{dz} = \left(\frac{\nu}{z}\right)\boldsymbol{E}_\nu(z) - \boldsymbol{E}_{\nu+1}(z) - \frac{[1-\cos(\pi\nu)]}{\pi z} \tag{2.6.18}$$

2.7 Interrelations, Recurrence Relations, Differential and Integral Formulas for the Kelvin $\mathrm{ber}_\nu(z)$, $\mathrm{bei}_\nu(z)$, the Modified Kelvin $\mathrm{ker}_\nu(z)$, $\mathrm{kei}_\nu(z)$ and $\mathrm{her}_\nu(z)$, $\mathrm{hei}_\nu(z)$ Functions

$$\begin{aligned} &R_n(z) = \mathrm{ber}_n(z)\ ;\ \mathrm{bei}_n(z)\ ;\ \mathrm{ker}_n(z)\ ;\ \mathrm{kei}_n(z) \quad ; \quad n = 0, \pm 1, \pm 2, \dots \\ &R_{-n}(z) = (-1)^n R_n(z) \end{aligned} \tag{2.7.1}$$

$$\mathrm{ber}_n(-z) = (-1)^n \mathrm{ber}_n(z) \quad ; \quad n = 0, \pm 1, \pm 2, \pm 3, \dots \tag{2.7.2}$$

$$\mathrm{ber}_{\nu+1}(z) = -\left\{\frac{\sqrt{2}\nu}{z}[\mathrm{ber}_\nu(z) - \mathrm{bei}_\nu(z)] + \mathrm{ber}_{\nu-1}(z)\right\} \tag{2.7.3}$$

$$\mathrm{ber}_{-\nu}(z) = \cos(\pi\nu)\ \mathrm{ber}_\nu(z) + \sin(\pi\nu)\ \mathrm{bei}_\nu(z) + \frac{2}{\pi}\sin(\pi\nu)\,\mathrm{ker}_\nu(z) \tag{2.7.4}$$

$$\mathrm{ber}_{-\nu}(z) = \cos(\pi\nu)\ \mathrm{ber}_\nu(z) - \sin(\pi\nu)\ [\mathrm{hei}_\nu(z) - \mathrm{bei}_\nu(z)] \tag{2.7.5}$$

$$\mathrm{bei}_n(-z) = (-1)^n \mathrm{bei}_n(z) \quad ; \quad n = 0, \pm 1, \pm 2, \pm 3, \dots \tag{2.7.6}$$

$$\mathrm{bei}_{\nu+1}(z) = -\left\{\frac{\sqrt{2}\nu}{z}[\mathrm{bei}_\nu(z) + \mathrm{ber}_\nu(z)] + \mathrm{bei}_{\nu-1}(z)\right\} \tag{2.7.7}$$

$$\mathrm{bei}_{-\nu}(z) = -\sin(\pi\nu)\,\mathrm{ber}_\nu(z) + \cos(\pi\nu)\ \mathrm{bei}_\nu(z) + \frac{2}{\pi}\sin(\pi\nu)\ \mathrm{kei}_\nu(z) \tag{2.7.8}$$

$$\mathrm{bei}_{-\nu}(z) = \cos(\pi\nu)\ \mathrm{bei}_\nu(z) + \sin(\pi\nu)\ [\mathrm{her}_\nu(z) - \mathrm{ber}_\nu(z)] \tag{2.7.9}$$

$$\frac{d\,\mathrm{ber}(z)}{dz} = \frac{1}{\sqrt{2}}[\mathrm{ber}_1(z) + \mathrm{bei}_1(z)] \tag{2.7.10}$$

$$\frac{d\,\mathrm{ber}_\nu(z)}{dz} = -\frac{\nu}{z}\,\mathrm{ber}_\nu(z) - \frac{1}{\sqrt{2}}[\mathrm{ber}_{\nu-1}(z) + \mathrm{bei}_{\nu-1}(z)] \tag{2.7.11}$$

$$\frac{d\,\mathrm{bei}(z)}{dz} = \frac{1}{\sqrt{2}}[\mathrm{bei}_1(z) - \mathrm{ber}_1(z)] \tag{2.7.12}$$

$$\frac{d\,\mathrm{bei}_\nu(z)}{dz} = \frac{1}{\sqrt{8}}[(\mathrm{bei}_{\nu+1}(z) - \mathrm{ber}_{\nu+1}(z)) - (\mathrm{bei}_{\nu-1}(z) - \mathrm{ber}_{\nu-1}(z))] \tag{2.7.13}$$

$$\frac{d\,\mathrm{bei}_\nu(z)}{dz} = \frac{\nu}{z}\,\mathrm{bei}_\nu(z) - \frac{1}{\sqrt{2}}[\mathrm{ber}_{\nu+1}(z) - \mathrm{bei}_{\nu+1}(z)] \tag{2.7.14}$$

$$\frac{d\,\mathrm{bei}_\nu(z)}{dz} = -\frac{\nu}{z}\,\mathrm{ber}_\nu(z) + \frac{1}{\sqrt{2}}[\mathrm{ber}_{\nu-1}(z) - \mathrm{bei}_{\nu-1}(z)] \tag{2.7.15}$$

$$\int \mathrm{ber}_1(z)\,dz = \frac{1}{\sqrt{2}}[\mathrm{ber}(z) - \mathrm{bei}(z) - 1] \tag{2.7.16}$$

$$\int z^n \mathrm{ber}_{n-1}(z)\,dz = -\frac{z^n}{\sqrt{2}}[\mathrm{ber}_n(z) - \mathrm{bei}_n(z)] \tag{2.7.17}$$

$$\int z^n \mathrm{bei}_{n-1}(z)\,dz = -\frac{z^n}{\sqrt{2}}[\mathrm{ber}_n(z) + \mathrm{bei}_n(z)] \tag{2.7.18}$$

$$\mathrm{ker}_{\nu+1}(z) = -\left\{\frac{\sqrt{2}\,\nu}{z}[\mathrm{ker}_\nu(z) - \mathrm{kei}_\nu(z)] + \mathrm{ker}_{\nu-1}(z)\right\} \tag{2.7.19}$$

$$\mathrm{ker}_{-\nu}(z) = \cos(\pi\nu)\,\mathrm{ker}_\nu(z) - \sin(\pi\nu)\,\mathrm{kei}_\nu(z) \tag{2.7.20}$$

$$\mathrm{kei}_{\nu+1}(z) = -\left\{\frac{\sqrt{2}\,\nu}{z}[\mathrm{kei}_\nu(z) + \mathrm{ker}_\nu(z)] + \mathrm{kei}_{\nu-1}(z)\right\} \tag{2.7.21}$$

$$\mathrm{kei}_{-\nu}(z) = \sin(\pi\nu)\,\mathrm{ker}_\nu(z) + \cos(\pi\nu)\,\mathrm{kei}_\nu(z) \tag{2.7.22}$$

$$\frac{d\,\mathrm{ker}_\nu(z)}{dz} = \frac{1}{\sqrt{8}}\{[\mathrm{ker}_{\nu+1}(z) + \mathrm{kei}_{\nu+1}(z)] - [\mathrm{ker}_{\nu-1}(z) + \mathrm{kei}_{\nu-1}(z)]\} \tag{2.7.23}$$

$$\frac{d\,\mathrm{kei}_\nu(z)}{dz} = \frac{1}{\sqrt{8}}\{[\mathrm{kei}_{\nu+1}(z) - \mathrm{ker}_{\nu+1}(z)] - [\mathrm{kei}_{\nu-1}(z) - \mathrm{ker}_{\nu-1}(z)]\} \tag{2.7.24}$$

$$\mathrm{her}_n(-z) = (-1)^n[\mathrm{her}_n(z) - 2\mathrm{ber}_n(z)] \quad ; \quad n = 0, \pm 1, \pm 2, \pm 3, \ldots \tag{2.7.25}$$

$$\mathrm{her}_{-\nu}(z) = \cos(\pi\nu)\,\mathrm{her}_\nu(z) - \sin(\pi\nu)\,\mathrm{hei}_\nu(z) \tag{2.7.26}$$

$$\mathrm{hei}_n(-z) = (-1)^n[\mathrm{hei}_n(z) - 2\mathrm{bei}_n(z)] \quad ; \quad n = 0, \pm 1, \pm 2, \pm 3, \ldots \tag{2.7.27}$$

$$\mathrm{hei}_{-\nu}(z) = \sin(\pi\nu)\,\mathrm{her}_\nu(z) + \cos(\pi\nu)\,\mathrm{hei}_\nu(z) \tag{2.7.28}$$

2.8 Interrelations, Recurrence Relations, Differential and Integral Formulas for the Bessel Integral Functions

$$\mathrm{Ji}_0(0) = ci(0) - \ln 2 \tag{2.8.1}$$

$$\mathrm{Ji}_0(z) = ci(z) - \ln 2 + \int_0^z \frac{J_0(x) - \cos x}{x}\,dx \tag{2.8.2}$$

$$\mathrm{Ji}_0(z) = \gamma + \ln\left(\frac{z}{2}\right) + \int_0^z \frac{J_0(x) - 1}{x}\,dx \tag{2.8.3}$$

$$\mathrm{Ji}_2(z) = -\frac{J_1(z)}{z} \tag{2.8.4}$$

$$\mathrm{Ji}_{2n}(z) = \mathrm{Ji}_{2n}(-z) \quad ; \quad n = 0, 1, 2, 3, \ldots \tag{2.8.5}$$

$$\mathrm{Ji}_{2n+1}(z) + \mathrm{Ji}_{2n+1}(-z) = -\frac{2}{2n+1} \tag{2.8.6}$$

$$\mathrm{Ji}_n(z) = \frac{z}{2n}[(n-1)\mathrm{Ji}_{n-1}(z) - (n+1)\mathrm{Ji}_{n+1}(z)] \tag{2.8.7}$$

$$\mathrm{Ji}_{2n-1}(z) = \frac{z}{(4n-2)}[(2n-2)\mathrm{Ji}_{2n-2}(z) - 2n\,\mathrm{Ji}_{2n}(z)] \tag{2.8.8}$$

$$\mathrm{Ji}_\nu(z) = \frac{1}{\nu}\left[\int J_{\nu-1}(z)dz - J_\nu(z) - 1\right] dz \tag{2.8.9}$$

$$\frac{d\mathrm{Ji}_0(z)}{dz} = \frac{J_0(z)}{z} \tag{2.8.10}$$

$$\frac{d\mathrm{Yi}_0(z)}{dz} = \frac{Y_0(z)}{z} \tag{2.8.11}$$

$$\frac{d\mathrm{Ji}_n(z)}{dz} = \frac{\mathrm{Ji}_n(z)}{z} \tag{2.8.12}$$

$$\frac{d\mathrm{Ji}_n(z)}{dz} = \frac{[(n-1)\mathrm{Ji}_{n-1}(z) - (n+1)\mathrm{Ji}_{n+1}(z)]}{2n} \tag{2.8.13}$$

$$\int_0^\infty si(t)\,\mathrm{Ji}_0(2\sqrt{xt})\,dt = \frac{\sin x}{2x} - \frac{ci(x)}{2} \tag{2.8.14}$$

$$\int_0^\infty ci(t)\,\mathrm{Ji}_0(2\sqrt{xt})\,dt = \frac{1-\cos x}{2x} - \frac{si(x)}{2} \tag{2.8.15}$$

$$\int_0^\infty si(t)\,\mathrm{Ji}_1(2\sqrt{xt})\,\frac{dt}{\sqrt{t}} = \frac{\frac{\pi}{2} + si(x) + 2\sin x}{\sqrt{x}} + 4C(\sqrt{x}) \tag{2.8.16}$$

$$\int_0^\infty ci(t)\,\mathrm{Ji}_1(2\sqrt{xt})\,\frac{dt}{\sqrt{t}} = \frac{\gamma + 2 + \ln x - ci(x) - 2\cos x}{\sqrt{x}} + 4S(\sqrt{x}) \tag{2.8.17}$$

2.9 Series Expansions of the Bessel and Related Functions

$$J_n(z) = \sum_{k=0}^{\infty} \frac{(-1)^{k+n}\left(\frac{z}{2}\right)^{2k+n}}{k!(k+n)!} \tag{2.9.1}$$

$$J_{\pm\nu}(z) = \sum_{k=0}^{\infty} \frac{(-1)^k \left(\frac{z}{2}\right)^{2k \pm \nu}}{k!\Gamma(k \pm \nu + 1)} \tag{2.9.2}$$

$$\begin{aligned} Y_n(z) = &-\frac{1}{\pi}\left(\frac{2}{z}\right)^n \sum_{k=0}^{n-1} \frac{(n-k-1)!}{k!}\left(\frac{z^2}{4}\right)^k + \frac{2\ln\left(\frac{z}{2}\right)}{\pi} J_n(z) \\ &-\frac{1}{\pi}\left(\frac{z}{2}\right)^n \sum_{k=0}^{\infty} \frac{\left(-\frac{z^2}{4}\right)^k}{k!(n+k)!}[\psi(k+1) + \psi(n+k+1)] \end{aligned} \tag{2.9.3}$$

$$I_n(z) = \sum_{k=0}^{\infty} \frac{\left(\frac{z}{2}\right)^{2k+n}}{k!(k+n)!} \tag{2.9.4}$$

$$I_\nu(z) = \sum_{k=0}^{\infty} \frac{\left(\frac{z}{2}\right)^{2k+\nu}}{k!\Gamma(k+\nu+1)} \tag{2.9.5}$$

$$\begin{aligned} K_n(z) = &(-1)^{n+1} I_n(z) \ln\left(\frac{z}{2}\right) + \frac{1}{2}\left(\frac{2}{z}\right)^n \sum_{k=0}^{n-1} (-1)^k \frac{(n-k-1)!}{k!}\left(\frac{z^2}{4}\right)^k \\ &+ \frac{(-1)^n}{2}\left(\frac{z}{2}\right)^n \sum_{k=0}^{\infty} \frac{[\psi(k+1) + \psi(n+k+1)]}{k!(n+k)!}\left(\frac{z^2}{4}\right)^k \end{aligned} \tag{2.9.6}$$

$$H_\nu(z) = \sum_{k=0}^{\infty} (-1)^k \frac{\left(\frac{z}{2}\right)^{2k+\nu+1}}{\Gamma\left(k+\frac{3}{2}\right)\Gamma\left(k+\nu+\frac{3}{2}\right)} \tag{2.9.7}$$

$$L_n(z) = \frac{1}{\pi}\sum_{k=0}^{n/2} \frac{\Gamma\left(k+\frac{1}{2}\right)}{\Gamma\left(n+\frac{1}{2}-k\right)}\left(\frac{z}{2}\right)^{n-2k-1} - H_n(z) \quad ; \quad n = 0, 1, 2, 3, \ldots \tag{2.9.8}$$

$$L_{-n}(z) = \frac{(-1)^{n+1}}{\pi}\sum_{k=0}^{n/2} \frac{\Gamma\left(n-k-\frac{1}{2}\right)}{\Gamma\left(k+\frac{3}{2}\right)}\left(\frac{2}{z}\right)^{n-2k-1} - H_{-n}(z) \tag{2.9.9}$$

$$L_\nu(z) = \sum_{k=0}^{\infty} \frac{\left(\frac{z}{2}\right)^{2k+\nu+1}}{\Gamma\left(k+\frac{3}{2}\right)\Gamma\left(k+\nu+\frac{3}{2}\right)} \tag{2.9.10}$$

$$s_{\mu,\nu}(z) = \frac{z^{\mu+1}}{4}\sum_{k=0}^{\infty} \frac{(-1)^k \left(\frac{z}{2}\right)^{2k} \Gamma\left(\frac{\mu-\nu+1}{2}\right)\Gamma\left(\frac{\mu+\nu+1}{2}\right)}{\Gamma\left(\frac{\mu-\nu+2k+3}{2}\right)\Gamma\left(\frac{\mu+\nu+2k+3}{2}\right)} \tag{2.9.11}$$

$$S_{-1,0}(z) = \frac{\pi^2}{8} J_0(z) + \sum_{k=1}^{\infty} \frac{(-1)^k \left(\frac{z}{2}\right)^{2k}}{2\,(k!)^2}\left\{\left[\ln\left(\frac{z}{2}\right) - \psi(k+1)\right]^2 - \frac{\psi'(k+1)}{2} + \frac{\pi^2}{2}\right\} \tag{2.9.12}$$

$$S_{\nu-1,\nu}(z) = -2^{\nu-2}\pi\,\Gamma(\nu)\,Y_\nu(z) + \frac{\Gamma(\nu)\,z^\nu}{4}\sum_{k=0}^{\infty}\frac{(-1)^k\left(\frac{z}{2}\right)^{2k}}{k!\Gamma(k+\nu+1)}\left[2\ln\left(\frac{z}{2}\right)-\psi(k+\nu+1)-\psi(k+1)\right] \tag{2.9.13}$$

$$J_\nu(z) = \cos\left(\frac{\pi\nu}{2}\right)\sum_{k=0}^{\infty}(-1)^k\frac{\left(\frac{z}{2}\right)^{2k}}{\Gamma\left(k+\frac{\nu}{2}+1\right)\Gamma\left(k-\frac{\nu}{2}+1\right)} + \sin\left(\frac{\pi\nu}{2}\right)\sum_{k=0}^{\infty}(-1)^k\frac{\left(\frac{z}{2}\right)^{2k+1}}{\Gamma\left(k+\frac{\nu+3}{2}\right)\Gamma\left(k+\frac{3-\nu}{2}\right)} \tag{2.9.14}$$

$$E_\nu(z) = \sin\left(\frac{\pi\nu}{2}\right)\sum_{k=0}^{\infty}(-1)^k\frac{\left(\frac{z}{2}\right)^{2k}}{\Gamma\left(k+\frac{\nu}{2}+1\right)\Gamma\left(k-\frac{\nu}{2}+1\right)} - \cos\left(\frac{\pi\nu}{2}\right)\sum_{k=0}^{\infty}(-1)^k\frac{\left(\frac{z}{2}\right)^{2k+1}}{\Gamma\left(k+\frac{\nu+3}{2}\right)\Gamma\left(k+\frac{3-\nu}{2}\right)} \tag{2.9.15}$$

$$\mathrm{Gi}(z) = \frac{1}{\pi}\sum_{k=0}^{\infty}\frac{3^{(k-2)/3}\Gamma\left(\frac{k+1}{3}\right)\cos\left[\frac{(2k-1)\pi}{3}\right]z^k}{k!} \tag{2.9.16}$$

$$\mathrm{Hi}(z) = \frac{1}{\pi}\sum_{k=0}^{\infty}\frac{3^{(k-2)/3}\Gamma\left(\frac{k+1}{3}\right)z^k}{k!} \tag{2.9.17}$$

$$\mathrm{ber}(z) = \sum_{k=0}^{\infty}\frac{(-1)^k}{[(2k)!]^2}\left(\frac{z^2}{2}\right)^{4k} \tag{2.9.18}$$

$$\mathrm{ber}_\nu(z) = \left(\frac{z}{2}\right)^\nu\sum_{k=0}^{\infty}\frac{\cos\left[\pi\left(\frac{3\nu+4k}{4}\right)\right]}{k!\Gamma(\nu+k+1)}\left(\frac{z^2}{2}\right)^k \quad ; \quad z\geq 0 \tag{2.9.19}$$

$$\mathrm{bei}(z) = \sum_{k=0}^{\infty}\frac{(-1)^k}{[(2k+1)!]^2}\left(\frac{z^2}{2}\right)^{4k+2} \tag{2.9.20}$$

$$\mathrm{bei}_\nu(z) = \left(\frac{z}{2}\right)^\nu\sum_{k=0}^{\infty}\frac{\sin\left[\pi\left(\frac{3\nu+4k}{4}\right)\right]}{k!\Gamma(\nu+k+1)}\left(\frac{z^2}{2}\right)^k \quad ; \quad z\geq 0 \tag{2.9.21}$$

$$\mathrm{ker}(z) = -\ln\left(\frac{z}{2}\right)\mathrm{ber}(z) + \frac{\pi}{4}\mathrm{bei}(z) + \sum_{k=0}^{\infty}(-1)^k\frac{\psi(2k+1)}{[(2k)!]^2}\left(\frac{z^2}{4}\right)^{2k} \tag{2.9.22}$$

$$\mathrm{kei}(z) = -\ln\left(\frac{z}{2}\right)\mathrm{bei}(z) - \frac{\pi}{4}\mathrm{ber}(z) + \sum_{k=0}^{\infty}(-1)^k\frac{\psi(2k+2)}{[(2k+1)!]^2}\left(\frac{z^2}{4}\right)^{2k} \tag{2.9.23}$$

$$\mathrm{Ji}_n(z) = \frac{1}{n}\left[-1+\frac{\left(\frac{z}{2}\right)^n}{n!}\right] + \sum_{k=1}^{\infty}\frac{(-1)^k\left(\frac{z}{2}\right)^{n+2k}}{(n+2k)k!(n+k)!} \tag{2.9.24}$$

$$\mathrm{Ji}_\nu(z) = \frac{1}{\nu}\left[-1 + \frac{\left(\frac{z}{2}\right)^\nu}{\Gamma(\nu+1)}\right] + \sum_{k=1}^{\infty} \frac{(-1)^k \left(\frac{z}{2}\right)^{\nu+2k}}{(\nu+2k)k!\Gamma(\nu+k+1)} \tag{2.9.25}$$

$$\mathrm{Ji}_0(z) = \gamma + \ln\left(\frac{z}{2}\right) + \sum_{k=1}^{\infty} \frac{(-1)^k \left(\frac{z}{2}\right)^{2k}}{2n\,(n!)^2} \tag{2.9.26}$$

$$\begin{aligned}\mathrm{Yi}_0(z) &= \frac{\gamma^2}{\pi} - \frac{\pi}{6} + \frac{2\gamma}{\pi}\ln\left(\frac{z}{2}\right) + \frac{1}{\pi}\left[\ln\left(\frac{z}{2}\right)\right]^2 \\ &\quad - \frac{2}{\pi}\sum_{k=1}^{\infty} \frac{(-1)^k \left(\frac{z}{2}\right)^{2k}}{2n\,(n!)^2}\left[\psi(n+1) + \frac{1}{2n} - \ln\left(\frac{z}{2}\right)\right]\end{aligned} \tag{2.9.27}$$

$$\begin{aligned}\mathrm{Ki}_0(z) &= -\frac{\gamma^2}{2} - \frac{\pi^2}{24} - \gamma\ln\left(\frac{z}{2}\right) - \frac{1}{2}\left[\ln\left(\frac{z}{2}\right)\right]^2 \\ &\quad + \frac{2}{\pi}\sum_{k=1}^{\infty} \frac{\left(\frac{z}{2}\right)^{2k}}{2n\,(n!)^2}\left[\psi(n+1) + \frac{1}{2n} - \ln\left(\frac{z}{2}\right)\right]\end{aligned} \tag{2.9.28}$$

$$J_0(z) = \left[J_0\left(\frac{z}{2}\right)\right]^2 + 2\sum_{k=1}^{\infty}(-1)^k\left[J_k\left(\frac{z}{2}\right)\right]^2 \tag{2.9.29}$$

$$Y_0(z) = \frac{2}{\pi}\left\{\left[\gamma + \ln\left(\frac{z}{2}\right)\right]J_0(z) + \sum_{k=0}^{\infty}\frac{\left(\frac{z}{2}\right)^k}{k!k}J_k(z)\right\} \tag{2.9.30}$$

$$Y_0(z) = \frac{2}{\pi}\left\{\left[\gamma + \ln\left(\frac{z}{2}\right)\right]J_0(z) - 2\sum_{k=0}^{\infty}\frac{(-1)^k}{k}J_{2k}(z)\right\} \tag{2.9.31}$$

$$I_0(z) = 1 - 2\sum_{k=1}^{\infty}(-1)^k I_{2k}(z) \tag{2.9.32}$$

$$I_0(z) = e^{-z} - 2\sum_{k=1}^{\infty}(-1)^k I_k(z) \tag{2.9.33}$$

$$I_0(z) = e^z - 2\sum_{k=1}^{\infty} I_k(z) \tag{2.9.34}$$

$$I_1(z) = \frac{2}{z}\sum_{k=1}^{\infty} k\,I_{2k}(z) \tag{2.9.35}$$

$$J_\nu(z) = \frac{1}{\Gamma(\nu)}\sum_{k=0}^{\infty}\frac{\left(\frac{z}{2}\right)^{k+\nu}}{k!(k+\nu)}J_k(z) \tag{2.9.36}$$

$$J_\nu(z) = \sum_{k=0}^{\infty}(-1)^k\frac{z^k}{k!}I_{k+\nu}(z) \tag{2.9.37}$$

$$I_\nu(z) = \sum_{k=0}^{\infty}\frac{z^k}{k!}J_{k+\nu}(z) \tag{2.9.38}$$

$$I_\nu(z) = \frac{1}{\Gamma(\nu)} \sum_{k=0}^{\infty} \frac{(-1)^k \left(\frac{z}{2}\right)^{k+\nu}}{k!(k+\nu)} I_k(z) \tag{2.9.39}$$

$$K_0(z) = -\left[\gamma + \ln\left(\frac{z}{2}\right)\right] I_0(z) - \sum_{k=0}^{\infty} \frac{(-1)^k \left(\frac{z}{2}\right)^k}{k!k} I_k(z)$$

$$K_0(z) = -\left[\gamma + \ln\left(\frac{z}{2}\right)\right] I_0(z) + 2\sum_{k=0}^{\infty} \frac{1}{k} I_{2k}(z) \tag{2.9.40}$$

$$\begin{aligned} s_{\lambda,\nu}(z) &= 2^{\lambda+1} \sum_{k=1}^{\infty} \frac{(2k+\lambda+1)\,\Gamma(k+\lambda+1)}{k!(2k+\lambda+\nu+1)(2k+\lambda-\nu+1)} J_{2k+\lambda+1}(z) \\ &+ \frac{2^{\lambda+1}\Gamma(\lambda+2)}{(\lambda+\nu+1)(\lambda-\nu+1)} J_{\lambda+1}(z) \quad ; \quad \lambda \pm \nu \neq n = -1, -3, -5, \ldots \end{aligned} \tag{2.9.41}$$

$$H_\nu(z) = \frac{4}{\sqrt{\pi}\,\Gamma\left(\nu+\frac{1}{2}\right)} \sum_{k=0}^{\infty} \frac{(2k+\nu+1)\,\Gamma(k+\nu+1)}{k!(2k+2\nu+1)(2k+1)} J_{2k+\nu+1}(z) \tag{2.9.42}$$

$$H_\nu(z) = \sqrt{\frac{z}{2\pi}} \sum_{k=0}^{\infty} \frac{\left(\frac{z}{2}\right)^k}{k!\left(k+\frac{1}{2}\right)} J_{k+\nu+1/2}(z) \tag{2.9.43}$$

$$H_0(z) = \sum_{k=0}^{\infty} (-1)^k \left[J_{k+1/2}(z)\right]^2 \tag{2.9.44}$$

$$H_0(z) = \frac{4}{\pi} \sum_{k=0}^{\infty} \frac{J_{2k+1}(z)}{2k+1} \tag{2.9.45}$$

$$H_1(z) = \frac{2}{\pi}\left[1 - J_0(z)\right] + \frac{4}{\pi} \sum_{k=1}^{\infty} \frac{J_{2k}(z)}{4k^2-1} \tag{2.9.46}$$

2.10 Bessel and Related Functions of Half Odd Integer Order

$$J_{-7/2}(z) = \sqrt{\frac{2}{\pi z}} \left[\left(1 - \frac{15}{z^2}\right)\sin z + \left(\frac{6}{z} - \frac{15}{z^3}\right)\cos z\right] \tag{2.10.1}$$

$$J_{-5/2}(z) = \sqrt{\frac{2}{\pi z}} \left[\frac{3\sin z}{z} + \left(\frac{3}{z^2} - 1\right)\cos z\right] \tag{2.10.2}$$

$$J_{-3/2}(z) = -\sqrt{\frac{2}{\pi z}} \left[\frac{\cos z}{z} + \sin z\right] \tag{2.10.3}$$

$$J_{-1/2}(z) = \sqrt{\frac{2}{\pi z}} \cos z \tag{2.10.4}$$

$$J_{1/2}(z) = \sqrt{\frac{2}{\pi z}} \sin z \tag{2.10.5}$$

$$J_{3/2}(z) = \sqrt{\frac{2}{\pi z}}\left[\frac{\sin z}{z} - \cos z\right] \tag{2.10.6}$$

$$J_{5/2}(z) = \sqrt{\frac{2}{\pi z}}\left[\left(\frac{3}{z^2} - 1\right)\sin z - \frac{3}{z}\cos z\right] \tag{2.10.7}$$

$$J_{7/2}(z) = \sqrt{\frac{2}{\pi z}}\left[\left(\frac{15}{z^3} - \frac{6}{z}\right)\sin z - \left(\frac{15}{z^2} - 1\right)\cos z\right] \tag{2.10.8}$$

$$Y_{-1/2}(z) = \sqrt{\frac{2}{\pi z}}\sin z \tag{2.10.9}$$

$$Y_{1/2}(z) = -\sqrt{\frac{2}{\pi z}}\cos z \tag{2.10.10}$$

$$Y_{3/2}(z) = -\sqrt{\frac{2}{\pi z}}\left(\sin z + \frac{\cos z}{z}\right) \tag{2.10.11}$$

$$Y_{5/2}(z) = \sqrt{\frac{2}{\pi z}}\left[\left(1 - \frac{3}{z}\right)\cos z - \frac{3}{z}\sin z\right] \tag{2.10.12}$$

$$J_{-n-1/2}(z) = (-1)^{n+1}Y_{n+1/2}(z) \quad ; \quad n = 0, 1, 2, 3, \ldots \tag{2.10.13}$$

$$Y_{-n-1/2}(z) = (-1)^{n}J_{n+1/2}(z) \quad ; \quad n = 0, 1, 2, 3, \ldots \tag{2.10.14}$$

$$I_{-7/2}(z) = \sqrt{\frac{2}{\pi z}}\left[\left(\frac{15}{z^2} + 1\right)\sinh z - \left(\frac{15}{z^3} + \frac{6}{z}\right)\cosh z\right] \tag{2.10.15}$$

$$I_{-5/2}(z) = \sqrt{\frac{2}{\pi z}}\left[\left(\frac{3}{z^2} + 1\right)\cosh z - \frac{3}{z}\sinh z\right] \tag{2.10.16}$$

$$I_{-3/2}(z) = \sqrt{\frac{2}{\pi z}}\left[\sinh z - \frac{\cosh z}{z}\right] \tag{2.10.17}$$

$$I_{-1/2}(z) = \sqrt{\frac{2}{\pi z}}\cosh z \tag{2.10.18}$$

$$I_{1/2}(z) = \sqrt{\frac{2}{\pi z}}\sinh z \tag{2.10.19}$$

$$I_{3/2}(z) = \sqrt{\frac{2}{\pi z}}\left[\cosh z - \frac{\sinh z}{z}\right] \tag{2.10.20}$$

$$I_{5/2}(z) = \sqrt{\frac{2}{\pi z}}\left[\left(\frac{3}{z^2} + 1\right)\sinh z - \frac{3}{z}\cosh z\right] \tag{2.10.21}$$

$$I_{7/2}(z) = \sqrt{\frac{2}{\pi z}}\left[\left(\frac{15}{z^2} + 1\right)\cosh z - \left(\frac{15}{z^3} - \frac{6}{z}\right)\sinh z\right] \tag{2.10.22}$$

$$K_{-1/2}(z) = \sqrt{\frac{\pi}{2z}}\, e^{-z} \tag{2.10.23}$$

$$K_{1/2}(z) = \sqrt{\frac{\pi}{2z}}\, e^{-z} \tag{2.10.24}$$

$$K_{3/2}(z) = \sqrt{\frac{\pi}{2z}}\, e^{-z}\left(1 + \frac{1}{z}\right) \tag{2.10.25}$$

$$K_{5/2}(z) = \sqrt{\frac{\pi}{2z}}\, e^{-z}\left(1 + \frac{3}{z} + \frac{3}{z^2}\right) \tag{2.10.26}$$

$$I_{-n-1/2}(z) = (-1)^n \frac{2}{\pi} K_{n+1/2}(z) + I_{n+1/2}(z) \quad ; \quad n = 0, 1, 2, 3, \ldots \tag{2.10.27}$$

$$K_{-n-1/2}(z) = K_{n+1/2}(z) \quad ; \quad n = 0, 1, 2, 3, \ldots \tag{2.10.28}$$

$$H_{-3/2}(z) = \sqrt{\frac{2}{\pi z}}\left(\cos z - \frac{\sin z}{z}\right) \tag{2.10.29}$$

$$H_{-1/2}(z) = \sqrt{\frac{2}{\pi z}}\, \sin z \tag{2.10.30}$$

$$H_{1/2}(z) = \sqrt{\frac{2}{\pi z}}\,(1 - \cos z) \tag{2.10.31}$$

$$H_{3/2}(z) = \sqrt{\frac{z}{2\pi}}\left(1 + \frac{2}{z^2}\right) - \sqrt{\frac{2}{\pi z}}\left(\sin z + \frac{\cos z}{z}\right) \tag{2.10.32}$$

$$I_{-n-1/2}(z) = (-1)^n \frac{2}{\pi} K_{n+1/2}(z) + I_{n+1/2}(z) \quad ; \quad n = 0, 1, 2, 3, \ldots \tag{2.10.33}$$

$$L_{-3/2}(z) = \sqrt{\frac{2}{\pi z}}\left(\cosh z - \frac{\sinh z}{z}\right) \tag{2.10.34}$$

$$L_{-1/2}(z) = \sqrt{\frac{2}{\pi z}}\, \sinh z \tag{2.10.35}$$

$$L_{1/2}(z) = \sqrt{\frac{2}{\pi z}}\,(1 - \cosh z) \tag{2.10.36}$$

$$L_{3/2}(z) = -\sqrt{\frac{z}{2\pi}}\left(1 - \frac{2}{z^2}\right) + \sqrt{\frac{2}{\pi z}}\left(\sinh z - \frac{\cosh z}{z}\right) \tag{2.10.37}$$

$$L_{-n-1/2}(z) = I_{n+1/2}(z) \quad ; \quad n = 0, 1, 2, 3, \ldots \tag{2.10.38}$$

$$\operatorname{ber}_{-1/2}(z) = \sqrt{\frac{2}{\pi z}}\left\{\cos\left(\frac{3\pi}{8}\right)\cos\left(\frac{z}{\sqrt{2}}\right)\cosh\left(\frac{z}{\sqrt{2}}\right) + \sin\left(\frac{3\pi}{8}\right)\sin\left(\frac{z}{\sqrt{2}}\right)\sinh\left(\frac{z}{\sqrt{2}}\right)\right\} \tag{2.10.39}$$

$$\operatorname{bei}_{-1/2}(z) = \sqrt{\frac{2}{\pi z}}\left\{\cos\left(\frac{3\pi}{8}\right)\sin\left(\frac{z}{\sqrt{2}}\right)\sinh\left(\frac{z}{\sqrt{2}}\right) - \sin\left(\frac{3\pi}{8}\right)\cos\left(\frac{z}{\sqrt{2}}\right)\cosh\left(\frac{z}{\sqrt{2}}\right)\right\} \tag{2.10.40}$$

$$\operatorname{ber}_{1/2}(z) = \frac{1}{\sqrt{2\pi z}}\left\{e^{z/\sqrt{2}}\cos\left(\frac{z}{\sqrt{2}} + \frac{\pi}{8}\right) - e^{-z/\sqrt{2}}\cos\left(\frac{z}{\sqrt{2}} - \frac{\pi}{8}\right)\right\} \tag{2.10.41}$$

$$\operatorname{bei}_{1/2}(z) = \frac{1}{\sqrt{2\pi z}}\left\{e^{z/\sqrt{2}}\sin\left(\frac{z}{\sqrt{2}} + \frac{\pi}{8}\right) - e^{-z/\sqrt{2}}\sin\left(\frac{z}{\sqrt{2}} - \frac{\pi}{8}\right)\right\} \tag{2.10.42}$$

$$\operatorname{ker}_{-1/2}(z) = \sqrt{\frac{\pi}{2z}}\,e^{z/\sqrt{2}}\cos\left(\frac{z}{\sqrt{2}} - \frac{\pi}{8}\right) \tag{2.10.43}$$

$$\operatorname{kei}_{-1/2}(z) = -\sqrt{\frac{\pi}{2z}}\,e^{-z/\sqrt{2}}\sin\left(\frac{z}{\sqrt{2}} - \frac{\pi}{8}\right) \tag{2.10.44}$$

$$\operatorname{ker}_{1/2}(z) = \sqrt{\frac{\pi}{2z}}\,e^{z/\sqrt{2}}\cos\left(\frac{z}{\sqrt{2}} + \frac{3\pi}{8}\right) \tag{2.10.45}$$

$$\operatorname{kei}_{1/2}(z) = -\sqrt{\frac{\pi}{2z}}\,e^{-z/\sqrt{2}}\sin\left(\frac{z}{\sqrt{2}} + \frac{3\pi}{8}\right) \tag{2.10.46}$$

$$\mathbf{J}_{-1/2}(z) = \sqrt{\frac{2}{\pi z}}\{\cos z[C(z) + S(z)] - \sin z[C(z) - S(z)]\} \tag{2.10.47}$$

$$\mathbf{J}_{1/2}(z) = \sqrt{\frac{2}{\pi z}}\{\cos z[C(z) - S(z)] + \sin z[C(z) + S(z)]\} \tag{2.10.48}$$

$$\mathbf{E}_{-1/2}(z) = -\sqrt{\frac{2}{\pi z}}\{\cos z[C(z) - S(z)] + \sin z[C(z) + S(z)]\} \tag{2.10.49}$$

$$\mathbf{E}_{1/2}(z) = \sqrt{\frac{2}{\pi z}}\{\cos z[C(z) + S(z)] - \sin z[C(z) - S(z)]\} \tag{2.10.50}$$

$$s_{-1,1/2}(z) = 2\sqrt{\frac{2\pi}{z}}\,[\sin z\,S(z) + \cos z\,C(z)] \tag{2.10.51}$$

$$s_{0,1/2}(z) = \sqrt{\frac{2\pi}{z}}\,[\sin z\,C(z) - \cos z\,S(z)] \tag{2.10.52}$$

$$s_{1/2,1/2}(z)=\frac{1-\cos z}{\sqrt{z}} \tag{2.10.53}$$

$$s_{3/2,1/2}(z)=\frac{z-\sin z}{\sqrt{z}} \tag{2.10.54}$$

$$s_{5/2,1/2}(z)=\frac{z^2+2\cos z-2}{\sqrt{z}} \tag{2.10.55}$$

$$S_{-3/2,1/2}(z)=-\frac{1}{\sqrt{z}}\,[\sin z\, si(z)+\cos z\, Ci(z)] \tag{2.10.56}$$

$$S_{-1,1/2}(z)=2\sqrt{\frac{2\pi}{z}}\left\{\cos z\left[\frac{1}{2}-C(z)\right]+\sin z\left[\frac{1}{2}-S(z)\right]\right\} \tag{2.10.57}$$

$$S_{-1/2,1/2}(z)=\frac{1}{\sqrt{z}}\,[\sin z\, Ci(z)-\cos z\, si(z)] \tag{2.10.58}$$

$$S_{0,1/2}(z)=\sqrt{\frac{2\pi}{z}}\left\{\cos z\left[\frac{1}{2}-S(z)\right]-\sin z\left[\frac{1}{2}-C(z)\right]\right\} \tag{2.10.59}$$

$$S_{3/2,1/2}(z)=\sqrt{z} \tag{2.10.60}$$

$$S_{1/2,1/2}(z)=\frac{1}{\sqrt{z}} \tag{2.10.61}$$

$$\mathrm{Ji}_{1/2}(z)=-\frac{2^{3/2}}{\sqrt{\pi}}\left[\frac{\sin z}{\sqrt{z}}+2C(\sqrt{z})\right] \tag{2.10.62}$$

2.11 Bessel and Related Functions Expressed in Terms of Special Functions

$$J_\nu(z)=\frac{\left(\frac{z}{2}\right)^\nu}{\Gamma(\nu+1)}\,{}_0F_1(;\nu+1;-\frac{z^2}{4}) \tag{2.11.1}$$

$$\begin{aligned}J_\nu(z)=&\frac{\left(\frac{z}{2}\right)^\nu}{\Gamma(\nu+1)}\,{}_0F_3\left(;\frac{1}{2}.\frac{\nu+1}{2},\frac{\nu+2}{2};\frac{z^4}{256}\right)\\&+\frac{\left(\frac{z}{2}\right)^{\nu+2}}{\Gamma(\nu+2)}\,{}_0F_3\left(;\frac{3}{2}.\frac{\nu+2}{2},\frac{\nu+3}{2};\frac{z^4}{256}\right)\end{aligned} \tag{2.11.2}$$

$$I_\nu(z)=\frac{\left(\frac{z}{2}\right)^\nu}{\Gamma(\nu+1)}\,{}_0F_1\left(;\nu+1;\frac{z^2}{4}\right) \tag{2.11.3}$$

$$I_\nu(z)=\frac{\left(\frac{z}{2}\right)^\nu e^{-z}}{\Gamma(\nu+1)}\,{}_1F_1\left(\nu+\frac{1}{2};1+2\nu;2z\right) \tag{2.11.4}$$

$$K_{1/4}(z)=\frac{\sqrt{\pi}}{z^{1/4}}D_{-1/2}(2\sqrt{z}) \tag{2.11.5}$$

$$K_\nu(z)=(2z)^\nu\sqrt{\pi}\,e^{-z}\,{}_1F_1\left(\nu+\frac{1}{2}2\nu+1;2z\right) \tag{2.11.6}$$

$$K_\nu(z)=\sqrt{\frac{\pi}{2z}}W_{0,\,\nu}(2z) \tag{2.11.7}$$

$$H_\nu(z)=\frac{2}{\sqrt{\pi}\,\Gamma\left(\nu+\frac{3}{2}\right)}\left(\frac{z}{2}\right)^{\nu+1}{}_1F_2\left(1;\frac{3}{2},\nu+\frac{3}{2};-\frac{z^2}{4}\right) \tag{2.11.8}$$

$$L_\nu(z)=\frac{2}{\sqrt{\pi}\,\Gamma\left(\nu+\frac{3}{2}\right)}\left(\frac{z}{2}\right)^{\nu+1}{}_1F_2\left(1;\nu+\frac{3}{2};\frac{z^2}{4}\right) \tag{2.11.9}$$

$$\mathrm{Ai}(z)=\frac{1}{3^{2/3}\Gamma\left(\frac{2}{3}\right)}\,{}_0F_1\left(;\frac{2}{3};\frac{z^3}{9}\right)-\frac{z}{3^{1/3}\Gamma\left(\frac{1}{3}\right)}\,{}_0F_1\left(;\frac{4}{3};\frac{z^3}{9}\right) \tag{2.11.10}$$

$$\mathrm{Ai}(-z)=\frac{1}{3^{2/3}\Gamma\left(\frac{2}{3}\right)}\,{}_0F_1\left(;\frac{2}{3};-\frac{z^3}{9}\right)+\frac{z}{3^{1/3}\Gamma\left(\frac{1}{3}\right)}\,{}_0F_1\left(;\frac{4}{3};-\frac{z^3}{9}\right) \tag{2.11.11}$$

$$\mathrm{Ai}'(z)=-\frac{1}{3^{1/3}\Gamma\left(\frac{1}{3}\right)}\,{}_0F_1\left(;\frac{1}{3};\frac{z^3}{9}\right)+\frac{z^2}{2\cdot 3^{2/3}\Gamma\left(\frac{2}{3}\right)}\,{}_0F_1\left(;\frac{5}{3};\frac{z^3}{9}\right) \tag{2.11.12}$$

$$\mathrm{Ai}'(-z)=-\frac{1}{3^{1/3}\Gamma\left(\frac{1}{3}\right)}\,{}_0F_1\left(;\frac{1}{3};-\frac{z^3}{9}\right)+\frac{z^2}{2\cdot 3^{2/3}\Gamma\left(\frac{2}{3}\right)}\,{}_0F_1\left(;\frac{5}{3};-\frac{z^3}{9}\right) \tag{2.11.13}$$

$$\mathrm{Bi}(z)=\frac{1}{3^{1/6}\Gamma\left(\frac{2}{3}\right)}\,{}_0F_1\left(;\frac{2}{3};\frac{z^3}{9}\right)+\frac{3^{1/6}z}{\Gamma\left(\frac{1}{3}\right)}\,{}_0F_1\left(;\frac{4}{3};\frac{z^3}{9}\right) \tag{2.11.14}$$

$$\mathrm{Bi}(-z)=\frac{1}{3^{1/6}\Gamma\left(\frac{2}{3}\right)}\,{}_0F_1\left(;\frac{2}{3};-\frac{z^3}{9}\right)-\frac{3^{1/6}z}{\Gamma\left(\frac{1}{3}\right)}\,{}_0F_1\left(;\frac{4}{3};-\frac{z^3}{9}\right) \tag{2.11.15}$$

$$\mathrm{Bi}'(z)=\frac{3^{1/6}}{\Gamma\left(\frac{1}{3}\right)}\,{}_0F_1\left(;\frac{1}{3};\frac{z^3}{9}\right)+\frac{z^2}{2\cdot 3^{1/6}\Gamma\left(\frac{2}{3}\right)}\,{}_0F_1\left(;\frac{5}{3};\frac{z^3}{9}\right) \tag{2.11.16}$$

$$\mathrm{Bi}'(-z)=\frac{3^{1/6}}{\Gamma\left(\frac{1}{3}\right)}\,{}_0F_1\left(;\frac{1}{3};-\frac{z^3}{9}\right)+\frac{z^2}{2\cdot 3^{1/6}\Gamma\left(\frac{2}{3}\right)}\,{}_0F_1\left(;\frac{5}{3};-\frac{z^3}{9}\right) \tag{2.11.17}$$

$$\mathrm{Gi}(-z)=\mathrm{Bi}(-z)-\frac{2\sqrt{z}}{3\pi}S_{0,1/3}\left(\frac{2}{3}z^{3/2}\right) \tag{2.11.18}$$

$$\mathrm{ber}(z)={}_0F_3\left(;\frac{1}{2},\frac{1}{2},1;-\frac{z^4}{256}\right) \tag{2.11.19}$$

$$\mathrm{ber}_\nu(z) = \frac{\left(\frac{z}{2}\right)^\nu \cos\left(\frac{3\pi\nu}{4}\right)}{\Gamma(\nu+1)}\, {}_0F_3\left(;\frac{1}{2}, \frac{\nu+1}{2}, \frac{\nu+2}{2}; -\frac{z^4}{256}\right) - \frac{\left(\frac{z}{2}\right)^{\nu+2} \sin\left(\frac{3\pi\nu}{4}\right)}{\Gamma(\nu+2)}\, {}_0F_3\left(;\frac{3}{2}, \frac{\nu+2}{2}, \frac{\nu+3}{2}; -\frac{z^4}{256}\right) \tag{2.11.20}$$

$$\mathrm{bei}(z) = \frac{z^2}{4}\, {}_0F_3\left(;\frac{3}{2}, \frac{3}{2}, 1; -\frac{z^4}{256}\right) \tag{2.11.21}$$

$$\mathrm{bei}_\nu(z) = \frac{\left(\frac{z}{2}\right)^\nu \sin\left(\frac{3\pi\nu}{4}\right)}{\Gamma(\nu+1)}\, {}_0F_3\left(;\frac{1}{2}, \frac{\nu+1}{2}, \frac{\nu+2}{2}; -\frac{z^4}{256}\right) + \frac{\left(\frac{z}{2}\right)^{\nu+2} \cos\left(\frac{3\pi\nu}{4}\right)}{\Gamma(\nu+2)}\, {}_0F_3\left(;\frac{3}{2}, \frac{\nu+2}{2}, \frac{\nu+3}{2}; -\frac{z^4}{256}\right) \tag{2.11.22}$$

$$\frac{d\,\mathrm{ber}(z)}{dz} = -\frac{z^3}{256}\, {}_0F_3\left(; 2, \frac{3}{2}, \frac{3}{2}; -\frac{z^4}{256}\right) \tag{2.11.23}$$

$$\frac{d\,\mathrm{bei}(z)}{dz} = \frac{z}{2}\, {}_0F_3\left(; 1, \frac{1}{2}, \frac{3}{2}; -\frac{z^4}{256}\right) \tag{2.11.24}$$

$$s_{\lambda,\nu}(z) = \frac{z^{\lambda+1}}{(\lambda+\nu+1)(\lambda-\nu+1)}\, {}_1F_2\left(1; \frac{\lambda-\nu+3}{2}, \frac{\lambda+\nu+3}{2}; -\frac{z^2}{4}\right) \qquad \lambda \pm \nu \neq -1, -2, -3, \ldots \tag{2.11.25}$$

$$\mathrm{Ji}_0(z) = \gamma + \ln\left(\frac{z}{2}\right) - \frac{z^2}{8}\, {}_2F_3\left(1, 1; 2, 2, 2; -\frac{z^2}{4}\right) \tag{2.11.26}$$

2.12 Integral Representation of the Bessel and Related Functions

$$J_0(z) = \frac{2}{\pi}\int_1^\infty \frac{\sin(zt)}{\sqrt{t^2-1}}\,dt \tag{2.12.1}$$

$$J_0(z) = \frac{1}{\pi}\int_{-1}^1 \frac{\cos(zt)}{\sqrt{1-t^2}}\,dt \tag{2.12.2}$$

$$J_0(z) = \frac{2}{\pi}\int_0^1 \frac{\cos(zt)}{\sqrt{1-t^2}}\,dt \tag{2.12.3}$$

$$J_0(z) = \frac{1}{\pi}\int_0^\pi \cos(z\cos t)\,dt$$

$$J_0(z) = \frac{2}{\pi^2}\int_0^{\pi} t\cos(z\sin t)\,dt \tag{2.12.4}$$

$$J_0(z) = \frac{2}{\pi}\int_0^{\infty} \sin(z\cosh t)\,dt \quad ; \quad z>0 \tag{2.12.5}$$

$$J_n(z) = \frac{1}{\pi}\int_0^{\pi} \cos z(\sin t - nt)\,dt \quad ; \quad n=0,1,2,3,\ldots \tag{2.12.6}$$

$$J_n(z) = \frac{1}{2\pi}\int_0^{2\pi} \cos z(nt - \sin t)\,dt \quad ; \quad n=0,1,2,3,\ldots \tag{2.12.7}$$

$$J_\nu(z) = \frac{2\left(\frac{z}{2}\right)^\nu}{\sqrt{\pi}\,\Gamma\left(\nu+\frac{1}{2}\right)}\int_0^{\pi/2} \cos(z\cos t)(\sin t)^{2\nu}dt \quad ; \quad \operatorname{Re}\nu > -\frac{1}{2} \tag{2.12.8}$$

$$J_\nu(z) = \frac{\left(\frac{z}{2}\right)^\nu}{\sqrt{\pi}\,\Gamma\left(\nu+\frac{1}{2}\right)}\int_0^{\pi} \cos(z\cos t)(\sin t)^{2\nu}dt \quad ; \quad \operatorname{Re}\nu > -\frac{1}{2} \tag{2.12.9}$$

$$J_\nu(z) = \frac{2\left(\frac{z}{2}\right)^\nu}{\sqrt{\pi}\,\Gamma\left(\nu+\frac{1}{2}\right)}\int_0^{\pi/2} \cos(z\sin t)(\cos t)^{2\nu}dt \quad ; \quad \operatorname{Re}\nu > -\frac{1}{2} \tag{2.12.10}$$

$$J_\nu(z) = \frac{\left(\frac{z}{2}\right)^\nu}{\sqrt{\pi}\,\Gamma\left(\nu+\frac{1}{2}\right)}\int_{-\pi/2}^{\pi/2} \cos(z\sin t)(\cos t)^{2\nu}dt \quad ; \quad \operatorname{Re}\nu > -\frac{1}{2} \tag{2.12.11}$$

$$J_\nu(z) = \frac{2}{\pi}\int_0^{\infty} \sin\left(z\cosh t - \frac{\pi\nu}{2}\right)\cosh(\nu t)\,dt \quad ; \quad |\operatorname{Re}\nu| < 1 \tag{2.12.12}$$

$$J_\nu(z) = \frac{2\left(\frac{z}{2}\right)^\nu}{\sqrt{\pi}\,\Gamma\left(\nu+\frac{1}{2}\right)}\int_0^{1} (1-t^2)^{\nu-1/2}\cos(zt)dt \quad ; \quad \operatorname{Re}\nu > -\frac{1}{2} \tag{2.12.13}$$

$$J_\nu(z) = \frac{\left(\frac{z}{2}\right)^\nu}{\sqrt{\pi}\,\Gamma\left(\nu+\frac{1}{2}\right)}\int_0^{1} \frac{t^{\nu-1/2}}{\sqrt{1-t}}\cos(z\sqrt{1-t})dt \quad ; \quad \operatorname{Re}\nu > -\frac{1}{2} \tag{2.12.14}$$

$$J_\nu(z) = \frac{\left(\frac{z}{2}\right)^\nu}{\sqrt{\pi}\,\Gamma\left(\nu+\frac{1}{2}\right)}\int_{-1}^{1} (1-t^2)^{\nu-1/2}\cos(zt)dt \quad ; \quad \operatorname{Re}\nu > -\frac{1}{2} \tag{2.12.15}$$

$$J_\nu(z) = \frac{1}{\pi}\int_0^{\pi} \cos(z\sin t - \nu t)\,dt - \frac{\sin(\pi\nu)}{\pi}\int_0^{\infty} e^{-(z\sinh t + \nu t)}dt \quad ; \quad \operatorname{Re} z > 0 \tag{2.12.16}$$

$$J_\nu(z) = \frac{1}{\pi}\int_0^{\pi/2} \cos(z\sin t - \nu t)\,dt + \frac{1}{\pi}\int_0^{\infty} e^{-\nu t}\sin\left(z\cosh t - \frac{\pi\nu}{2}\right)dt \tag{2.12.17}$$

$z > 0 \quad ; \quad \mathrm{Re}\,\nu \geq 0$

$$J_{-\nu}(z) = \frac{2\left(\frac{z}{2}\right)^\nu}{\sqrt{\pi}\,\Gamma\left(\nu+\frac{1}{2}\right)}\left[\int_0^1 (1-t^2)\cos(zt+\pi\nu)\,dt + \sin(\pi\nu)\int_0^\infty e^{-zt}(1+t^2)^{\nu-1/2}dt\right] \quad ; \quad \mathrm{Re}\,z > 0 \quad ; \quad \mathrm{Re}\,\nu > -\frac{1}{2} \tag{2.12.18}$$

$$Y_0(z) = -\frac{2}{\pi}\int_0^\infty \cos(z\cosh t)\,dt \quad ; \quad z>0 \quad ; \quad |\mathrm{Re}\,\nu| < 1 \tag{2.12.19}$$

$$Y_0(z) = \frac{2}{\pi}\int_0^{\pi/2} \cos(z\cos t)\,\ln[4z(\sin t)^2]\,dt \quad ; \quad z>0 \tag{2.12.20}$$

$$Y_0(z) = \frac{4}{\pi^2}\int_0^1 \sin(zt)\frac{\sin^{-1}t}{\sqrt{1-t^2}}dt - \frac{4}{\pi^2}\int_1^\infty \sin(zt)\frac{\ln(t+\sqrt{t^2-1})}{\sqrt{t^2-1}}dt \quad z>0 \tag{2.12.21}$$

$$Y_\nu(z) = \frac{1}{\pi}\int_0^\pi \sin(z\sin t - \nu t)\,dt - \frac{1}{\pi}\int_0^\infty e^{-z\sinh t}\,[e^{\nu t} + \cos(\pi\nu)\,e^{-\nu t}]\,dt \quad ; \quad \mathrm{Re}\,z > 0 \tag{2.12.22}$$

$$Y_\nu(z) = \frac{2\left(\frac{z}{2}\right)^\nu}{\sqrt{\pi}\,\Gamma\left(\nu+\frac{1}{2}\right)}\left[\int_0^{\pi/2} \sin(z\,\sin t)(\cos t)^{2\nu}dt - \int_0^\infty e^{-z\sinh t}(\cosh t)^{2\nu}dt\right]$$

$$\mathrm{Re}\,\nu > -\frac{1}{2} \quad ; \quad \mathrm{Re}\,z > 0 \tag{2.12.23}$$

$$Y_\nu(z) = \frac{2\left(\frac{z}{2}\right)^\nu}{\sqrt{\pi}\,\Gamma\left(\nu+\frac{1}{2}\right)}\left[\int_0^1 (1-t^2)\sin(zt)\,dt - \int_0^\infty e^{-zt}(1+t^2)^{\nu-1/2}dt\right] \tag{2.12.24}$$

$\mathrm{Re}\,z > 0 \quad ; \quad \mathrm{Re}\,\nu > -\frac{1}{2}$

$$I_0(z)=\frac{2}{\pi}\int_0^{\pi/2}\cosh(z\cos t)\,dt \tag{2.12.25}$$

$$I_0(z)=\frac{1}{\pi}\int_0^{\pi}\cosh(z\cos t)\,dt \tag{2.12.26}$$

$$I_0(z)=\frac{2}{\pi}\int_0^{1}\frac{\cosh(zt)}{\sqrt{1-t^2}}\,dt \tag{2.12.27}$$

$$I_0(z)=\frac{1}{\pi}\int_{-1}^{1}\frac{e^{-zt}}{\sqrt{1-t^2}}\,dt \tag{2.12.28}$$

$$I_0(z)=\frac{2z}{\pi}\int_0^{1}\sqrt{1-t^2}\cosh(zt)\,dt \tag{2.12.29}$$

$$I_n(z)=\frac{1}{\pi}\int_0^{\pi}e^{z\cos t}\cos(nt)\,dt \quad ; \quad n=0,1,2,3,\ldots \tag{2.12.30}$$

$$I_\nu(z)=\frac{2\left(\frac{z}{2}\right)^\nu}{\sqrt{\pi}\,\Gamma\left(\nu+\frac{1}{2}\right)}\int_0^{\pi/2}\cosh(z\cos t)(\sin t)^{2\nu}dt \qquad \operatorname{Re}\nu>-\frac{1}{2} \tag{2.12.31}$$

$$I_\nu(z)=\frac{\left(\frac{z}{2}\right)^\nu}{\sqrt{\pi}\,\Gamma\left(\nu+\frac{1}{2}\right)}\int_0^{\pi}\cosh(z\cos t)(\sin t)^{2\nu}dt \quad ; \quad \operatorname{Re}\nu>-\frac{1}{2} \tag{2.12.32}$$

$$I_\nu(z)=\frac{1}{\pi}\int_0^{\pi}e^{z\cos t}\cos(\nu t)\,dt-\frac{\sin(\pi\nu)}{\pi}\int_0^{\infty}e^{-(z\cosh t+\nu t)}dt \qquad \operatorname{Re}\nu>0 \tag{2.12.33}$$

$$I_\nu(z)=\frac{\left(\frac{z}{2}\right)^\nu}{\sqrt{\pi}\,\Gamma\left(\nu+\frac{1}{2}\right)}\int_{-1}^{1}e^{-zt}(1-t^2)^{\nu-1/2}dt \quad ; \qquad \operatorname{Re}\nu>-\frac{1}{2} \tag{2.12.34}$$

$$I_\nu(z)=\frac{\left(\frac{z}{2}\right)^\nu}{\sqrt{\pi}\,\Gamma\left(\nu+\frac{1}{2}\right)}\int_{-1}^{1}e^{-zt}(1-t^2)^{\nu-1/2}\cosh(zt)dt \qquad \operatorname{Re}\nu>-\frac{1}{2} \tag{2.12.35}$$

$$I_{-\nu}(z) = I_\nu(z) + \frac{2\left(\frac{z}{2}\right)^\nu \sin(\pi\nu)}{\sqrt{\pi}\,\Gamma\left(\nu+\frac{1}{2}\right)} \int_1^\infty \left(t^2-1\right)^{\nu-1/2} e^{-zt}\,dt$$

$$\operatorname{Re}\nu > -\frac{1}{2} \tag{2.12.36}$$

$$I_{-\nu}(z) = I_\nu(z) + \frac{2\sin(\pi\nu)}{\pi} \int_0^\infty e^{-z\cosh t} \cosh(\nu t)\,dt \quad ; \quad \operatorname{Re}\nu > 0 \tag{2.12.37}$$

$$I_{-\nu}(z) = \frac{2\left(\frac{z}{2}\right)^\nu}{\sqrt{\pi}\,\Gamma\left(\nu+\frac{1}{2}\right)} \left[\frac{1}{2}\int_{-1}^{1} e^{zt}\left(1-t^2\right)^{\nu-1/2} dt + \sin(\pi\nu)\int_1^\infty e^{-zt}\left(t^2-1\right)^{\nu-1/2} dt\right]$$

$$\operatorname{Re} z > 0 \quad ; \quad \operatorname{Re}\nu > -\frac{1}{2} \tag{2.12.38}$$

$$K_0(z) = \int_0^\infty \frac{e^{-z\sqrt{1+t^2}}}{\sqrt{1+t^2}}\,dt \quad ; \quad z > 0 \tag{2.12.39}$$

$$K_0(z) = \int_0^\infty \frac{\cos(zt)}{\sqrt{1+t^2}}\,dt \tag{2.12.40}$$

$$K_0(z) = \int_0^\infty \cos(z\sinh t)\,dt \quad ; \quad z > 0 \tag{2.12.41}$$

$$K_0(z) = \int_0^\infty \frac{t\,J_0(zt)}{1+t^2}\,dt \tag{2.12.42}$$

$$K_0(z) = \int_0^\infty \frac{Y_0(zt)}{1+t^2}\,dt \tag{2.12.43}$$

$$K_0(z) = z\int_0^\infty t\,e^{-z\cosh t} \sinh t\,dt \tag{2.12.44}$$

$$K_\nu(z) = \int_0^\infty e^{-z\cosh t} \cosh(\nu t)\,dt \quad ; \quad \operatorname{Re} z > 0 \tag{2.12.45}$$

$$K_\nu(z) = \frac{1}{2}\int_{-\infty}^\infty e^{-z\cosh t - \nu t}\,dt \quad ; \quad \operatorname{Re} z > 0 \tag{2.12.46}$$

$$K_\nu(z) = \frac{\sqrt{\pi}\,\left(\frac{z}{2}\right)^\nu}{\Gamma\left(\nu+\frac{1}{2}\right)} \int_0^\infty e^{-z\cosh t}(\sinh t)^{2\nu}\,dt \tag{2.12.47}$$

$$\operatorname{Re} z > 0 \quad ; \quad \operatorname{Re}\nu > -\frac{1}{2}$$

$$K_\nu(z) = \frac{\Gamma\left(\nu+\frac{1}{2}\right)\left(\frac{2}{z}\right)^\nu}{\sqrt{\pi}} \int_0^\infty \frac{\cos(z\sinh t)}{(\cosh t)^{2\nu}}\,dt \tag{2.12.48}$$

$$\operatorname{Re} z > 0 \quad ; \quad \operatorname{Re}\nu > -\frac{1}{2}$$

$$K_\nu(z) = \frac{\Gamma\left(\nu+\frac{1}{2}\right)(2z)^\nu}{\sqrt{\pi}} \int_0^\infty \frac{\cos t}{(1+t^2)^{\nu+1/2}}\,dt \tag{2.12.49}$$

$$\operatorname{Re} z > 0 \quad ; \quad \operatorname{Re}\nu > -\frac{1}{2}$$

$$K_\nu(z) = \frac{1}{2}\int_0^\infty t^{\nu-1} e^{-z(t+1/t)/2}\,dt \quad ; \quad \operatorname{Re} z > 0 \quad ; \quad \operatorname{Re}\nu > -\frac{1}{2} \tag{2.12.50}$$

$$K_\nu(z) = \frac{1}{2}\left(\frac{z}{2}\right)^\nu \int_0^\infty t^{-(\nu+1)}\, e^{-(t+z^2/4t)}\,dt \tag{2.12.51}$$

$$\operatorname{Re} z > 0 \quad ; \quad \operatorname{Re}\nu > -\frac{1}{2}$$

$$K_\nu(z) = \frac{\sqrt{\pi}\,\left(\frac{z}{2}\right)^\nu}{\Gamma\left(\nu+\frac{1}{2}\right)} \int_1^\infty (t^2-1)^{\nu-1/2} e^{-zt}\,dt \tag{2.12.52}$$

$$\operatorname{Re} z > 0 \quad ; \quad \operatorname{Re}\nu > -\frac{1}{2}$$

$$K_\nu(z) = \frac{\sqrt{\frac{\pi}{2z}}\, e^{-z}}{\Gamma\left(\nu+\frac{1}{2}\right)} \int_0^\infty t^{\nu-1/2}\left(1+\frac{t}{2z}\right) e^{-t}\,dt \tag{2.12.53}$$

$$\operatorname{Re} z > 0 \quad ; \quad \operatorname{Re}\nu > -\frac{1}{2}$$

$$K_\nu(z) = \frac{\sqrt{\frac{\pi}{2}}\, z^\nu e^{-z}}{\Gamma\left(\nu+\frac{1}{2}\right)} \int_0^\infty \left(t+\frac{t^2}{2}\right)^{\nu-1/2} e^{-zt}\,dt \tag{2.12.54}$$

$$\operatorname{Re} z > 0 \quad ; \quad \operatorname{Re}\nu > -\frac{1}{2}$$

$$K_\nu(z) = \frac{\sqrt{\pi}\,\left(\frac{z}{2}\right)^\nu e^{-z}}{\Gamma\left(\nu+\frac{1}{2}\right)} \int_0^1 t^{z-1}[(\ln t)^2 - 2\ln t]^{\nu-1/2}\,dt \tag{2.12.55}$$

$$\mathbf{H}_0(z)=\frac{2}{\pi}\int_0^1\frac{\sin(zt)}{\sqrt{1-t^2}}dt \tag{2.12.56}$$

$$\mathbf{H}_0(z)=\frac{1}{\pi}Y_0(z)+\frac{2}{\pi}\int_0^\infty e^{-z\sinh t}\,dt \tag{2.12.57}$$

$$\mathbf{H}_\nu(z)=\frac{2\left(\frac{z}{2}\right)^\nu}{\sqrt{\pi}\,\Gamma\left(\nu+\frac{1}{2}\right)}\int_0^{\pi/2}\sin(z\cos t)(\sin t)^{2\nu}dt \quad ; \quad \operatorname{Re}\nu>-\frac{1}{2} \tag{2.12.58}$$

$$\mathbf{H}_\nu(z)=\frac{2\left(\frac{z}{2}\right)^\nu}{\sqrt{\pi}\,\Gamma\left(\nu+\frac{1}{2}\right)}\int_0^1(1-t^2)^{\nu-1/2}\sin(zt)dt \quad ; \quad \operatorname{Re}\nu>-\frac{1}{2} \tag{2.12.59}$$

$$\mathbf{H}_\nu(z)=Y_\nu(z)+\frac{2\left(\frac{z}{2}\right)^\nu}{\sqrt{\pi}\,\Gamma\left(\nu+\frac{1}{2}\right)}\int_0^\infty e^{-zt}(1+t^2)^{\nu-1/2}dt \quad ; \quad \operatorname{Re}\nu>0 \tag{2.12.60}$$

$$\mathbf{L}_0(z)=I_0(z)-\frac{2}{\pi}\int_0^\infty\sin(z\sinh t)\,dt \tag{2.12.61}$$

$$\mathbf{L}_0(z)=I_0(z)-\frac{2}{\pi}\int_0^\infty\frac{\sin(zt)}{\sqrt{1+t^2}}dt \tag{2.12.62}$$

$$\mathbf{L}_\nu(z)=\frac{2\left(\frac{z}{2}\right)^\nu}{\sqrt{\pi}\,\Gamma\left(\nu+\frac{1}{2}\right)}\int_0^{\pi/2}\sinh(z\cos t)(\sin t)^{2\nu}dt$$

$$\operatorname{Re}\nu>-\frac{1}{2} \tag{2.12.63}$$

$$\mathbf{L}_\nu(z)=I_\nu(z)-\frac{2\left(\frac{z}{2}\right)^\nu}{\sqrt{\pi}\,\Gamma\left(\nu+\frac{1}{2}\right)}\int_0^1 e^{-zt}(1-t^2)^{\nu-1/2}dt$$

$$\operatorname{Re}\nu>-\frac{1}{2} \tag{2.12.64}$$

$$\mathbf{J}_\nu(z)=J_\nu(z)+\frac{\sin(\pi\nu)}{\pi}\int_0^\infty e^{-z\sinh t-\nu t}dt \quad ; \quad \operatorname{Re}z>0 \tag{2.12.65}$$

$$\mathbf{E}_\nu(z)=-Y_\nu(z)-\frac{1}{\pi}\int_0^\infty\left(e^{\nu t}+e^{-\nu t}\cos(\pi\nu)\right)e^{-z\sinh t}dt$$

$$\operatorname{Re}z>0 \tag{2.12.66}$$

$$s_{-1,\nu}(z) = -\frac{1}{\nu\sin(\pi\nu)}\int_0^{\pi}\cos(z\sin t)\cos(\nu t)\,dt \tag{2.12.67}$$

$$s_{-1,\nu}(z) = -\frac{1}{\nu[1-\cos(\pi\nu)]}\int_0^{\pi}\cos(z\sin t)\sin(\nu t)\,dt \tag{2.12.68}$$

$$s_{-1,\nu}(z) = -\frac{1}{\nu\sin\left(\frac{\pi\nu}{2}\right)}\int_0^{\pi/2}\cos(z\cos t)\cos(\nu t)\,dt \tag{2.12.69}$$

$$s_{0,\nu}(z) = \frac{1}{\sin(\pi\nu)}\int_0^{\pi}\sin(z\sin t)\cos(\nu t)\,dt \tag{2.12.70}$$

$$s_{0,\nu}(z) = \frac{1}{\nu\cos\left(\frac{\pi\nu}{2}\right)}\int_0^{\pi/2}\cos(z\cos t)\cos(\nu t)\,dt \tag{2.12.71}$$

$$s_{0,\nu}(z) = -\frac{1}{[1+\cos(\pi\nu)]}\int_0^{\pi}\sin(z\sin t)\cos(\nu t)\,dt \tag{2.12.72}$$

$$s_{\nu,\nu}(z) = \sqrt{\pi}\,z^{\nu}\int_0^{\pi/2}(\cos t)^{2\nu}\sin(z\sin t)\,dt \quad ; \quad \mathrm{Re}\nu > -\frac{1}{2} \tag{2.12.73}$$

$$s_{\mu-1/2,1/2}(z) = \frac{1}{\sqrt{z}}\int_0^{z}t^{\mu-1}\sin(z-t)\,dt \quad ; \quad \mu > 0 \tag{2.12.74}$$

$$s_{\mu,\nu}(z) = 2^{\mu}\Gamma\left(\frac{\mu-\nu+1}{2}\right)\left(\frac{z}{2}\right)^{(\mu+\nu+1)/2} I(z)$$

$$I(z) = \int_0^{\pi/2}(\cos t)^{\lambda+\nu}(\sin t)^{(\nu-\mu+1)/2}J_{(\mu-\nu+1)/2}(z\sin t)\,dt \tag{2.12.75}$$

$$\mathrm{Re}(\mu+\nu) > -1$$

$$S_{-2,1}(z) = \frac{1}{2}\int_0^{\infty}t e^{-z\sinh t}\sinh t\,dt = \frac{1}{2}\int_0^{\infty}\frac{t e^{-zt}\sinh^{-1}t}{\sqrt{1+t^2}}\,dt \tag{2.12.76}$$

$$\begin{aligned} S_{-1,0}(z) &= \int_0^{\infty}t e^{-z\sinh t}\,dt \\ &= \int_0^{\infty}\frac{e^{-zt}\sinh^{-1}t}{\sqrt{1+t^2}}\,dt = \frac{z}{2}\int_0^{\infty}e^{-zt}[\sinh^{-1}t]^2\,dt \end{aligned} \tag{2.12.77}$$

$$S_{0,1/3}(z) = \frac{\sqrt{3}z}{\pi} \int_0^\infty \frac{K_{1/3}(t)}{z^2+t^2}\, dt \tag{2.12.78}$$

$$S_{0,\nu}(z) = \int_0^\infty e^{-z\sinh t} \cosh(\nu t)\, dt \quad ; \quad \operatorname{Re} z > 0 \tag{2.12.79}$$

$$S_{0,\nu}(z) = \frac{z}{\nu} \int_0^\infty e^{-z\sinh t} \sinh(\nu t)\, \cosh t\, dt \quad ; \quad \operatorname{Re} z > 0 \tag{2.12.80}$$

$$S_{1,\nu}(z) = z \int_0^\infty e^{-z\sinh t} \cosh(\nu t)\, \cosh t\, dt \quad ; \quad \operatorname{Re} z > 0 \tag{2.12.81}$$

$$S_{\nu,\nu}(z) = z^\nu \int_0^\infty e^{-z\sinh t} (\cosh t)^{2\nu} dt \quad ; \quad \operatorname{Re} z > 0 \tag{2.12.82}$$

$$S_{\mu,\nu}(z) = \frac{(2z)^{\mu+1}}{\Gamma\left(\frac{1-\mu-\nu}{2}\right)\Gamma\left(\frac{1-\mu+\nu}{2}\right)} \int_0^\infty \frac{x^{-\mu} K_\nu(x)}{z^2+t^2}\, dt \qquad \operatorname{Re}(\mu \pm \nu) < 1 \tag{2.12.83}$$

$$S_{\mu,\nu}(z) = z^{\mu+1} \int_0^\infty t e^{-zt} {}_2F_1\left(\frac{1-\mu+\nu}{2}, \frac{1-\mu-\nu}{2}; \frac{3}{2}; -t^2\right) dt \qquad \operatorname{Re} z > 0 \tag{2.12.84}$$

$$\operatorname{ber}(\sqrt{2}z) = \frac{2}{\pi} \int_0^1 \frac{\cos(zt)\cosh(zt)}{\sqrt{1-t^2}}\, dt \tag{2.12.85}$$

$$\operatorname{ber}_n(\sqrt{2}z) = \frac{(-1)^n}{\pi} \int_0^\pi \cos(z\sin t - nt) \cosh(z \sin t)\, dt \qquad n = 0, 1, 2, 3, \ldots \tag{2.12.86}$$

$$\begin{aligned} \operatorname{ber}_\nu(\sqrt{2}z) = \frac{1}{\pi} \int_0^\pi & [\cos(\pi\nu)\, \cos(z\sin t - \nu t)\, \cosh(z\sin t) \\ & - \sin(\pi\nu)\, \sin(z\sin t - \nu t)\, \sinh(z\sin t)]\, dt \\ & - \frac{\sin(\pi\nu)}{\pi} \int_0^\infty e^{-\nu t - z\sinh t} \cos(z\sinh t + \pi\nu)\, dt \end{aligned} \tag{2.12.87}$$

$$\mathrm{bei}(\sqrt{2}z) = \frac{2}{\pi}\int_0^1 \frac{\sin(zt)\sinh(zt)}{\sqrt{1-t^2}}\,dt \tag{2.12.88}$$

$$\mathrm{bei}_n(\sqrt{2}z) = \frac{(-1)^n}{\pi}\int_0^\pi \sin(z\sin t - nt)\sinh(z\sin t)\,dt \qquad n = 0, 1, 2, 3, \ldots \tag{2.12.89}$$

$$\begin{aligned}\mathrm{bei}_\nu(\sqrt{2}z) &= \frac{1}{\pi}\int_0^\pi [\cos(\pi\nu)\ \sin(z\sin t - \nu t)\ \sinh(z\sin t) \\ &\quad + \sin(\pi\nu)\ \cos(z\sin t - \nu t)\ \cosh(z\sin t)]\,dt \\ &\quad - \frac{\sin(\pi\nu)}{\pi}\int_0^\infty e^{-\nu t - z\sinh t}\sin(z\sinh t + \pi\nu)\,dt\end{aligned} \tag{2.12.90}$$

$$\mathrm{ker}(\sqrt{2}z) = \int_0^\infty e^{-z\cosh t}\cos(z\cosh t)dt \tag{2.12.91}$$

$$\mathrm{ker}(z) = -\frac{z^2}{8}\int_0^\infty e^{-z^2t/4}\,Ci\left(\frac{1}{t}\right)\,dt \tag{2.12.92}$$

$$\mathrm{ker}(z) = \int_0^\infty \frac{t^3 J_0(zt)}{1+t^4}\,dt \tag{2.12.93}$$

$$\mathrm{ker}(z) = \frac{z}{4}\int_0^\infty \ln(1+t^4)\,J_1(zt)\,dt \tag{2.12.94}$$

$$\mathrm{ker}_\nu(\sqrt{2}z) = \int_0^\infty e^{-z\cosh t}\cos\left(z\cosh t + \frac{\pi\nu}{2}\right)\cosh(\nu t)dt \tag{2.12.95}$$

$$\mathrm{kei}(\sqrt{2}z) = -\int_0^\infty e^{-z\cosh t}\sin(z\cosh t)dt \tag{2.12.96}$$

$$\mathrm{kei}(z) = \frac{z^2}{8}\int_0^\infty e^{-z^2t/4}\left[Si\left(\frac{1}{t}\right) - \frac{\pi}{2}\right]dt \tag{2.12.97}$$

$$\mathrm{kei}(z) = -\int_0^\infty \frac{t\,J_0(zt)}{1+t^4}\,dt \tag{2.12.98}$$

$$\mathrm{kei}(z) = -\frac{z}{2}\int_0^\infty \tan^{-1}(t^2)\,J_1(zt)\,dt \tag{2.12.99}$$

$$\mathrm{kei}_\nu(\sqrt{2}z) = -\int_0^\infty e^{-z\cosh t}\sin\left(z\cosh t + \frac{\pi\nu}{2}\right)\cosh(\nu t)dt \tag{2.12.100}$$

$$\mathrm{Ji}_0(z) = \frac{2}{\pi}\int_0^\infty si(z\cosh t)dt \tag{2.12.101}$$

$$\mathrm{Ji}_0(z) = \gamma + \ln\left(\frac{z}{2}\right) + \int_0^z \frac{J_0(t)-1}{t}dt \tag{2.12.102}$$

$$\mathrm{Ji}_0(z) = \gamma + J_0(z)\ln z - \ln 2 + \int_0^z J_0(t)\ln t\,dt \tag{2.12.103}$$

$$\mathrm{Ji}_{2n}(z) = \frac{1}{\pi}\int_0^\pi \cos(2nt)\,ci(z\sin t)\,dt \tag{2.12.104}$$

$$\mathrm{Ji}_{2n+1}(z) = \frac{1}{\pi}\int_0^\pi \sin[(2n+1)t]\,si(z\sin t)\,dt \tag{2.12.105}$$

$$\mathrm{Yi}_0(z) = -\frac{2}{\pi}\int_0^\infty ci(z\cosh t)dt \tag{2.12.106}$$

$$\mathrm{Ki}_\nu(z) = \frac{1}{\Gamma(\nu)}\int_z^\infty (t-z)^{\nu-1}K_0(t)\,dt \qquad \mathrm{Re}\,z>0 \;;\; \mathrm{Re}\,\nu>0 \tag{2.12.107}$$

$$\mathrm{Ki}_\nu(z) = \int_0^\infty \frac{e^{-z\cosh t}}{(\cosh t)^\nu}dt \;;\; \mathrm{Re}\,z>0 \;;\; \mathrm{Re}\,\nu>0 \tag{2.12.108}$$

2.13 Differential Equations Reducible to the Bessel Differential Equations

Differential equations presented here were collected from many sources in the literature, mainly from [7–9, 13, 14, 16, 18, 20, 21, 23, 24, 30–32]. For the readers convenience, this list of differential equations includes not only these which are of less or more general character, but also particular cases having only different coefficients. Most of the solutions expressible in terms of the Bessel functions are for differential equations that contain power functions, much less exist for exponential or other elementary functions.

The following notation is used to express solutions of differential equations $C_\nu(z) = A_1 J_\nu(z) + A_2 Y_\nu(z)$ and $Z_\nu(z) = A_1 I_\nu(z) + A_2 K_\nu(z)$

$$\frac{d^2w(z)}{dz^2} + z\,w(z) = 0 \quad ; \quad w(z) = \sqrt{z}\, C_{1/3}\left(\frac{2}{3} z^{3/2}\right) \tag{2.13.1}$$

$$\frac{d^2w(z)}{dz^2} - z\,w(z) = 0 \quad ; \quad w(z) = \sqrt{z}\, C_{1/3}\left(\frac{2}{3} i z^{3/2}\right)$$
$$w(z) = A_1 \mathrm{Ai}(z) + A_2 \mathrm{Bi}(z) \tag{2.13.2}$$

$$\frac{d^2w(z)}{dz^2} + \alpha^2 z^2 w(z) = 0 \quad ; \quad w(z) = \sqrt{z}\, C_{1/4}\left(\frac{\alpha}{2} z^2\right) \tag{2.13.3}$$

$$\frac{d^2w(z)}{dz^2} + \alpha^4 z^4 w(z) = 0 \quad ; \quad w(z) = \sqrt{z}\, C_{1/6}\left(\frac{\alpha^2 z^3}{3}\right) \tag{2.13.4}$$

$$\frac{d^2w(z)}{dz^2} + \frac{1}{\sqrt{z}} w(z) = 0 \quad ; \quad w(z) = \sqrt{z}\, C_{2/3}\left(\frac{4}{3} z^{3/4}\right) \tag{2.13.5}$$

$$\frac{d^2w(z)}{dz^2} - \frac{1}{\sqrt{z}} w(z) = 0 \quad ; \quad w(z) = \sqrt{z}\, C_{2/3}\left(\frac{4}{3} i z^{3/4}\right) \tag{2.13.6}$$

$$\frac{d^2w(z)}{dz^2} + \frac{1}{z} w(z) = 0 \quad ; \quad w(z) = \sqrt{z}\, C_1(2\sqrt{z}) \tag{2.13.7}$$

$$\frac{d^2w(z)}{dz^2} - \frac{1}{z} w(z) = 0 \quad ; \quad w(z) = \sqrt{z}\, Z_1(2\sqrt{z}) \tag{2.13.8}$$

$$\frac{d^2w(z)}{dz^2} + \alpha^2 z^{-3/2} w(z) = 0 \quad ; \quad w(z) = \sqrt{z}\, C_2(4\alpha z^{1/4}) \tag{2.13.9}$$

$$\frac{d^2w(z)}{dz^2} + \alpha^2 z^{\beta+2} w(z) = 0 \quad ; \quad w(z) = \sqrt{z}\, C_{1/(\beta+2)}\left(\frac{2\alpha z^{\beta/2+1}}{\beta+2}\right) \tag{2.13.10}$$

$$\frac{d^2w(z)}{dz^2} + \alpha^2 z^{2-\beta} w(z) = 0 \quad ; \quad w(z) = \sqrt{z}\, C_{1/(2-\beta)}\left(\frac{2\alpha z^{1-\beta/2}}{2-\beta}\right) \tag{2.13.11}$$

$$\frac{d^2w(z)}{dz^2} - \alpha^2 z^{2\beta-2} w(z) = 0 \quad ; \quad w(z) = \sqrt{z}\, C_{1/(2\beta-2)}\left(\frac{i\alpha z^{\beta-1/2}}{\beta-1}\right) \tag{2.13.12}$$

$$z^2 \frac{d^2w(z)}{dz^2} + \left(\alpha^2 z + \frac{1}{4}\right) w(z) = 0 \quad ; \quad w(z) = \sqrt{z}\, C_0(2\alpha\sqrt{z}) \tag{2.13.13}$$

$$z^2 \frac{d^2w(z)}{dz^2} + \left(\alpha^2 z^2 - \frac{3}{4}\right) w(z) = 0 \quad ; \quad w(z) = z\, C_1(\alpha z) \tag{2.13.14}$$

$$z^2 \frac{d^2w(z)}{dz^2} + (\alpha^2 z^2 - 2) w(z) = 0 \quad ; \quad w(z) = \frac{1}{z}\, C_{3/2}(\alpha z) \tag{2.13.15}$$

$$z^2 \frac{d^2w(z)}{dz^2} + \left(\alpha^2 z^2 - \frac{15}{4}\right) w(z) = 0 \quad ; \quad w(z) = \sqrt{z}\, C_2(\alpha z) \tag{2.13.16}$$

$$z^2 \frac{d^2w(z)}{dz^2} + (\alpha^2 z + \beta) w(z) = 0 \quad ; \quad w(z) = \sqrt{z}\, C_{\sqrt{1-4\beta}}(2\alpha\sqrt{z}) \tag{2.13.17}$$

$$z^2 \frac{d^2w(z)}{dz^2} + \left(\alpha^2 z^2 - \alpha^2 + \frac{1}{4}\right) w(z) = 0 \quad ; \quad w(z) = \sqrt{z}\, C_\beta(\alpha z) \tag{2.13.18}$$

$$z^2 \frac{d^2w(z)}{dz^2} - [\alpha^2 z^2 + \beta(1-\beta)]\, w(z) = 0$$

$$w(z) = \sqrt{z}\, C_{\beta-1/2}(i\alpha z) \tag{2.13.19}$$

$$z^2 \frac{d^2w(z)}{dz^2} + \left(\alpha^2 z^3 + \frac{1}{4}\right) w(z) = 0 \quad ; \quad w(z) = \sqrt{z}\, C_0\left(\frac{2\alpha}{3} z^{3/2}\right) \tag{2.13.20}$$

$$z^2 \frac{d^2w(z)}{dz^2} + (\alpha^2 z^3 - \beta) w(z) = 0 \quad ; \quad w(z) = \sqrt{z}\, C_{\frac{1}{3}\sqrt{1+4\beta}}\left(\frac{2\alpha}{3} z^{3/2}\right) \tag{2.13.21}$$

$$z^2 \frac{d^2w(z)}{dz^2} + (\alpha^2 z^4 + 4) w(z) = 0 \quad ; \quad w(z) = \sqrt{z}\, C_0\left(\frac{\alpha}{2} z^2\right) \tag{2.13.22}$$

$$z^2 \frac{d^2w(z)}{dz^2} + \left(\alpha^2 z^4 - \frac{\beta}{4}\right) w(z) = 0 \quad ; \quad w(z) = \sqrt{z}\, C_{\frac{1}{2}\sqrt{1+\beta}}\left(\frac{\alpha}{2} z^2\right) \tag{2.13.23}$$

$$z^2 \frac{d^2w(z)}{dz^2} + (\alpha^2 z^5 - 2) w(z) = 0 \quad ; \quad w(z) = \sqrt{z}\, C_{3/5}\left(\frac{2\alpha}{5} z^{5/2}\right) \tag{2.13.24}$$

$$z^2 \frac{d^2w(z)}{dz^2} + \left(\frac{\alpha^2}{z} + \frac{1}{4}\right) w(z) = 0 \quad ; \quad w(z) = \sqrt{z}\, C_0\left(\frac{2\alpha}{\sqrt{z}}\right) \tag{2.13.25}$$

$$z^2 \frac{d^2w(z)}{dz^2} + \left(\alpha^2 z^3 + \frac{1}{4}\right) w(z) = 0 \quad ; \quad w(z) = \sqrt{z}\, C_0\left(\frac{2\alpha}{3} z^{3/2}\right) \tag{2.13.26}$$

$$z^2 \frac{d^2w(z)}{dz^2} + \left(\alpha^2 z^{n+2} - \frac{n^2 + 2n}{4}\right) w(z) = 0$$

$$w(z) = \sqrt{z}\, C_{(n+1)/(n+2)}\left(\frac{2\alpha}{n+2} z^{n/2+1}\right) \tag{2.13.27}$$

$$z^2 \frac{d^2w(z)}{dz^2} + \left(\alpha^2 z^{n+2} - \frac{n^2 + 4n + 3}{4}\right) w(z) = 0$$

$$w(z) = \sqrt{z}\, C_1\left(\frac{2\alpha}{n+2} z^{n/2+1}\right) \tag{2.13.28}$$

$$z^2\frac{d^2w(z)}{dz^2}+\left(\alpha^2 z^{\beta+2}+\frac{1}{4}\right)w(z)=0$$
$$w(z)=\sqrt{z}\,C_0\left(\frac{2\alpha}{\beta+2}z^{\beta/2+1}\right) \tag{2.13.29}$$

$$z^2\frac{d^2w(z)}{dz^2}+(\alpha^2 z^{\beta+2}+\lambda^2)w(z)=0 \quad ; \quad \alpha\neq 0 \quad ; \quad \beta\neq -2$$
$$w(z)=\sqrt{z}\,C_{\sqrt{1-4\lambda^2}/(\beta+2)}\left(\frac{2\alpha}{\beta+2}z^{(\beta+2)/2}\right) \tag{2.13.30}$$

$$z^2\frac{d^2w(z)}{dz^2}+\left(\alpha^2\beta^2 z^{\alpha}-\alpha^2\lambda^2+\frac{1}{4}\right)w(z)=0$$
$$w(z)=\sqrt{z}\,C_\lambda(\beta z^{\alpha}) \tag{2.13.31}$$

$$z^2\frac{d^2w(z)}{dz^2}+\left(\alpha^2\lambda^2 z^{\alpha}+\frac{1-\alpha^2\beta^2}{4}\right)w(z)=0$$
$$w(z)=\sqrt{z}\,C_\lambda(2\lambda z^{\alpha/2}) \tag{2.13.32}$$

$$\frac{d^2w(z)}{dz^2}+\frac{dw(z)}{dz}+\left(\alpha^2 z^n+\frac{1}{4}\right)w(z)=0$$
$$w(z)=\sqrt{z}\,e^{-z/2}C_{1/(n+2)}\left(\frac{2\alpha}{n+2}z^{n/2+1}\right) \tag{2.13.33}$$

$$\frac{d^2w(z)}{dz^2}-\frac{dw(z)}{dz}+\left(\alpha^2 z^n+\frac{1}{4}\right)w(z)=0$$
$$w(z)=\sqrt{z}\,e^{z/2}C_{1/(n+2)}\left(\frac{2\alpha}{n+2}z^{n/2+1}\right) \tag{2.13.34}$$

$$\frac{d^2w(z)}{dz^2}+2\frac{dw(z)}{dz}+(\alpha^2 z^n+1)w(z)=0$$
$$w(z)=\sqrt{z}\,e^{-z}C_{1/(n+2)}\left(\frac{2\alpha}{n+2}z^{n/2+1}\right) \tag{2.13.35}$$

$$\frac{d^2w(z)}{dz^2}-2\frac{dw(z)}{dz}+(\alpha^2 z^n+1)w(z)=0$$
$$w(z)=\sqrt{z}\,e^{z}C_{1/(n+2)}\left(\frac{2\alpha}{n+2}z^{n/2+1}\right) \tag{2.13.36}$$

$$\frac{d^2w(z)}{dz^2}+\alpha\frac{dw(z)}{dz}+\left(\frac{\alpha^2}{4}+\beta^2\right)w(z)=0$$
$$w(z)=\sqrt{z}\,e^{-\alpha z/2}C_{1/2}(\beta z) \tag{2.13.37}$$

$$\frac{d^2w(z)}{dz^2} + \alpha\frac{dw(z)}{dz} + \left(\frac{\alpha^2}{4} + \beta^2 z\right)w(z) = 0$$
$$w(z) = \sqrt{z}\,e^{-\alpha z/2}C_{1/3}\left(\frac{2\beta}{3}z^{3/2}\right) \qquad (2.13.38)$$

$$\frac{d^2w(z)}{dz^2} + \alpha\frac{dw(z)}{dz} + \left(\beta^2 z^n + \frac{\alpha^2}{4}\right)w(z) = 0$$
$$w(z) = z\,e^{-\alpha z/2}C_0\left(\frac{2\beta}{n+2}z^{n/2+1}\right) \qquad (2.13.39)$$

$$z\frac{d^2w(z)}{dz^2} + \frac{1}{2}\frac{dw(z)}{dz} + \frac{1}{4}w(z) = 0 \quad ; \quad w(z) = z^{1/4}\,C_{1/2}(\sqrt{z}) \qquad (2.13.40)$$

$$z\frac{d^2w(z)}{dz^2} - \frac{dw(z)}{dz} + z\,w(z) = 0 \quad ; \quad w(z) = z\,C_1(\sqrt{z}) \qquad (2.13.41)$$

$$z\frac{d^2w(z)}{dz^2} + \beta\frac{dw(z)}{dz} + \frac{\alpha^2}{4}w(z) = 0$$
$$w(z) = z^{(1-\beta)/2}C_{1-\beta}(\alpha\sqrt{z}) \qquad (2.13.42)$$

$$z\frac{d^2w(z)}{dz^2} - 2\beta\frac{dw(z)}{dz} - \alpha^2 z\,w(z) = 0$$
$$w(z) = z^{\beta+1/2}Z_{\beta+1/2}(\alpha z) \qquad (2.13.43)$$

$$z\frac{d^2w(z)}{dz^2} + (1+\beta)\frac{dw(z)}{dz} + w(z) = 0$$
$$w(z) = z^{-\beta/2}C_\beta(2\sqrt{z}) \qquad (2.13.44)$$

$$z\frac{d^2w(z)}{dz^2} + (1+\beta)\frac{dw(z)}{dz} - w(z) = 0$$
$$w(z) = z^{-\beta/2}Z_\beta(2\sqrt{z}) \qquad (2.13.45)$$

$$z\frac{d^2w(z)}{dz^2} + (1-\beta)\frac{dw(z)}{dz} + \frac{1}{4}w(z) = 0$$
$$w(z) = z^{\beta/2}C_\beta(\sqrt{z}) \qquad (2.13.46)$$

$$z\frac{d^2w(z)}{dz^2} + (1+2\beta)\frac{dw(z)}{dz} - z\,w(z) = 0$$
$$w(z) = z^{-\beta}Z_\beta(z) \qquad (2.13.47)$$

$$z\frac{d^2w(z)}{dz^2} + (1-2\beta)\frac{dw(z)}{dz} + z\,w(z) = 0$$
$$w(z) = z^{\beta}C_\beta(z) \qquad (2.13.48)$$

$$z^2\frac{d^2w(z)}{dz^2}+2z\frac{dw(z)}{dz}+\alpha^2 z w(z)=0 \quad ; \quad w(z)=C_1(2\alpha\sqrt{z}) \tag{2.13.49}$$

$$z^2\frac{d^2w(z)}{dz^2}+(\beta+1)z\frac{dw(z)}{dz}+\alpha^2 z w(z)=0$$
$$w(z)=z^{-\beta/2}C_\beta(2\alpha\sqrt{z}) \tag{2.13.50}$$

$$z^2\frac{d^2w(z)}{dz^2}+\frac{z}{2}\frac{dw(z)}{dz}+\left(\alpha^2 z-\frac{\beta^2}{4}+\frac{1}{16}\right)w(z)=0$$
$$w(z)=z^{1/4}\,C_\beta(2\alpha\sqrt{z}) \tag{2.13.51}$$

$$z^2\frac{d^2w(z)}{dz^2}+z\frac{dw(z)}{dz}+\left(\alpha^2 z-\frac{1}{4}\right)w(z)=0 \quad ; \quad w(z)=C_1(2\alpha\sqrt{z}) \tag{2.13.52}$$

$$z^2\frac{d^2w(z)}{dz^2}+z\frac{dw(z)}{dz}+\frac{1}{4}(z-\beta^2)w(z)=0 \quad ; \quad w(z)=C_\beta(\sqrt{z}) \tag{2.13.53}$$

$$z^2\frac{d^2w(z)}{dz^2}+z\frac{dw(z)}{dz}+\left(\alpha^2 z-\frac{\beta^2}{4}\right)w(z)=0$$
$$w(z)=C_\beta(2\alpha\sqrt{z}) \tag{2.13.54}$$

$$z^2\frac{d^2w(z)}{dz^2}-z\frac{dw(z)}{dz}+(\alpha^2 z+\beta^2)w(z)=0$$
$$w(z)=z\,C_{2\sqrt{1-\beta^2}}(2\alpha\sqrt{z}) \tag{2.13.55}$$

$$z^2\frac{d^2w(z)}{dz^2}-z\frac{dw(z)}{dz}+(\alpha^2 z+1)w(z)=0$$
$$w(z)=z\,C_0(2\alpha\sqrt{z}) \tag{2.13.56}$$

$$z^2\frac{d^2w(z)}{dz^2}-z\frac{dw(z)}{dz}+\left(\alpha^2 z+1-\frac{\beta^2}{4}\right)w(z)=0$$
$$w(z)=z\,C_\beta(2\alpha\sqrt{z}) \tag{2.13.57}$$

$$z^2\frac{d^2w(z)}{dz^2}+2z\frac{dw(z)}{dz}+\left(\alpha^2 z+\frac{1}{4}\right)w(z)=0$$
$$w(z)=\frac{1}{\sqrt{z}}\,C_0(2\alpha\sqrt{z}) \tag{2.13.58}$$

$$z^2\frac{d^2w(z)}{dz^2}+2z\frac{dw(z)}{dz}+(\alpha^2 z+\beta^2)w(z)=0$$
$$w(z)=\frac{1}{\sqrt{z}}C_{\sqrt{1-4\beta^2}}(2\alpha\sqrt{z}) \tag{2.13.59}$$

$$z^2\frac{d^2w(z)}{dz^2}+2z\frac{dw(z)}{dz}+\left(\alpha^2 z+\frac{1-\beta^2}{4}\right)w(z)=0$$
$$w(z)=\frac{1}{\sqrt{z}}C_{\beta}(2\alpha\sqrt{z}) \tag{2.13.60}$$

$$z^2\frac{d^2w(z)}{dz^2}-2z\frac{dw(z)}{dz}+\left(\alpha^2 z+\frac{9}{4}\right)w(z)=0$$
$$w(z)=z^{3/2}C_0(2\alpha\sqrt{z}) \tag{2.13.61}$$

$$z^2\frac{d^2w(z)}{dz^2}-2z\frac{dw(z)}{dz}+(\alpha^2 z+2)w(z)=0$$
$$w(z)=z^{3/2}C_1(2\alpha\sqrt{z}) \tag{2.13.62}$$

$$z^2\frac{d^2w(z)}{dz^2}-2z\frac{dw(z)}{dz}+\left(\alpha^2 z+\frac{9-\beta^2}{4}\right)w(z)=0$$
$$w(z)=z^{3/2}C_{\beta}(2\alpha\sqrt{z}) \tag{2.13.63}$$

$$z^2\frac{d^2w(z)}{dz^2}-\beta z\frac{dw(z)}{dz}+\left[\alpha^2 z+\frac{(\beta+1)^2}{4}\right]w(z)=0$$
$$w(z)=z^{(\beta+1)/2}C_0(2\alpha\sqrt{z}) \tag{2.13.64}$$

$$z^2\frac{d^2w(z)}{dz^2}-\beta z\frac{dw(z)}{dz}+(\alpha^2 z+\beta)w(z)=0$$
$$w(z)=z^{(\beta+1)/2}C_{\beta-1}(2\alpha\sqrt{z}) \tag{2.13.65}$$

$$z^2\frac{d^2w(z)}{dz^2}+(\beta+1)z\frac{dw(z)}{dz}+\left(\alpha^2 z+\frac{\beta^2}{4}\right)w(z)=0$$
$$w(z)=z^{-\beta/2}C_0(2\alpha\sqrt{z}) \tag{2.13.66}$$

$$z^2\frac{d^2w(z)}{dz^2}+(\beta+1)z\frac{dw(z)}{dz}+\left(z-\frac{3\beta^2}{4}\right)w(z)=0$$
$$w(z)=z^{-\beta/2}C_{2\beta}(2\sqrt{z}) \tag{2.13.67}$$

$$z^2 \frac{d^2w(z)}{dz^2} + (\beta+1)z\frac{dw(z)}{dz} + \left(\alpha^2 z + \frac{\beta^2-1}{4}\right)w(z) = 0$$

$$w(z) = z^{-\beta/2} C_1(2\alpha\sqrt{z}) \tag{2.13.68}$$

$$z^2 \frac{d^2w(z)}{dz^2} - (\beta+1)z\frac{dw(z)}{dz} + (\alpha^2 z + \lambda^2)w(z) = 0$$

$$w(z) = z^{\beta/2+1} C_{\sqrt{(\beta+2)^2-4\lambda^2}}(2\alpha\sqrt{z}) \tag{2.13.69}$$

$$z^2 \frac{d^2w(z)}{dz^2} - (\beta+1)z\frac{dw(z)}{dz} + (\alpha^2 z + \beta + 1)w(z) = 0$$

$$w(z) = z^{\beta/2+1} C_\nu(2\alpha\sqrt{z}) \tag{2.13.70}$$

$$z^2 \frac{d^2w(z)}{dz^2} - (\beta+1)z\frac{dw(z)}{dz} + \left[\alpha^2 z + \frac{(\beta+2)^2}{4}\right]w(z) = 0$$

$$w(z) = z^{\beta/2+1} C_0(2\alpha\sqrt{z}) \tag{2.13.71}$$

$$z^2 \frac{d^2w(z)}{dz^2} + (\beta+1)\, z\frac{dw(z)}{dz} + \alpha^2 z^2 w(z) = 0$$

$$w(z) = z^{-\beta} C_{\beta/2}(\alpha z) \tag{2.13.72}$$

$$z^2 \frac{d^2w(z)}{dz^2} + (2\beta+1)z\frac{dw(z)}{dz} + \alpha^2 z^2 w(z) = 0 \quad ; \quad w(z) = C_\beta(\alpha z) \tag{2.13.73}$$

$$z^2 \frac{d^2w(z)}{dz^2} + \frac{z}{2}\frac{dw(z)}{dz} + \left(\alpha^2 z^2 + \frac{1}{16}\right)w(z) = 0$$

$$w(z) = z^{1/4}\, C_0(\alpha z) \tag{2.13.74}$$

$$z^2 \frac{d^2w(z)}{dz^2} + \frac{z}{2}\frac{dw(z)}{dz} + \left(\alpha^2 z^2 - \beta^2 + \frac{1}{16}\right)w(z) = 0$$

$$w(z) = z^{1/4}\, C_\beta(\alpha z) \tag{2.13.75}$$

$$z^2 \frac{d^2w(z)}{dz^2} + z\frac{dw(z)}{dz} + 4\,(z^2 - 1)w(z) = 0$$

$$w(z) = C_2(2z) \tag{2.13.76}$$

$$z^2 \frac{d^2w(z)}{dz^2} + z\frac{dw(z)}{dz} + \left(z^2 - \frac{1}{16}\right)w(z) = 0$$

$$w(z) = C_{1/4}(z) \tag{2.13.77}$$

$$z^2 \frac{d^2w(z)}{dz^2} + z\frac{dw(z)}{dz} - \left(3z^2 + \frac{1}{4}\right)w(z) = 0 \qquad (2.13.78)$$

$$w(z) = Z_{1/3}(\sqrt{3}z)$$

$$z^2 \frac{d^2w(z)}{dz^2} + z\frac{dw(z)}{dz} + \left(3z^2 - \frac{1}{4}\right)w(z) = 0 \quad ; \quad w(z) = C_{1/3}(\sqrt{3}z) \qquad (2.13.79)$$

$$z^2 \frac{d^2w(z)}{dz^2} + z\frac{dw(z)}{dz} + (\alpha^2 z^2 - 1)w(z) = 0 \quad ; \quad w(z) = C_1(\alpha z) \qquad (2.13.80)$$

$$z^2 \frac{d^2w(z)}{dz^2} + z\frac{dw(z)}{dz} + \left[\alpha^2 z^2 - \frac{(2n+1)^2}{4}\right]w(z) = 0 \quad ; \quad w(z) = C_{n+1/2}(\alpha z) \qquad (2.13.81)$$

$$z^2 \frac{d^2w(z)}{dz^2} + z\frac{dw(z)}{dz} + (\alpha^2 z^2 - \nu^2)w(z) = 0 \quad ; \quad w(z) = C_\nu(\alpha z) \qquad (2.13.82)$$

$$z^2 \frac{d^2w(z)}{dz^2} - z\frac{dw(z)}{dz} + (\alpha^2 z^2 + 1)\,w(z) = 0 \quad ; \quad w(z) = z\,C_0(\alpha z) \qquad (2.13.83)$$

$$z^2 \frac{d^2w(z)}{dz^2} - z\frac{dw(z)}{dz} + (\alpha^2 z^2 + \beta^2)\,w(z) = 0 \qquad (2.13.84)$$

$$w(z) = C_{\sqrt{1-\beta^2}}(\alpha z)$$

$$z^2 \frac{d^2w(z)}{dz^2} - z\frac{dw(z)}{dz} + (\alpha^2 z^2 - \beta^2 + 1)\,w(z) = 0 \qquad (2.13.85)$$

$$w(z) = z\,C_\beta(\alpha z)$$

$$z^2 \frac{d^2w(z)}{dz^2} + 2z\frac{dw(z)}{dz} + \left(\alpha^2 z^2 - \frac{1}{4}\right)w(z) = 0 \qquad (2.13.86)$$

$$w(z) = \frac{1}{\sqrt{z}}\,C_0(\alpha z)$$

$$z^2 \frac{d^2w(z)}{dz^2} + 2z\frac{dw(z)}{dz} + (\alpha^2 z^2 - 2)\,w(z) = 0 \qquad (2.13.87)$$

$$w(z) = \frac{1}{\sqrt{z}}\,C_{3/2}(\alpha z)$$

$$z^2 \frac{d^2w(z)}{dz^2} + 2z\frac{dw(z)}{dz} + (\alpha^2 z^2 - \beta^2)\,w(z) = 0 \qquad (2.13.88)$$

$$w(z) = \frac{1}{\sqrt{z}}\,C_{\frac{1}{2}\sqrt{1-4\beta^2}}(\alpha z)$$

$$z^2\frac{d^2w(z)}{dz^2}+2z\frac{dw(z)}{dz}+(\alpha^2z^2-\beta^2)\,w(z)=0$$
$$w(z)=\frac{1}{\sqrt{z}}\,C_{\frac{1}{2}\sqrt{1-4\beta^2}}(\alpha z) \tag{2.13.89}$$

$$z^2\frac{d^2w(z)}{dz^2}+2z\frac{dw(z)}{dz}+(\alpha^2z^2-4\beta^2+1)\,w(z)=0$$
$$w(z)=\frac{1}{\sqrt{z}}\,C_\beta(\alpha z) \tag{2.13.90}$$

$$z^2\frac{d^2w(z)}{dz^2}-2z\frac{dw(z)}{dz}+\beta^2(\alpha^2z^2+\beta^2)w(z)=0$$
$$w(z)=z^{3/2}\,C_{\frac{1}{2}\sqrt{9-4\beta^2}}(\alpha z) \tag{2.13.91}$$

$$z^2\frac{d^2w(z)}{dz^2}-2z\frac{dw(z)}{dz}+\beta^2\left(\alpha^2z^2-\beta^2+\frac{9}{4}\right)w(z)=0$$
$$w(z)=z^{3/2}\,C_\beta(\alpha z) \tag{2.13.92}$$

$$z^2\frac{d^2w(z)}{dz^2}+3z\frac{dw(z)}{dz}+(\alpha^2z^2+1)w(z)=0$$
$$w(z)=\frac{1}{z}C_0(\alpha z) \tag{2.13.93}$$

$$z^2\frac{d^2w(z)}{dz^2}+(\beta+1)z\frac{dw(z)}{dz}+\left(\alpha^2z^2+\frac{\beta^2}{4}\right)w(z)=0$$
$$w(z)=z^{-\beta/2}C_0(\alpha z) \tag{2.13.94}$$

$$z^2\frac{d^2w(z)}{dz^2}+(\beta+1)z\frac{dw(z)}{dz}+\left(\alpha^2z^2-\frac{3\beta^2}{4}\right)w(z)=0$$
$$w(z)=z^{-\beta/2}C_\beta(\alpha z) \tag{2.13.95}$$

$$z^2\frac{d^2w(z)}{dz^2}+(\beta+1)z\frac{dw(z)}{dz}+(\alpha^2z^2+\lambda^2)w(z)=0$$
$$w(z)=z^{-\beta/2}C_{\frac{1}{2}\sqrt{\beta^2-4\lambda^2}}(\alpha z) \tag{2.13.96}$$

$$z^2\frac{d^2w(z)}{dz^2}-(\beta+1)z\frac{dw(z)}{dz}+\left[\alpha^2z^2+\frac{(\beta+2)^2}{4}\right]w(z)=0$$
$$w(z)=z^{\beta/2+1}C_0(\alpha z) \tag{2.13.97}$$

$$z^2\frac{d^2w(z)}{dz^2}-(\beta+1)z\frac{dw(z)}{dz}+(\alpha^2z^2+\lambda^2)w(z)=0$$
$$w(z)=z^{\beta/2+1}C_{\frac{1}{2}\sqrt{(\beta+2)^2-4\lambda^2}}(\alpha z) \tag{2.13.98}$$

$$z^2\frac{d^2w(z)}{dz^2}+(1+2\beta)z\frac{dw(z)}{dz}+(\beta^2+z^2)w(z)=0$$
$$w(z)=z^{-\beta}C_0(z) \tag{2.13.99}$$

$$z^2\frac{d^2w(z)}{dz^2}+z\frac{dw(z)}{dz}+\alpha^2z^3w(z)=0\quad;\quad w(z)=C_0\left(\frac{2\alpha}{3}z^{3/2}\right) \tag{2.13.100}$$

$$z^2\frac{d^2w(z)}{dz^2}-2z\frac{dw(z)}{dz}+\alpha^2z^3w(z)=0$$
$$w(z)=z^{3/2}C_1\left(\frac{2\alpha}{3}z^{3/2}\right) \tag{2.13.101}$$

$$z^2\frac{d^2w(z)}{dz^2}+\frac{z}{2}\frac{dw(z)}{dz}+\left(\alpha^2z^3+\frac{1}{16}\right)w(z)=0$$
$$w(z)=z^{1/4}C_0\left(\frac{2\alpha}{3}z^{3/2}\right) \tag{2.13.102}$$

$$z^2\frac{d^2w(z)}{dz^2}+\frac{z}{2}\frac{dw(z)}{dz}+\left(\alpha^2z^3+\frac{1-36\beta^2}{16}\right)w(z)=0$$
$$w(z)=z^{1/4}C_\beta\left(\frac{2\alpha}{3}z^{3/2}\right) \tag{2.13.103}$$

$$z^2\frac{d^2w(z)}{dz^2}+z\frac{dw(z)}{dz}+\left(\alpha^2z^3-\frac{9}{4}\right)w(z)=0$$
$$w(z)=C_1\left(\frac{2\alpha}{3}z^{3/2}\right) \tag{2.13.104}$$

$$z^2\frac{d^2w(z)}{dz^2}+z\frac{dw(z)}{dz}+(\alpha^2z^3-\beta^2)w(z)=0$$
$$w(z)=C_{2\beta/3}\left(\frac{2\alpha}{3}z^{3/2}\right) \tag{2.13.105}$$

$$z^2\frac{d^2w(z)}{dz^2}-z\frac{dw(z)}{dz}+(\alpha^2z^3-1)w(z)=0$$
$$w(z)=z\,C_0\left(\frac{2\alpha}{3}z^{3/2}\right) \tag{2.13.106}$$

$$z^2\frac{d^2w(z)}{dz^2}-z\frac{dw(z)}{dz}+\left(\alpha^2z^3-\frac{5}{4}\right)w(z)=0$$
$$w(z)=z\,C_1\left(\frac{2\alpha}{3}z^{3/2}\right) \tag{2.13.107}$$

$$z^2\frac{d^2w(z)}{dz^2}-z\frac{dw(z)}{dz}+(\alpha^2z^3+1-\beta^2)w(z)=0$$
$$w(z)=z\,C_{2\beta/3}\left(\frac{2\alpha}{3}z^{3/2}\right) \tag{2.13.108}$$

$$z^2\frac{d^2w(z)}{dz^2}-z\frac{dw(z)}{dz}+(\alpha^2z^3+\beta^2)w(z)=0$$
$$w(z)=C_{\frac{2}{3}\sqrt{1-\beta^2}}\left(\frac{2\alpha}{3}z^{3/2}\right) \tag{2.13.109}$$

$$z^2\frac{d^2w(z)}{dz^2}+2z\frac{dw(z)}{dz}+\left(\alpha^2z^3+\frac{1}{4}\right)w(z)=0$$
$$w(z)=\frac{1}{\sqrt{z}}\,C_0\left(\frac{2\alpha}{3}z^{3/2}\right) \tag{2.13.110}$$

$$z^2\frac{d^2w(z)}{dz^2}+2z\frac{dw(z)}{dz}+(\alpha^2z^3-2)w(z)=0$$
$$w(z)=\frac{1}{\sqrt{z}}\,C_1\left(\frac{2\alpha}{3}z^{3/2}\right) \tag{2.13.111}$$

$$z^2\frac{d^2w(z)}{dz^2}+2z\frac{dw(z)}{dz}+(\alpha^2z^3+\beta^2)w(z)=0$$
$$w(z)=C_{\frac{1}{3}\sqrt{1-4\beta^2}}\left(\frac{2\alpha}{3}z^{3/2}\right) \tag{2.13.112}$$

$$z^2\frac{d^2w(z)}{dz^2}+2z\frac{dw(z)}{dz}+\left(\alpha^2z^3+\frac{1-9\beta^2}{4}\right)w(z)=0$$
$$w(z)=\frac{1}{\sqrt{z}}\,C_\beta\left(\frac{2\alpha}{3}z^{3/2}\right) \tag{2.13.113}$$

$$z^2\frac{d^2w(z)}{dz^2}-2z\frac{dw(z)}{dz}+(\alpha^2z^3+\beta^2)w(z)=0$$
$$w(z)=\sqrt{z}\,C_{\frac{1}{3}\sqrt{9-\beta^2}}\left(\frac{2\alpha}{3}z^{3/2}\right) \tag{2.13.114}$$

$$z^2\frac{d^2w(z)}{dz^2}-2z\frac{dw(z)}{dz}+\frac{9}{4}\left[\alpha^2z^3+\frac{9}{4}(1-\beta^2)\right]w(z)=0$$
$$w(z)=z^{3/2}\,C_\beta\left(\frac{2\alpha}{3}z^{3/2}\right) \tag{2.13.115}$$

$$z^2\frac{d^2w(z)}{dz^2}+(\beta+1)\,z\frac{dw(z)}{dz}+\alpha^2z^3w(z)=0$$
$$w(z)=z^{-\beta/2}C_{\beta/3}\left(\frac{2\alpha}{3}z^{3/2}\right) \tag{2.13.116}$$

$$z^2\frac{d^2w(z)}{dz^2}+(\beta+1)z\frac{dw(z)}{dz}+(\alpha^2z^3+\beta^2)w(z)=0$$
$$w(z)=z^{-\nu/2}C_{\frac{1}{3}\sqrt{\nu^2-4\beta^2}}\left(\frac{2\alpha}{3}z^{3/2}\right) \tag{2.13.117}$$

$$z^2\frac{d^2w(z)}{dz^2}+(\beta+1)z\frac{dw(z)}{dz}+(\alpha^2z^3-2\beta^2)w(z)=0$$
$$w(z)=z^{-1/2}C_\beta\left(\frac{2\alpha}{3}z^{3/2}\right) \tag{2.13.118}$$

$$z^2\frac{d^2w(z)}{dz^2}+(\beta+1)z\frac{dw(z)}{dz}+\left(\alpha^2z^3+\frac{\beta^2}{4}\right)w(z)=0$$
$$w(z)=z^{-\beta/2}C_0\left(\frac{2\alpha}{3}z^{3/2}\right) \tag{2.13.119}$$

$$z^2\frac{d^2w(z)}{dz^2}-(\beta+1)z\frac{dw(z)}{dz}+(\alpha^2z^3+\beta^2)w(z)=0$$
$$w(z)=z^{\nu/2+1}C_{\frac{1}{3}\sqrt{(\nu+2)^2-4\beta^2}}\left(\frac{2\alpha}{3}z^{3/2}\right) \tag{2.13.120}$$

$$z^2\frac{d^2w(z)}{dz^2}+z\frac{dw(z)}{dz}+4z^4w(z)=0\quad;\quad w(z)=C_0(z^2) \tag{2.13.121}$$

$$z^2\frac{d^2w(z)}{dz^2}+(\beta+1)\,z\frac{dw(z)}{dz}+\alpha^2z^4w(z)=0$$
$$w(z)=z^{-\beta/2}C_{\beta/4}\left(\frac{\alpha z^2}{2}\right) \tag{2.13.122}$$

$$z^2\frac{d^2w(z)}{dz^2}+\frac{z}{2}\frac{dw(z)}{dz}+\left(\alpha^2z^4+\frac{1}{16}\right)w(z)=0$$
$$w(z)=z^{1/4}\,C_0\left(\frac{\alpha}{2}z^2\right) \tag{2.13.123}$$

$$z^2\frac{d^2w(z)}{dz^2}+\frac{z}{2}\frac{dw(z)}{dz}+\left(\alpha^2z^4-4\beta^2+\frac{1}{16}\right)w(z)=0$$
$$w(z)=z^{1/4}\,C_\beta\left(\frac{\alpha}{2}z^2\right) \tag{2.13.124}$$

$$z^2\frac{d^2w(z)}{dz^2}+z\frac{dw(z)}{dz}+4(z^4-\beta^2)w(z)=0$$
$$w(z)=C_\beta(z^2) \tag{2.13.125}$$

$$z^2\frac{d^2w(z)}{dz^2}+z\frac{dw(z)}{dz}+(\alpha^2z^4-4\beta^2)w(z)=0$$
$$w(z)=C_\beta\left(\frac{\alpha z^2}{2}\right) \tag{2.13.126}$$

$$z^2\frac{d^2w(z)}{dz^2}+z\frac{dw(z)}{dz}+[\alpha^2z^4-(2n+1)^2]w(z)=0$$
$$w(z)=C_{n+1/2}\left(\frac{z^2}{2}\right) \tag{2.13.127}$$

$$z^2\frac{d^2w(z)}{dz^2}-z\frac{dw(z)}{dz}+(\alpha^2z^4+1)w(z)=0$$
$$w(z)=z\,C_0\left(\frac{\alpha}{2}z^2\right) \tag{2.13.128}$$

$$z^2\frac{d^2w(z)}{dz^2}-z\frac{dw(z)}{dz}+(\alpha^2z^4+\beta^2)w(z)=0$$
$$w(z)=z\,C_{\frac{1}{2}\sqrt{1-\beta^2}}\left(\frac{\alpha z^2}{2}\right) \tag{2.13.129}$$

$$z^2\frac{d^2w(z)}{dz^2}-z\frac{dw(z)}{dz}+(\alpha^2z^4-\beta^2+1)w(z)=0$$
$$w(z)=z\,C_\beta\left(\frac{\alpha}{2}z^2\right) \tag{2.13.130}$$

$$z^2\frac{d^2w(z)}{dz^2}+2z\frac{dw(z)}{dz}+\left(\alpha^2z^4+\frac{1}{4}\right)w(z)=0$$
$$w(z)=\frac{1}{\sqrt{z}}\,C_0\left(\frac{\alpha}{2}z^2\right) \tag{2.13.131}$$

$$z^2\frac{d^2w(z)}{dz^2}+2z\frac{dw(z)}{dz}+\left(\alpha^2z^4-\frac{15}{4}\right)w(z)=0$$
$$w(z)=\frac{1}{\sqrt{z}}\,C_1\left(\frac{\alpha}{2}z^2\right) \tag{2.13.132}$$

$$z^2\frac{d^2w(z)}{dz^2}+2z\frac{dw(z)}{dz}+\left(\alpha^2z^4-4\beta^2+\frac{1}{4}\right)w(z)=0 \tag{2.13.133}$$
$$w(z)=\frac{1}{\sqrt{z}}\,C_\beta\left(\frac{\alpha z^2}{2}\right)$$

$$z^2\frac{d^2w(z)}{dz^2}-2z\frac{dw(z)}{dz}+\left(\alpha^2z^4+\frac{9}{4}\right)w(z)=0 \tag{2.13.134}$$
$$w(z)=z^{3/2}\,C_0\left(\frac{\alpha}{2}z^2\right)$$

$$z^2\frac{d^2w(z)}{dz^2}-2z\frac{dw(z)}{dz}+\left(\alpha^2z^4+\frac{9-16\beta^2}{4}\right)w(z)=0 \tag{2.13.135}$$
$$w(z)=z^{3/2}\,C_\beta\left(\frac{\alpha}{2}z^2\right)$$

$$z^2\frac{d^2w(z)}{dz^2}+(\beta+1)z\frac{dw(z)}{dz}+\left(\alpha^2z^4+\frac{\beta^2}{4}\right)w(z)=0 \tag{2.13.136}$$
$$w(z)=z^{-\beta/2}C_0(\alpha z^2)$$

$$z^2\frac{d^2w(z)}{dz^2}+(\beta+1)z\frac{dw(z)}{dz}+(\alpha^2z^4+\lambda^2)w(z)=0 \tag{2.13.137}$$
$$w(z)=z^{-\beta/2}C_{\frac{1}{4}\sqrt{\beta^2-4\lambda^2}}(\alpha z^2)$$

$$z^2\frac{d^2w(z)}{dz^2}-(\beta+1)z\frac{dw(z)}{dz}+\left[\alpha^2z^4+\frac{(\beta+2)^2}{4}\right]w(z)=0 \tag{2.13.138}$$
$$w(z)=z^{\beta/2+1}C_0\left(\frac{\alpha z^2}{2}\right)$$

$$z^2\frac{d^2w(z)}{dz^2}+(\beta+1)z\frac{dw(z)}{dz}+\left(\alpha^2z^4+\frac{\beta^2-16}{4}\right)w(z)=0 \tag{2.13.139}$$
$$w(z)=z^{-\beta/2}C_1(\alpha z^2)$$

$$z\frac{d^2w(z)}{dz^2}-3\frac{dw(z)}{dz}-9z^5w(z)=0 \tag{2.13.140}$$
$$w(z)=z^2Z_{2/3}(z^3)$$

$$z^2\frac{d^2w(z)}{dz^2}+z\frac{dw(z)}{dz}+16z^8w(z)=0 \quad ; \quad w(z)=C_0(z^4) \tag{2.13.141}$$

$$z^2\frac{d^2w(z)}{dz^2}+z\frac{dw(z)}{dz}+\left(\frac{\alpha^2}{z^2}-\beta^2\right)w(z)=0 \quad ; \quad w(z)=C_\beta\left(\frac{\alpha}{z}\right) \tag{2.13.142}$$

$$z^2\frac{d^2w(z)}{dz^2}+\frac{z}{3}\frac{dw(z)}{dz}+\left(\alpha^2z^{n+2}+\frac{1}{9}\right)w(z)=0$$
$$w(z)=z^{1/3}C_0\left(\frac{2\alpha}{n+2}z^{n/2+1}\right) \tag{2.13.143}$$

$$z^2\frac{d^2w(z)}{dz^2}+\frac{z}{2}\frac{dw(z)}{dz}+(\alpha^2z^{n+2}+\beta^2)w(z)=0$$
$$w(z)=z^{1/4}\,C_{\frac{1}{2}\sqrt{1-16\beta^2}/(n+2)}\left(\frac{2\alpha}{n+2}z^{n/2+1}\right) \tag{2.13.144}$$

$$z^2\frac{d^2w(z)}{dz^2}-\frac{z}{2}\frac{dw(z)}{dz}+\left(\alpha^2z^{n+2}+\frac{9-4\beta^2}{16}\right)w(z)=0$$
$$w(z)=z^{3/4}\,C_{\beta/(n+2)}\left(\frac{2\alpha}{n+2}z^{n/2+1}\right) \tag{2.13.145}$$

$$z^2\frac{d^2w(z)}{dz^2}+z\frac{dw(z)}{dz}+n^2z^{2n-2}w(z)=0$$
$$w(z)=C_0(z^n) \tag{2.13.146}$$

$$z^2\frac{d^2w(z)}{dz^2}+z\frac{dw(z)}{dz}+\alpha^2z^{n+2}w(z)=0$$
$$w(z)=C_0\left(\frac{2\alpha}{n+2}z^{n/2+1}\right) \tag{2.13.147}$$

$$z^2\frac{d^2w(z)}{dz^2}+z\frac{dw(z)}{dz}+(\alpha^2z^2-\beta^2)w(z)=0$$
$$w(z)=C_\beta(\alpha z) \tag{2.13.148}$$

$$z^2\frac{d^2w(z)}{dz^2}+z\frac{dw(z)}{dz}+[(n+2)^2z^{n+2}-n^2)]w(z)=0$$
$$w(z)=C_{2n/(n+2)}(2z^{n/2+1}) \tag{2.13.149}$$

$$z^2\frac{d^2w(z)}{dz^2}+z\frac{dw(z)}{dz}+\left[\alpha^2z^{n+2}-\frac{(n+2)^2}{4}\right]w(z)=0$$
$$w(z)=C_{2n/(n+2)}\left(\frac{2\alpha}{n+2}z^{n/2+1}\right) \tag{2.13.150}$$

$$z^2\frac{d^2w(z)}{dz^2}+z\frac{dw(z)}{dz}+(\alpha^2z^{n+2}-\beta^2)w(z)=0$$
$$w(z)=C_{2\beta/(n+2)}\left(\frac{2\alpha}{n+2}z^{n/2+1}\right) \tag{2.13.151}$$

$$z^2\frac{d^2w(z)}{dz^2}+z\frac{dw(z)}{dz}+\beta^2(\alpha^2z^{2\beta}-1)w(z)=0$$
$$w(z)=C_1(\alpha z^{\beta}) \tag{2.13.152}$$

$$z^2\frac{d^2w(z)}{dz^2}+z\frac{dw(z)}{dz}-(iz^2+\nu^2)w(z)=0$$
$$w(z)=\mathrm{ber}_\nu(z)+i\,\mathrm{bei}_\nu(z)$$
$$w(z)=\mathrm{ber}_{-\nu}(z)+i\,\mathrm{bei}_{-\nu}(z) \tag{2.13.153}$$
$$w(z)=\mathrm{ker}_\nu(z)+i\,\mathrm{kei}_\nu(z)$$
$$w(z)=\mathrm{ker}_{-\nu}(z)+i\,\mathrm{kei}_{-\nu}(z)$$

$$z^2\frac{d^2w(z)}{dz^2}-z\frac{dw(z)}{dz}+\alpha^2z^{n+2}w(z)=0$$
$$w(z)=z\,C_{2/(n+2)}\left(\frac{2\alpha}{n+2}z^{n/2+1}\right) \tag{2.13.154}$$

$$z^2\frac{d^2w(z)}{dz^2}-z\frac{dw(z)}{dz}+(\alpha^2z^{n+2}+1)w(z)=0$$
$$w(z)=z\,C_0\left(\frac{2\alpha}{n+2}z^{n/2+1}\right) \tag{2.13.155}$$

$$z^2\frac{d^2w(z)}{dz^2}-z\frac{dw(z)}{dz}+\left(\alpha^2z^{n+2}-\frac{n(n+4)}{4}\right)w(z)=0$$
$$w(z)=z\,C_1\left(\frac{2\alpha}{n+2}z^{n/2+1}\right) \tag{2.13.156}$$

$$z^2\frac{d^2w(z)}{dz^2}-z\frac{dw(z)}{dz}+(\alpha^2z^{n+2}+\beta^2)w(z)=0$$
$$w(z)=z\,C_{2\sqrt{1-\beta^2}/(n+2)}\left(\frac{2\alpha}{n+2}z^{n/2+1}\right) \tag{2.13.157}$$

$$z^2\frac{d^2w(z)}{dz^2}+2z\frac{dw(z)}{dz}+\alpha^2z^{n+2}w(z)=0$$
$$w(z)=\frac{1}{\sqrt{z}}C_{1/(n+2)}\left(\frac{2\alpha}{n+2}z^{n/2+1}\right) \tag{2.13.158}$$

$$z^2\frac{d^2w(z)}{dz^2}-2z\frac{dw(z)}{dz}+\alpha^2z^{n+2}w(z)=0$$
$$w(z)=z^{5/2}C_{1/(n+2)}\left(\frac{2\alpha}{n+2}z^{n/2+1}\right) \tag{2.13.159}$$

$$z^2\frac{d^2w(z)}{dz^2}-2z\frac{dw(z)}{dz}+(\alpha^2z^{n+2}+2)w(z)=0$$
$$w(z)=z^{3/2}C_{1/(n+2)}\left(\frac{2\alpha}{n+2}z^{n/2+1}\right) \tag{2.13.160}$$

$$z^2\frac{d^2w(z)}{dz^2}-2z\frac{dw(z)}{dz}+(\alpha^2z^{n+2}+\beta^2)w(z)=0$$
$$w(z)=z^{3/2}C_{\sqrt{9-4\beta^2}/(n+2)}\left(\frac{2\alpha}{n+2}z^{n/2+1}\right) \tag{2.13.161}$$

$$z^2\frac{d^2w(z)}{dz^2}+\beta z\frac{dw(z)}{dz}+\alpha^2z^{\lambda+2}w(z)=0$$
$$w(z)=z^{(1-\beta)/2}C_{(1-\beta)/(\lambda+2)}\left(\frac{2\alpha}{\lambda+2}z^{(\lambda+2)/2}\right) \tag{2.13.162}$$

$$z^2\frac{d^2w(z)}{dz^2}-\beta z\frac{dw(z)}{dz}+(\alpha^2z^{n+2}+\beta)w(z)=0$$
$$w(z)=z^{(\beta+1)/2}C_{(\beta-1)/(n+2)}\left(\frac{2\alpha}{n+2}z^{n/2+1}\right) \tag{2.13.163}$$

$$z^2\frac{d^2w(z)}{dz^2}-\beta z\frac{dw(z)}{dz}+(\alpha z^{\lambda+2}+\mu^2)w(z)=0$$
$$w(z)=z^{(\beta+1)/2}C_{\sqrt{(\beta+1)^2-4\mu^2}/(\lambda+2)}\left(\frac{2\beta}{\lambda+2}z^{\lambda/2+1}\right) \tag{2.13.164}$$

$$z^2\frac{d^2w(z)}{dz^2}-2\beta z\frac{dw(z)}{dz}+\alpha^2z^2w(z)=0$$
$$w(z)=z^{\beta+1/2}C_{\beta+1/2}(\alpha z) \tag{2.13.165}$$

$$z^2\frac{d^2w(z)}{dz^2}+\frac{z}{\beta}\frac{dw(z)}{dz}+\left[\alpha^2z^{n+2}+\left(\frac{\beta-1}{2\beta}\right)^2\right]w(z)=0$$
$$w(z)=z^{(\beta-1)/2\beta}C_0\left(\frac{2\alpha}{n+2}z^{n/2+1}\right) \tag{2.13.166}$$

$$z^2\frac{d^2w(z)}{dz^2}+(\beta+1)z\frac{dw(z)}{dz}+\alpha^2z^{n+2}w(z)=0$$
$$w(z)=z^{-\beta/2}C_{\beta/(n+2)}\left(\frac{2\alpha}{n+2}z^{n/2+1}\right) \tag{2.13.167}$$

$$z^2\frac{d^2w(z)}{dz^2}+(\beta+1)z\frac{dw(z)}{dz}+\alpha^2z^\beta w(z)=0$$
$$w(z)=z^{-\beta/2}C_1\left(\frac{2\alpha}{\beta}z^{\beta/2}\right) \tag{2.13.168}$$

$$z^2\frac{d^2w(z)}{dz^2}+(\beta+1)z\frac{dw(z)}{dz}+\beta^2z^\beta w(z)=0$$
$$w(z)=z^{-\beta/2}C_1(2z^{\beta/2}) \tag{2.13.169}$$

$$z^2\frac{d^2w(z)}{dz^2}+(\beta+1)z\frac{dw(z)}{dz}+\left(\alpha^2z^{n+2}+\frac{\beta^2}{4}\right)w(z)=0$$
$$w(z)=z^{-\beta/2}C_0\left(\frac{2\alpha}{n+2}z^{n/2+1}\right) \tag{2.13.170}$$

$$z^2\frac{d^2w(z)}{dz^2}+(\beta+1)z\frac{dw(z)}{dz}+\left(\alpha^2z^{n+2}-\frac{3\beta^2}{4}\right)w(z)=0$$
$$w(z)=z^{-\beta/2}C_{2\nu/(n+2)}\left(\frac{2\alpha}{n+2}z^{n/2+1}\right) \tag{2.13.171}$$

$$z^2\frac{d^2w(z)}{dz^2}+(\beta+1)z\frac{dw(z)}{dz}+\left(\alpha^2z^{\beta}-\frac{3\beta^2}{4}\right)w(z)=0$$
$$w(z)=z^{-\beta/2}C_2\left(\frac{2\alpha}{\beta}z^{n/2+1}\right) \tag{2.13.172}$$

$$z^2\frac{d^2w(z)}{dz^2}-(\beta+1)z\frac{dw(z)}{dz}+\alpha^2z^{n+2}w(z)=0$$
$$w(z)=z^{(\beta+2)/2}C_{(\beta+2)/(n+2)}\left(\frac{2\alpha}{n+2}z^{n/2+1}\right) \tag{2.13.173}$$

$$z^2\frac{d^2w(z)}{dz^2}+(\beta-1)z\frac{dw(z)}{dz}+(\alpha^2z^{n+2}-2\beta)w(z)=0$$
$$w(z)=z^{1-\beta/2}C_{(\beta+2)/(n+2)}\left(\frac{2\alpha}{n+2}z^{n/2+1}\right) \tag{2.13.174}$$

$$z^2\frac{d^2w(z)}{dz^2}-(\beta+1)z\frac{dw(z)}{dz}+\left[\alpha^2z^{n+2}+\frac{(\beta+2)^2}{4}\right]w(z)=0$$
$$w(z)=z^{\beta/2+1}C_0\left(\frac{2\alpha}{n+2}z^{n/2+1}\right) \tag{2.13.175}$$

$$z^2\frac{d^2w(z)}{dz^2}+(1+2\beta)\,z\frac{dw(z)}{dz}+(\alpha^2z^{2\lambda}+\mu^2)w(z)=0$$
$$w(z)=z^{-\beta}C_{\frac{1}{\lambda}\sqrt{\beta^2-\mu^2}}\left(\frac{\alpha}{\lambda}z^{\lambda}\right) \tag{2.13.176}$$

$$z^2\frac{d^2w(z)}{dz^2}+(1-2\beta)\,z\frac{dw(z)}{dz}+\beta^2(z^{2\beta}+1-\beta^2)w(z)=0$$
$$w(z)=z^{\beta}C_{\beta}(z^{\beta}) \tag{2.13.177}$$

$$z^2\frac{d^2w(z)}{dz^2}+(1-2\beta)\,z\frac{dw(z)}{dz}+(\alpha^2z^2+\beta^2-\lambda^2)w(z)=0$$
$$w(z)=z^{\beta}C_{\lambda}(\alpha z) \tag{2.13.178}$$

$$z^2\frac{d^2w(z)}{dz^2}+(1-2\beta)\,z\frac{dw(z)}{dz}+(\alpha^2\lambda^2z^{\lambda}+\beta^2-\lambda^2\mu^2)w(z)=0$$
$$w(z)=z^{\beta}C_{\mu}(\alpha z^{\lambda}) \tag{2.13.179}$$

$$z^2\frac{d^2w(z)}{dz^2}-(2\beta\lambda-1)\,z\frac{dw(z)}{dz}+\alpha^2\beta^2z^{2\beta}w(z)=0$$
$$w(z)=z^{\beta\lambda}\,C_{\lambda}(\alpha z^{\beta}) \tag{2.13.180}$$

$$z^2\frac{d^2w(z)}{dz^2}+(2\alpha-2\beta\mu+1)z\frac{dw(z)}{dz}+[\beta^2\lambda^2z^{2\beta}+\alpha(\alpha-2\beta\mu)]w(z)=0\quad;$$
$$w(z)=z^{\beta\mu-\alpha}C_{\mu}(\lambda z^{\beta}) \tag{2.13.181}$$

$$z^2\frac{d^2w(z)}{dz^2}-z^2\frac{dw(z)}{dz}+\left(\frac{1-\alpha^2+z^2}{4}+\alpha^2z^{\beta}\right)w(z)=0$$
$$w(z)=\sqrt{z}\,e^{z/2}C_{\alpha/\beta}\left(\frac{2\alpha}{\beta}z^{\beta/2}\right) \tag{2.13.182}$$

$$z^2\frac{d^2w(z)}{dz^2}+\alpha z^2\frac{dw(z)}{dz}+\left[\left(\frac{\alpha^2}{4}+\beta^2\right)z^2+\lambda^2\right]w(z)=0$$
$$w(z)=\sqrt{z}\,e^{-\alpha z/2}C_{\frac{1}{2}\sqrt{1-4\lambda^2}}(\beta z) \tag{2.13.183}$$

$$z^2\frac{d^2w(z)}{dz^2}-2\alpha z^2\frac{dw(z)}{dz}+\left(\alpha^2z^2+\lambda^2z+\frac{1-\beta^2}{4}\right)w(z)=0$$
$$w(z)=\sqrt{z}\,e^{\alpha z}C_{\beta}(2\lambda\sqrt{z}) \tag{2.13.184}$$

$$z^2\frac{d^2w(z)}{dz^2}+\alpha z^2\frac{dw(z)}{dz}+\left(\frac{\alpha^2z^2}{4}+\beta^2z^{n+2}+\frac{1}{4}\right)w(z)=0$$
$$w(z)=\sqrt{z}\,e^{-\alpha z/2}C_0\left(\frac{2\beta}{n+2}z^{n/2+1}\right) \tag{2.13.185}$$

$$z^2\frac{d^2w(z)}{dz^2}+\alpha z^2\frac{dw(z)}{dz}+\left(\frac{\alpha^2z^2}{4}+\beta^2z^{n+2}+\lambda^2\right)w(z)=0$$
$$w(z)=\sqrt{z}\,e^{-\alpha z/2}C_{\sqrt{1-4\lambda^2}/(n+2)}\left(\frac{2\beta}{n+2}z^{n/2+1}\right) \tag{2.13.186}$$

$$z^2\frac{d^2w(z)}{dz^2}+\alpha z^2\frac{dw(z)}{dz}+\left(\frac{\alpha^2z^2}{4}+\beta^2z\right)w(z)=0$$
$$w(z)=\sqrt{z}\,e^{-\alpha z/2}C_1(2\beta\sqrt{z}) \tag{2.13.187}$$

$$z^2\frac{d^2w(z)}{dz^2}+\alpha z^2\frac{dw(z)}{dz}+\left(\frac{\alpha^2z^2}{4}+\beta^2z+\lambda^2\right)w(z)=0$$
$$w(z)=\sqrt{z}\,e^{-\alpha z/2}C_{\sqrt{1-4\lambda^2}}(2\beta\sqrt{z}) \tag{2.13.188}$$

$$z^2\frac{d^2w(z)}{dz^2}+\alpha z^2\frac{dw(z)}{dz}+\left(\frac{\alpha^2z^2}{4}+\beta^2z^3-2\right)w(z)=0$$
$$w(z)=\sqrt{z}\,e^{-\alpha z/2}C_1\left(\frac{2\beta}{3}z^{3/2}\right) \tag{2.13.189}$$

$$z^2\frac{d^2w(z)}{dz^2}+\alpha z^2\frac{dw(z)}{dz}+\left(\frac{\alpha^2}{4}+\beta^2z^2\right)z^2w(z)=0$$
$$w(z)=\sqrt{z}\,e^{-\alpha z/2}C_{1/4}\left(\frac{\beta}{2}z^2\right) \tag{2.13.190}$$

$$z^2\frac{d^2w(z)}{dz^2}+\alpha z^2\frac{dw(z)}{dz}+\left(\frac{\alpha^2z^2}{4}+\beta^2z^3+\lambda^2\right)w(z)=0$$
$$w(z)=\sqrt{z}\,e^{-\alpha z/2}C_{\frac{1}{3}\sqrt{1-4\lambda^2}}\left(\frac{2\beta}{3}z^{3/2}\right) \tag{2.13.191}$$

$$z^2\frac{d^2w(z)}{dz^2}+\alpha z^2\frac{dw(z)}{dz}+\left(\lambda^2z^2+\frac{1-\beta^2+\alpha^2z^2}{4}\right)w(z)=0$$
$$w(z)=\sqrt{z}\,e^{-\alpha z/2}C_\beta(2\lambda\sqrt{z}) \tag{2.13.192}$$

$$z^2\frac{d^2w(z)}{dz^2}+\alpha z^2\frac{dw(z)}{dz}+\left(\lambda^2z^3+\frac{1-9\beta^2+\alpha^2z^2}{4}\right)w(z)=0$$
$$w(z)=\sqrt{z}\,e^{-\alpha z/2}C_\beta\left(\frac{2\lambda}{3}z^{3/2}\right) \tag{2.13.193}$$

$$z^2\frac{d^2w(z)}{dz^2}+\alpha z^2\frac{dw(z)}{dz}+\left(\lambda^2z^4+\frac{1-16\beta^2+\alpha^2z^2}{4}\right)w(z)=0$$
$$w(z)=\sqrt{z}\,e^{-\alpha z/2}C_\beta\left(\frac{\lambda}{2}z^2\right) \tag{2.13.194}$$

$$z^2\frac{d^2w(z)}{dz^2}+z(1+2\alpha z)\frac{dw(z)}{dz}+(\alpha z+\beta^2-\lambda^2)w(z)=0$$
$$w(z)=e^{-\alpha z}C_\lambda(\sqrt{\beta^2-\alpha^2}z) \tag{2.13.195}$$

$$z^2\frac{d^2w(z)}{dz^2}+z(\beta+1-2\alpha z)\frac{dw(z)}{dz}+z[\alpha(\beta+1)-1-\alpha^2z]w(z)=0$$
$$w(z)=z^{-\beta/2}e^{\alpha z}C_\beta(2\sqrt{z}) \tag{2.13.196}$$

$$z^2\frac{d^2w(z)}{dz^2}+\left[\alpha^2\beta^2z^{2\beta}+\frac{1}{4}-\beta^2\lambda^2\right]w(z)=\alpha^{\mu+1}\beta^2z^{\beta\mu+\beta+1/2}$$
$$w(z)=z^{1/2}[A_1J_\lambda(\alpha z^\beta)+A_2Y_\lambda(\alpha z^\beta)+s_{\mu,\lambda}(\alpha z^\beta)] \tag{2.13.197}$$

$$z^2\frac{d^2w(z)}{dz^2}+\alpha z\frac{dw(z)}{dz}+\left[z^2-\lambda^2+\frac{(\alpha-1)^2}{4}\right]w(z)=\beta z^{\mu-(\alpha+1)/2}$$
$$w(z)=z^{-(\alpha-1)/2}[A_1J_\lambda(z)+A_2Y_\lambda(z)+\beta s_{\mu,\lambda}(z)] \tag{2.13.198}$$

$$z^2\frac{d^2w(z)}{dz^2}+(1-2\alpha)z\frac{dw(z)}{dz}+[\beta^2z^2+\alpha^2-\lambda^2]w(z)=\beta^{\mu+1}z^{\alpha+\mu-1}$$
$$w(z)=z^\alpha[A_1J_\lambda(\beta z)+A_2Y_\lambda(\beta z)+s_{\mu,\lambda}(\beta z)] \tag{2.13.199}$$

$$z^2\frac{d^2w(z)}{dz^2}+(1-2\alpha)z\frac{dw(z)}{dz}+[\beta^2\lambda^2z^{2\lambda}+\alpha^2-\lambda^2v^2]w(z)=\beta^{\mu+1}\lambda^2z^{\alpha+\lambda\mu+\lambda}$$
$$w(z)=z^\alpha[A_1J_v(\beta z^\lambda)+A_2Y_v(\beta z^\lambda)+s_{\mu,v}(\beta z^\lambda)] \tag{2.13.200}$$

$$z^2\frac{d^2w(z)}{dz^2}+(1-2\alpha\lambda)z\frac{dw(z)}{dz}+\alpha^2\beta^2z^{2\alpha}w(z)=\alpha^2\beta^{\mu+1}z^{\alpha(\lambda+\mu+1)}$$
$$w(z)=z^{\alpha\lambda}[A_1J_\lambda(\beta z^\alpha)+A_2Y_\lambda(\beta z^\alpha)+s_{\mu,\lambda}(\beta z^\alpha)] \tag{2.13.201}$$

$$\frac{d^3w(z)}{dz^3} - 2z\frac{dw(z)}{dz}2w(z) = 0$$
$$w(z) = A_1Ai^2(z) + A_2\mathrm{Ai}(z)\mathrm{Bi}(z) + A_3A_1Bi^2(z) \tag{2.13.202}$$

$$\frac{d^3w(z)}{dz^3} + z^{2\alpha-2}\frac{dw(z)}{dz} + (\alpha-1)z^{2\alpha-3}w(z) = 0 \quad ; \quad \zeta = \frac{z^\nu}{2\alpha}$$
$$w(z) = z\left\{A_1\left[J_{1/2\alpha}(\zeta)\right]^2 + A_2J_{1/2\alpha}(\zeta)Y_\nu(\zeta) + A_1\left[Y_{1/2\alpha}(\zeta)\right]^2\right\} \tag{2.13.203}$$

$$z\frac{d^3w(z)}{dz^3} - (z+2\alpha)\frac{d^2w(z)}{dz^2} + (z-2\alpha-1)\frac{dw(z)}{dz} + (z-1)w(z) = 0$$
$$w(z) = A_1e^z + z^{\alpha+1}[A_2I_{\alpha+1}(z) + A_3K_{\alpha+1}(z)] \tag{2.13.204}$$

$$2z\frac{d^3w(z)}{dz^3} - 4(z+\alpha-1)\frac{d^2w(z)}{dz^2} + (2z+6\alpha-5)\frac{dw(z)}{dz} + (1-2\alpha)w(z) = 0$$
$$w(z) = A_1e^z + z^\alpha e^{z/2}\left[A_2J_\alpha\left(\frac{iz}{2}\right) + A_3Y_\alpha\left(\frac{iz}{2}\right)\right] \tag{2.13.205}$$

$$z^2\frac{d^3w(z)}{dz^3} + 3z\frac{d^2w(z)}{dz^2} + (1+z^2)\frac{dw(z)}{dz} = 0$$
$$w(z) = A_1 + A_2\mathrm{Ji}_0(z) + A_3\mathrm{Yi}_0(z) \tag{2.13.206}$$

$$z^2\frac{d^3w(z)}{dz^3} + 3z\frac{d^2w(z)}{dz^2} + [4z^2 - 4n^2 + 1]\frac{dw(z)}{dz} + 4zw(z) = 0$$
$$w(z) = A_1J_n^2(z) + A_2J_n(z)Y_n(z) + A_3Y_n^2(z) \tag{2.13.207}$$

$$z^2\frac{d^3w(z)}{dz^3} + 3z\frac{d^2w(z)}{dz^2} + [4\alpha^2z^{2\alpha} - 4\alpha^2\beta^2 + 1]\frac{dw(z)}{dz} + 4\alpha^2z^{2\alpha-1}w(z) = 0$$
$$w(z) = A_1J_\beta^2(z^\alpha) + A_2J_\beta(z^\alpha)Y_\beta(z^\alpha) + A_3Y_\beta^2(z^\alpha) \tag{2.13.208}$$

$$z^2\frac{d^3w(z)}{dz^3} + 2z(1-z)\frac{d^2w(z)}{dz^2} + \left(z^2 - 2z - \alpha^2 + \frac{1}{4}\right)\frac{dw(z)}{dz} + \left(\alpha^2 - \frac{1}{4}\right)w(z) = 0 \tag{2.13.209}$$
$$w(z) = A_1e^z + \sqrt{z}e^{z/2}\left[A_2J_\alpha\left(\frac{iz}{2}\right) + A_3Y_\alpha\left(i\frac{z}{2}\right)\right]$$

$$z^2\frac{d^3w(z)}{dz^3} - 2z\,(1-z)\frac{d^2w(z)}{dz^2} + \left(z^2 - 2z - \alpha^2 + \frac{1}{4}\right)\frac{dw(z)}{dz} + \left(\alpha^2 - \frac{1}{4}\right)w(z) = 0 \tag{2.13.210}$$

$$w(z) = A_1e^z + \sqrt{z}e^{z/2}\left[A_2I_\alpha\left(\frac{z}{2}\right) + A_3K_\alpha\left(\frac{z}{2}\right)\right]$$

$$z^2\frac{d^3w(z)}{dz^3} - z\,(z+\alpha)\frac{d^2w(z)}{dz^2} + \alpha\,(1+2z)\frac{dw(z)}{dz} - \alpha\,(1+z)w(z) = 0 \tag{2.13.211}$$

$$w(z) = A_1e^z + z^{(\alpha+1)/2}\,[A_2J_{\alpha+1}(2\sqrt{\alpha z}) + A_3Y_{\alpha+1}(2\sqrt{\alpha z})]$$

$$z^2\frac{d^3w(z)}{dz^3} - z\,(z-2)\frac{d^2w(z)}{dz^2} - \left(z^2 + \alpha^2 - \frac{1}{4}\right)\frac{dw(z)}{dz} + \left(z^2 - 2z + \alpha^2 - \frac{1}{4}\right)w(z) = 0 \tag{2.13.212}$$

$$w(z) = A_1e^z + \sqrt{z}\,[A_2I_\alpha(z) + A_3K_\alpha(z)]$$

$$z^3\frac{d^3w(z)}{dz^3} + (4z^3 + \alpha z)\,\frac{d^2w(z)}{dz^2} - \alpha w(z) = 0 \tag{2.13.213}$$

$$w(z) = A_1J_\alpha^2(z) + A_2J_\alpha(z)\,Y_\alpha(z) + A_3Y_\alpha^2(z)$$

$$z^3\frac{d^3w(z)}{dz^3} + [4\,z^3 + (1 - 4\,\alpha^2)z]\frac{dw(z)}{dz} - (1 - 4\,\alpha^2)\,w(z) = 0 \tag{2.13.214}$$

$$w(z) = z\,[A_1J_\alpha^2(z) + A_2J_\alpha(z)Y_\alpha(z) + A_3Y_\alpha^2(z)]$$

$$z\frac{d^4w(z)}{dz^4} - \frac{d^3w(z)}{dz^3} + 4z^3w(z) = 0$$

$$w(z) = [A_1\,\mathrm{Ai}(z)\,\mathrm{Ai}(-z) + A_2\mathrm{Ai}(z)\,\mathrm{Bi}(-z) + A_3\mathrm{Ai}(-z)\,\mathrm{Bi}(z) + A_4\mathrm{Bi}(z)\,\mathrm{Bi}(-z)] \tag{2.13.215}$$

$$\frac{d^4w(z)}{dz^4} - 10z\frac{d^3w(z)}{dz^3} - 10z\frac{dw(z)}{dz} + 9z^2w(z) = 0 \tag{2.13.216}$$

$$w(z) = A_1\,[\mathrm{Ai}(z)]^3 + A_2[\mathrm{Ai}(z)]^2\mathrm{Bi}(z) + A_3\mathrm{Ai}(z)[\mathrm{Bi}(z)]^2 + A_4[\mathrm{Bi}(z)]^3$$

$$z^2\frac{d^4w(z)}{dz^4} + 6z\frac{d^3w(z)}{dz^3} + 6\frac{d^2w(z)}{dz^2} - \alpha^2w(z) = 0 \tag{2.13.217}$$

$$w(z) = \frac{1}{\sqrt{z}}[A_1J_1(2\sqrt{\alpha z}) + A_2Y_1(2\sqrt{\alpha z}) + A_3J_1(2i\sqrt{\alpha z}) + A_4Y_1(2i\sqrt{\alpha z})]$$

$$z^2\frac{d^4w(z)}{dz^4}+8z\frac{d^3w(z)}{dz^3}+12\frac{d^2w(z)}{dz^2}-\alpha^2w(z)=0$$

$$w(z)=\frac{1}{z}[A_1J_1(2\sqrt{\alpha z})+A_2Y_1(2\sqrt{\alpha z})+A_3J_1(2i\sqrt{\alpha z})+A_4Y_1(2i\sqrt{\alpha z})] \tag{2.13.218}$$

$$z^2\frac{d^4w(z)}{dz^4}+2(\alpha+2)z\frac{d^3w(z)}{dz^3}+(\alpha+1)(\alpha+2)\frac{d^2w(z)}{dz^2}-\beta^4w(z)=0 \tag{2.13.219}$$

$$w(z)=z^{-\alpha/2}[A_1J_\alpha(2\sqrt{\beta z})+A_2Y_\alpha(2\sqrt{\beta z})+A_3I_\alpha(2\sqrt{\beta z})+A_4K_\alpha(2\sqrt{\beta z})]$$

$$z^3\frac{d^4w(z)}{dz^4}+2z^2\frac{d^3w(z)}{dz^3}-z\frac{d^2w(z)}{dz^2}+\frac{dw(z)}{dz}-\alpha^4z^3w(z)=0 \tag{2.13.220}$$

$$w(z)=A_1J_0(\alpha z)+A_2Y_0(\alpha z)+A_3I_0(\alpha z)+A_4K_0(\alpha z)$$

$$z^4\frac{d^4w(z)}{dz^4}+6z^2\frac{d^3w(z)}{dz^3}+6z^2\frac{d^2w(z)}{dz^2}+4\alpha^4z^2w(z)=0$$

$$w(z)=\frac{1}{\sqrt{z}}[A_1\,\mathrm{ber}_1(2^{3/2}\alpha\sqrt{z})+A_2\mathrm{bei}_1(2^{3/2}\alpha\sqrt{z}) \tag{2.13.221}$$

$$+A_3\mathrm{ker}_1(2^{3/2}\alpha\sqrt{z})+A_4\mathrm{kei}_1(2^{3/2}\alpha\sqrt{z})]$$

$$z^4\frac{d^4w(z)}{dz^4}+2(2-n)z^3\frac{d^3w(z)}{dz^3}+(1-n)(2-n)z^2\frac{d^2w(z)}{dz^2}$$

$$-\alpha^4z^{2n}w(z)=0\quad;\quad w(z)=A_1J_{1/n}\left(\frac{2\alpha}{n}z^{n/2}\right)+A_2Y_{1/n}\left(\frac{2\alpha}{n}z^{n/2}\right) \tag{2.13.222}$$

$$+A_3I_{1/n}\left(\frac{2\alpha}{n}z^{n/2}\right)+A_4K_{1/n}\left(\frac{2\alpha}{n}z^{n/2}\right)$$

$$z^4\frac{d^4w(z)}{dz^4}+2z^3\frac{d^3w(z)}{dz^3}-(1+2\alpha^2)\left[z^2\frac{d^2w(z)}{dz^2}-z\frac{dw(z)}{dz}\right]$$

$$+(\beta^4\alpha^4-\alpha^2+z^4)w(z)=0 \tag{2.13.223}$$

$$w(z)=A_1J_\alpha(\beta z)+A_2Y_\alpha(\beta z)+A_3I_\alpha(\beta z)+A_4K_\alpha(\beta z)$$

$$w(z)=\mathrm{ber}_{\pm\alpha}(\beta z)\quad;\quad\mathrm{bei}_{\pm\alpha}(\beta z);\quad\mathrm{ker}_{\pm\alpha}(\beta z);\quad\mathrm{kei}_{\pm\alpha}(\beta z)$$

$$z^4\frac{d^4w(z)}{dz^4}+6z^3\frac{d^3w(z)}{dz^3}+[4z^4+(7-\alpha^2-\beta^2)z^2]\frac{d^2w(z)}{dz^2}$$

$$+z(16z^2+1-\alpha^2-\beta^2)\frac{dw(z)}{dz}+(8z^2+\alpha^2\beta^2)w(z)=0 \tag{2.13.224}$$

$$w(z)=A_1J_{(\alpha+\beta)/2}(z)J_{(\alpha-\beta)/2}(z)+A_2Y_{(\alpha+\beta)/2}(z)J_{(\alpha-\beta)/2}(z)$$

$$+A_3J_{(\alpha+\beta)/2}(z)\,Y_{(\alpha-\beta)/2}(z)+A_4Y_{(\alpha+\beta)/2}(z)\,Y_{(\alpha-\beta)/2}(z)$$

$$\frac{d^2w(z)}{dz^2} + e^{2z} w(z) = 0 \quad ; \quad w(z) = C_0(e^z) \tag{2.13.225}$$

$$\frac{d^2w(z)}{dz^2} + \alpha^2 e^{\alpha z} w(z) = 0 \quad ; \quad w(z) = C_0(2e^{\alpha z/2}) \tag{2.13.226}$$

$$\frac{d^2w(z)}{dz^2} + \alpha^2 e^{\beta z} w(z) = 0 \quad ; \quad w(z) = C_0\left(\frac{2\alpha}{\beta} e^{\beta z/2}\right) \tag{2.13.227}$$

$$\frac{d^2w(z)}{dz^2} + \alpha^2\beta^2 e^{\alpha z} w(z) = 0 \quad ; \quad w(z) = C_0(2\beta e^{\alpha z/2}) \tag{2.13.228}$$

$$\frac{d^2w(z)}{dz^2} + \left(\alpha^2 e^{\beta z} - \frac{1}{4}\right) w(z) = 0 \quad ; \quad w(z) = C_\beta\left(2\frac{\alpha}{\beta}\beta e^{\beta z}\right) \tag{2.13.229}$$

$$\frac{d^2w(z)}{dz^2} + (\alpha^2 e^{\alpha z} - 4) w(z) = 0 \quad ; \quad w(z) = C_{4/\alpha}(2e^{\alpha z/2}) \tag{2.13.230}$$

$$\frac{d^2w(z)}{dz^2} + (4e^{2z} - \alpha^2) w(z) = 0 \quad ; \quad w(z) = C_\alpha(2e^z) \tag{2.13.231}$$

$$\frac{d^2w(z)}{dz^2} + (4\beta^2 e^{2z} - \alpha^2) w(z) = 0 \quad ; \quad w(z) = C_\alpha(2\beta e^z) \tag{2.13.232}$$

$$\frac{d^2w(z)}{dz^2} + \alpha^2(4e^{2\alpha z} - 1) w(z) = 0 \quad ; \quad w(z) = C_1(2e^{\alpha z}) \tag{2.13.233}$$

$$\frac{d^2w(z)}{dz^2} + \alpha^2(4e^{2\alpha z} - \alpha^2) w(z) = 0 \quad ; \quad w(z) = C_\alpha(2e^{\alpha z}) \tag{2.13.234}$$

$$\frac{d^2w(z)}{dz^2} + (\alpha^2 e^{2z} - \beta^2) w(z) = 0 \quad ; \quad w(z) = C_{2\beta}(2\alpha e^z) \tag{2.13.235}$$

$$\frac{d^2w(z)}{dz^2} + (\alpha^2 e^{2\beta z} - \beta^2) w(z) = 0 \quad ; \quad w(z) = C_1\left(\frac{\alpha}{\beta} e^{\beta z}\right) \tag{2.13.236}$$

$$\frac{d^2w(z)}{dz^2} + (\alpha^2 e^{2\beta z} - \beta^4) w(z) = 0 \quad ; \quad w(z) = C_\beta\left(\frac{\alpha}{\beta} e^{\beta z}\right) \tag{2.13.237}$$

$$\frac{d^2w(z)}{dz^2} + (\alpha^2 e^{2\beta^{n-1} z} - \beta^{2n}) w(z) = 0$$
$$w(z) = C_\beta\left(\frac{\alpha}{\beta^{n-1}} e^{\beta^{n-1} z}\right) \tag{2.13.238}$$

$$\frac{d^2w(z)}{dz^2} + \left(\alpha^2 e^{\beta/z} - \frac{1}{4}\right) w(z) = 0 \quad ; \quad w(z) = C_{1/\beta}\left(2\frac{\alpha}{\beta} e^{\beta/2z}\right) \tag{2.13.239}$$

$$z^4 \frac{d^2w(z)}{dz^2} + \alpha^2 e^{\alpha/z} w(z) = 0 \quad ; \quad w(z) = z\, C_0(e^{\alpha/2z}) \tag{2.13.240}$$

$$z^4 \frac{d^2w(z)}{dz^2} + \alpha^2 e^{\alpha/z} w(z) = 0 \quad ; \quad w(z) = z\, C_0(e^{\alpha/2z}) \tag{2.13.241}$$

$$z^4 \frac{d^2w(z)}{dz^2} + (e^{2/z} - \alpha^2) w(z) = 0 \quad ; \quad w(z) = z\, C_\alpha(e^{1/z}) \tag{2.13.242}$$

$$\begin{aligned} &z^4 \frac{d^2w(z)}{dz^2} + (\alpha^2 e^{\beta/z} - \lambda^2) w(z) = 0 \\ &w(z) = z\, C_{2\lambda/\beta}\left(2\frac{\alpha}{\beta} e^{\lambda/z}\right) \end{aligned} \tag{2.13.243}$$

$$\begin{aligned} &\frac{d^2w(z)}{dz^2} + \alpha\beta \frac{dw(z)}{dz} + \alpha^2 \lambda^2 e^{\alpha z}\, w(z) = 0 \\ &w(z) = e^{-\alpha\beta z/2} C_\beta(2\lambda e^{\alpha z/2}) \end{aligned} \tag{2.13.244}$$

$$\begin{aligned} &\frac{d^2w(z)}{dz^2} + \frac{dw(z)}{dz} + \left(\alpha^2 e^{\beta z} + \frac{1}{4}\right) w(z) = 0 \\ &w(z) = e^{-\alpha z/2} C_0\left(\frac{2\alpha}{\beta} e^{\beta z/2}\right) \end{aligned} \tag{2.13.245}$$

$$\begin{aligned} &\frac{d^2w(z)}{dz^2} + 2\frac{dw(z)}{dz} + (\alpha^2 e^{\beta z} + 1)\, w(z) = 0 \\ &w(z) = e^{-z} C_0\left(\frac{2\alpha}{\beta} e^{\beta z/2}\right) \end{aligned} \tag{2.13.246}$$

$$\begin{aligned} &\frac{d^2w(z)}{dz^2} - 2\frac{dw(z)}{dz} + (\alpha^2 e^{\beta z} + 1)\, w(z) = 0 \\ &w(z) = e^{z} C_0\left(\frac{2\alpha}{\beta} e^{\beta z/2}\right) \end{aligned} \tag{2.13.247}$$

$$\begin{aligned} &\frac{d^2w(z)}{dz^2} + \alpha\frac{dw(z)}{dz} + \left(\beta^2 e^{\lambda z} + \frac{\alpha^2}{4}\right) w(z) = 0 \\ &w(z) = e^{-\alpha z/2} C_0\left(\frac{2\beta}{\lambda} e^{\lambda z/2}\right) \end{aligned} \tag{2.13.248}$$

$$\begin{aligned} &\frac{d^2w(z)}{dz^2} + \alpha\frac{dw(z)}{dz} + \left(\beta^2 e^{\lambda z} + \frac{\alpha^2 - 4\mu^2}{4}\right) w(z) = 0 \\ &w(z) = e^{-\alpha z/2} C_{2\mu/\lambda}\left(\frac{2\beta}{\lambda} e^{\lambda z/2}\right) \end{aligned} \tag{2.13.249}$$

$$\frac{d^2w(z)}{dz^2}+\alpha\frac{dw(z)}{dz}+\left(\beta^2e^{2z}+\frac{\alpha^2-4\lambda^2}{4}\right)w(z)=0 \tag{2.13.250}$$
$$w(z)=e^{-\alpha z/2}C_\lambda(\beta e^z)$$

$$\frac{d^2w(z)}{dz^2}+\alpha\frac{dw(z)}{dz}+(\beta e^{\lambda z}+\delta)w(z)=0$$
$$w(z)=e^{-\alpha z/2}C_{\frac{\sqrt{\alpha^2-4\delta}}{\lambda}}\left(\frac{2\sqrt{\beta}}{\lambda}z^{\lambda z/2}\right) \tag{2.13.251}$$

$$\frac{d^2w(z)}{dz^2}+\alpha(\beta-2)\frac{dw(z)}{dz}+[\alpha^2(1-\beta)+\beta^2\lambda^2e^{\beta z}]w(z)=0 \tag{2.13.252}$$
$$w(z)=e^{(\alpha-\beta/2)z}C_\alpha(2\lambda e^{\beta z/2})$$

$$z^2\frac{d^2w(z)}{dz^2}+\frac{z}{2}\frac{dw(z)}{dz}+\left[z^2(\alpha^2e^{\beta z}-\lambda^2)-\frac{3}{16}\right]w(z)=0$$
$$w(z)=z^{-1/4}C_{2\lambda/\beta}\left(\frac{2\alpha}{\beta}e^{\beta z/2}\right) \tag{2.13.253}$$

$$z^2\frac{d^2w(z)}{dz^2}+\frac{z}{2}\frac{dw(z)}{dz}+\left[\frac{1}{z^2}(\alpha^2e^{\beta/z}-\lambda^2)-\frac{3}{16}\right]w(z)=0$$
$$w(z)=z^{3/4}C_{2\lambda/\beta}\left(\frac{2\alpha}{\beta}e^{\beta/2z}\right) \tag{2.13.254}$$

$$z^2\frac{d^2w(z)}{dz^2}+-\frac{z}{2}\frac{dw(z)}{dz}+\left[\frac{1}{z^2}(\alpha^2e^{\beta/z}-\lambda^2)+\frac{5}{16}\right]w(z)=0$$
$$w(z)=z^{5/4}C_{2\lambda/\beta}\left(\frac{2\alpha}{\beta}e^{\beta/2z}\right) \tag{2.13.255}$$

$$z^2\frac{d^2w(z)}{dz^2}-\frac{z}{2}\frac{dw(z)}{dz}+\left[z^2(\alpha^2e^{\beta z}-\lambda^2)-\frac{5}{16}\right]w(z)=0$$
$$w(z)=z^{1/4}C_{2\lambda/\beta}\left(\frac{2\alpha}{\beta}e^{\beta z/2}\right) \tag{2.13.256}$$

$$z^2\frac{d^2w(z)}{dz^2}+z\frac{dw(z)}{dz}+(\alpha^2z^2e^z-\frac{1}{4})w(z)=0$$
$$w(z)=\frac{1}{\sqrt{z}}C_0(2\alpha e^{z/2}) \tag{2.13.257}$$

$$z^2\frac{d^2w(z)}{dz^2}+z\frac{dw(z)}{dz}+\left[z^2\left((\alpha^2 e^z-\frac{\beta^2}{4}\right)-\frac{1}{4}\right]w(z)=0 \tag{2.13.258}$$

$$w(z)=\frac{1}{\sqrt{z}}C_\beta(2\alpha e^{z/2})$$

$$z^2\frac{d^2w(z)}{dz^2}+z\frac{dw(z)}{dz}+\left[z^2\left(\alpha^2 e^{\alpha z}-\frac{\beta^2}{4}\right)-\frac{1}{4}\right]w(z)=0 \tag{2.13.259}$$

$$w(z)=\frac{1}{\sqrt{z}}C_{2\beta/\alpha}(2\alpha e^{\alpha z/2})$$

$$z^2\frac{d^2w(z)}{dz^2}+z\frac{dw(z)}{dz}+\left[z^2(\alpha^2 e^{2z}-\beta^2)-\frac{1}{4}\right]w(z)=0 \tag{2.13.260}$$

$$w(z)=\frac{1}{\sqrt{z}}C_\beta(\alpha e^{z/2})$$

$$z^2\frac{d^2w(z)}{dz^2}+z\frac{dw(z)}{dz}+\left[z^2(\alpha^2 e^{\beta z}-\lambda^2)-\frac{1}{4}\right]w(z)=0 \tag{2.13.261}$$

$$w(z)=\frac{1}{\sqrt{z}}C_{2\lambda/\beta}\left(\frac{2\alpha}{\beta}e^{\beta z/2}\right)$$

$$z^2\frac{d^2w(z)}{dz^2}+z\frac{dw(z)}{dz}+\left[z^2(\alpha^2 e^{2\beta z}-\lambda^2)-\frac{1}{4}\right]w(z)=0 \tag{2.13.262}$$

$$w(z)=\frac{1}{\sqrt{z}}C_{\lambda/\beta}\left(\frac{\alpha}{\beta}e^{\beta z}\right)$$

$$z^2\frac{d^2w(z)}{dz^2}+z\frac{dw(z)}{dz}+\left(\frac{\alpha^2}{z^2}e^{2/z}-\frac{1}{4}\right)w(z)=0 \tag{2.13.263}$$

$$w(z)=\sqrt{z}\,C_0(\alpha e^{1/z})$$

$$z^2\frac{d^2w(z)}{dz^2}+z\frac{dw(z)}{dz}+\left[\frac{1}{z^2}(\alpha^2 e^{2/z}-\beta^2)-\frac{1}{4}\right]w(z)=0 \tag{2.13.264}$$

$$w(z)=\sqrt{z}\,C_\beta(\alpha e^{1/z})$$

$$z^2\frac{d^2w(z)}{dz^2}+z\frac{dw(z)}{dz}+\left[\frac{1}{z^2}\left(\alpha^2 e^{1/z}-\frac{\beta^2}{4}\right)-\frac{1}{4}\right]w(z)=0 \tag{2.13.265}$$

$$w(z)=\sqrt{z}\,C_\beta(2\alpha e^{1/2z})$$

$$z^2 \frac{d^2 w(z)}{dz^2} + z \frac{dw(z)}{dz} + \left[\frac{1}{z^2} \left(\alpha^2 e^{\beta/z} - \frac{\lambda^2}{4} \right) - \frac{1}{4} \right] w(z) = 0$$
$$w(z) = \sqrt{z}\, C_{\lambda/\beta} \left(\frac{2\alpha}{\beta} e^{\beta/2z} \right) \tag{2.13.266}$$

$$z^2 \frac{d^2 w(z)}{dz^2} - z \frac{dw(z)}{dz} + \left(\alpha^2 z^2 e^{2z} + \frac{3}{4} \right) w(z) = 0$$
$$w(z) = \sqrt{z}\, C_0(\alpha e^z) \tag{2.13.267}$$

$$z^2 \frac{d^2 w(z)}{dz^2} - z \frac{dw(z)}{dz} + \left[z^2 (\alpha^2 e^{2z} - \beta^2) + \frac{3}{4} \right] w(z) = 0$$
$$w(z) = \sqrt{z}\, C_\beta(\alpha e^z) \tag{2.13.268}$$

$$z^2 \frac{d^2 w(z)}{dz^2} - z \frac{dw(z)}{dz} + \left[z^2 \left(\alpha^2 e^{\beta z} - \frac{\lambda^2}{4} \right) + \frac{3}{4} \right] w(z) = 0$$
$$w(z) = \sqrt{z}\, C_{\lambda/\beta} \left(\frac{2\alpha}{\beta} e^{\beta z/2} \right) \tag{2.13.269}$$

$$z^2 \frac{d^2 w(z)}{dz^2} - z \frac{dw(z)}{dz} + \left(\frac{\alpha^2}{z^2} e^{2/z} + \frac{3}{4} \right) w(z) = 0$$
$$w(z) = z^{3/2} C_0(\alpha e^{1/z}) \tag{2.13.270}$$

$$z^2 \frac{d^2 w(z)}{dz^2} - z \frac{dw(z)}{dz} + \left[\frac{1}{z^2} \left(\alpha^2 e^{1/z} - \frac{\beta^2}{4} \right) + \frac{3}{4} \right] w(z) = 0$$
$$w(z) = z^{3/2} C_\beta(2\alpha e^{1/2z}) \tag{2.13.271}$$

$$z^2 \frac{d^2 w(z)}{dz^2} - z \frac{dw(z)}{dz} + \left[\frac{1}{z^2} (\alpha^2 e^{2/z} - \beta^2) + \frac{3}{4} \right] w(z) = 0$$
$$w(z) = z^{3/2} C_\beta(2\alpha e^{1/z}) \tag{2.13.272}$$

$$z^2 \frac{d^2 w(z)}{dz^2} - z \frac{dw(z)}{dz} + \left[\frac{1}{z^2} \left(\alpha^2 e^{\beta/z} - \frac{\lambda^2}{4} \right) + \frac{3}{4} \right] w(z) = 0$$
$$w(z) = z^{3/2} C_{\lambda/\beta} \left(\frac{2\alpha}{\beta} e^{\beta/2z} \right) \tag{2.13.273}$$

$$z^2 \frac{d^2 w(z)}{dz^2} + 2z \frac{dw(z)}{dz} + \alpha^2 (e^{2\alpha z} - \alpha^2) z^2 w(z) = 0$$
$$w(z) = \frac{1}{z} C_\alpha(e^{\alpha z}) \tag{2.13.274}$$

$$z^2\frac{d^2w(z)}{dz^2}+2z\frac{dw(z)}{dz}+(\alpha^2e^{2z}-\beta^2)z^2w(z)=0$$
$$w(z)=\frac{1}{z}C_\beta(\alpha e^z) \tag{2.13.275}$$

$$z^2\frac{d^2w(z)}{dz^2}+2z\frac{dw(z)}{dz}+(\alpha^2e^{\beta z}-\lambda^2)z^2w(z)=0$$
$$w(z)=\frac{1}{z}C_{2\lambda/\beta}\left(\frac{2\alpha}{\beta}e^{\beta z/2}\right) \tag{2.13.276}$$

$$z^2\frac{d^2w(z)}{dz^2}+2z\frac{dw(z)}{dz}+(\alpha^2e^{2\beta z}-\lambda^2)z^2w(z)=0$$
$$w(z)=\frac{1}{z}C_{\lambda/\beta}\left(\frac{\alpha}{\beta}e^{\beta z}\right) \tag{2.13.277}$$

$$z^2\frac{d^2w(z)}{dz^2}+2z\frac{dw(z)}{dz}+(\alpha^2e^{2\beta z}-\beta^2\lambda^2)z^2w(z)=0$$
$$w(z)=\frac{1}{z}C_\lambda\left(\frac{\alpha}{\beta}e^{\beta z}\right) \tag{2.13.278}$$

$$z^2\frac{d^2w(z)}{dz^2}+2z\frac{dw(z)}{dz}+\frac{\alpha^2}{z^2}(e^{\alpha/z}-\beta^2)w(z)=0$$
$$w(z)=C_{2\beta}(\alpha e^{\alpha/2z}) \tag{2.13.279}$$

$$z^2\frac{d^2w(z)}{dz^2}+2z\frac{dw(z)}{dz}+\frac{1}{z^2}(\alpha^2e^{2/z}-\beta^2)w(z)=0$$
$$w(z)=C_\beta(\alpha e^{1/z}) \tag{2.13.280}$$

$$z^2\frac{d^2w(z)}{dz^2}+2z\frac{dw(z)}{dz}+\frac{1}{z^2}(\alpha^2e^{\beta/z}-\lambda^2)w(z)=0$$
$$w(z)=C_{2\lambda/\beta}\left(\frac{2\alpha}{\beta}e^{\beta/2z}\right) \tag{2.13.281}$$

$$z^2\frac{d^2w(z)}{dz^2}-2z\frac{dw(z)}{dz}+[z^2(\alpha^2e^{2z}-\beta^2)+2]\,w(z)=0$$
$$w(z)=z\,C_\beta(\alpha e^z) \tag{2.13.282}$$

$$z^2\frac{d^2w(z)}{dz^2}-2z\frac{dw(z)}{dz}+[z^2(\alpha^2e^{\beta z}-\lambda^2)+2]\,w(z)=0$$
$$w(z)=z\,C_{2\lambda/\beta}\left(\frac{2\alpha}{\beta}e^{\beta z/2}\right) \tag{2.13.283}$$

$$z^2\frac{d^2w(z)}{dz^2}-2z\frac{dw(z)}{dz}+\left[\frac{1}{z^2}(\alpha^2 e^{2/z}-\beta^2)+2\right]w(z)=0 \tag{2.13.284}$$
$$w(z)=z^2\,C_\beta(\alpha e^{1/z})$$

$$z^2\frac{d^2w(z)}{dz^2}-2z\frac{dw(z)}{dz}+\left[\frac{1}{z^2}(\alpha^2 e^{\beta/z}-\lambda^2)+2\right]w(z)=0 \tag{2.13.285}$$
$$w(z)=z^2\,C_{2\lambda/\beta}\left(\frac{2\alpha}{\beta}\,e^{\beta/2z}\right)$$

$$z^2\frac{d^2w(z)}{dz^2}-2z\frac{dw(z)}{dz}+\left[\frac{1}{z^2}(\alpha^2 e^{2/z}-\beta^2)+2\right]w(z)=0 \tag{2.13.286}$$
$$w(z)=z^2\,C_\beta(\alpha e^{1/z})$$

$$z^2\frac{d^2w(z)}{dz^2}-2z\frac{dw(z)}{dz}+\left[\frac{1}{z^2}(\alpha^2 e^{\beta/z}-\lambda^2)+2\right]w(z)=0 \tag{2.13.287}$$
$$w(z)=z^2\,C_{2\lambda/\beta}\left(\frac{2\alpha}{\beta}\,e^{\beta/2z}\right)$$

$$z^2\frac{d^2w(z)}{dz^2}+2\alpha z\frac{dw(z)}{dz}+[(\beta^2 e^{2\mu z}-\lambda^2)\mu^2 z^2+\alpha(\alpha-1)]\,w(z)=0 \tag{2.13.288}$$
$$w(z)=z^{-\alpha}C_\lambda(\beta e^{\mu z})$$

$$z^2\frac{d^2w(z)}{dz^2}-2\alpha z\frac{dw(z)}{dz}+[\alpha(\alpha+1)-z^2(1-4\beta^2\lambda^2 e^{2\beta z})]\,w(z)=0 \tag{2.13.289}$$
$$w(z)=z^{\alpha}\,C_{1/\beta}(2\lambda e^{\beta z})$$

$$z^2\frac{d^2w(z)}{dz^2}+(\beta+1)z\frac{dw(z)}{dz}+\left(\alpha^2 z^2 e^z-\frac{\beta^2}{4}\right)w(z)=0 \tag{2.13.290}$$
$$w(z)=z^{-(\beta+1)}\,C_0(2\alpha e^{z/2})$$

$$z^2\frac{d^2w(z)}{dz^2}+(\beta+1)z\frac{dw(z)}{dz}+\left(\alpha^2 z^2 e^{\beta z}+\frac{\beta^2-1}{4}\right)w(z)=0 \tag{2.13.291}$$
$$w(z)=z^{-(\beta+1)/2}\,C_0\left(\frac{2\alpha}{\beta}\,e^{\beta z/2}\right)$$

$$z^2\frac{d^2w(z)}{dz^2}+(\beta+1)z\frac{dw(z)}{dz}+\left[z^2(4e^{2z}-\beta^2)+\frac{\beta^2-1}{4}\right]w(z)=0 \tag{2.13.292}$$
$$w(z)=z^{-(\beta+1)/2}\,C_\beta(2e^{z})$$

$$z^2 \frac{d^2w(z)}{dz^2} + (\beta+1) z \frac{dw(z)}{dz} + \left[z^2 \left(\alpha^2 e^{\lambda z} - \frac{\beta^2}{4} \right) + \frac{\beta^2 - 1}{4} \right] w(z) = 0 \tag{2.13.290}$$

$$w(z) = z^{-(\beta+1)/2} C_{\beta/\lambda} \left(\frac{2\alpha}{\beta} e^{\lambda z/2} \right)$$

$$z^2 \frac{d^2w(z)}{dz^2} + (\beta+1) z \frac{dw(z)}{dz} + \left[z^2 (\alpha^2 e^{2\lambda z} - \mu^2) + \frac{\beta^2 - 1}{4} \right] w(z) = 0 \tag{2.13.291}$$

$$w(z) = z^{-(\beta+1)/2} C_{\mu/\lambda} \left(\frac{\alpha}{\lambda} e^{\lambda z} \right)$$

$$z^2 \frac{d^2w(z)}{dz^2} + (\beta+1) z \frac{dw(z)}{dz} + \left(\frac{e^{1/z}}{z^2} + \frac{\beta^2 - 1}{4} \right) w(z) = 0 \tag{2.13.292}$$

$$w(z) = z^{(1-\beta)/2} C_0 (2 e^{1/2z})$$

$$z^2 \frac{d^2w(z)}{dz^2} + (\beta+1) z \frac{dw(z)}{dz} + \left[\frac{1}{z^2} (4 e^{2/z} - \beta^2) + \frac{\beta^2 - 1}{4} \right] \tag{2.13.293}$$

$$w(z) = 0 \quad ; \quad w(z) = z^{(1-\beta)/2} C_\beta (2 e^{1/z})$$

$$z^2 \frac{d^2w(z)}{dz^2} + (\beta+1) z \frac{dw(z)}{dz} + \left[\frac{1}{z^2} (\alpha^2 e^{2/z} - \lambda^2) + \frac{\beta^2 - 1}{4} \right] \tag{2.13.294}$$

$$w(z) = 0 \quad ; \quad w(z) = z^{(1-\beta)/2} C_\lambda (\alpha e^{1/z})$$

$$z^2 \frac{d^2w(z)}{dz^2} + (\beta+1) z \frac{dw(z)}{dz} + \left[\frac{1}{z^2} (\alpha^2 e^{2\lambda/z} - \mu^2) + \frac{\beta^2 - 1}{4} \right] \tag{2.13.295}$$

$$w(z) = 0 \quad ; \quad w(z) = z^{(1-\beta)/2} C_{\mu/\lambda} \left(\frac{\alpha}{\lambda} e^{\lambda/z} \right)$$

$$z^2 \frac{d^2w(z)}{dz^2} + z(\beta z - 2\alpha) \frac{dw(z)}{dz} + [\alpha(\alpha+1) - \alpha\beta z + \lambda^2 e^z] w(z) = 0 \tag{2.13.296}$$

$$w(z) = z^\alpha e^{-\beta z/2} C_\beta (2\lambda e^{z/2})$$

$$z \frac{d^2w(z)}{dz^2} - (2\alpha z^2 + 1) \frac{dw(z)}{dz} + 4\beta^2 z^3 e^{2\lambda z^2} w(z) = 0 \tag{2.13.297}$$

$$w(z) = e^{\alpha^2 z/2} C_{\alpha/2\lambda} \left(\frac{\sqrt{\beta}}{\lambda} z^{\lambda z^2} \right)$$

$$z^2 \frac{d^2w(z)}{dz^2} - 2\alpha z \frac{dw(z)}{dz} + [\alpha(\alpha+1) - z^2(1 - 4\beta^2 \lambda^2 e^{2\beta z})] w(z) = 0 \tag{2.13.298}$$

$$w(z) = z^\alpha e^z C_{1/\beta} (2\lambda e^{\beta z})$$

$$z^4 \frac{d^2w(z)}{dz^2} + z^5 \frac{dw(z)}{dz} + (z^2 - \alpha^2) w(z) = 0 \quad ; \quad w(z) = C_\alpha\left(\frac{1}{z}\right) \tag{2.13.299}$$

$$z^4 \frac{d^2w(z)}{dz^2} + z^5 \frac{dw(z)}{dz} - (z^2 + \alpha^2) w(z) = 0 \quad ; \quad w(z) = Z_\alpha\left(\frac{1}{z}\right) \tag{2.13.300}$$

$$\begin{gathered} z^2 \frac{d^2w(z)}{dz^2} + z\frac{dw(z)}{dz} + \alpha^2[\ln(\beta z)]^n w(z) = 0 \\ w(z) = \sqrt{\ln(\beta z)}\, C_{1/2n}\left(\frac{\alpha}{n}[\ln(\beta z)]^n\right) \end{gathered} \tag{2.13.301}$$

$$\begin{gathered} z^2 \frac{d^2w(z)}{dz^2} + (z - 2z^2 \tan z)\frac{dw(z)}{dz} - (z \tan z + \alpha^2)\, w(z) = 0 \\ w(z) = \frac{1}{\cos z} C_\alpha(z) \end{gathered} \tag{2.13.302}$$

$$\begin{gathered} z^2 \frac{d^2w(z)}{dz^2} + (z + 2z^2 \cot z)\frac{dw(z)}{dz} + (z \cot z - \alpha^2)\, w(z) = 0 \\ w(z) = \frac{1}{\sin z} C_\alpha(z) \end{gathered} \tag{2.13.303}$$

$$\begin{gathered} z^2 \frac{d^2w(z)}{dz^2} + z(\alpha + 1 - 2z \tan z)\frac{dw(z)}{dz} - \\ \left[(\alpha+1) z(\tan z - 1) - (z \tan z)^2 - \left(\frac{z}{\cos z}\right)^2\right] w(z) = 0 \\ w(z) = \frac{z^{-\alpha/2}}{\cos z} C_\alpha(2\sqrt{z}) \end{gathered} \tag{2.13.304}$$

$$\begin{gathered} z^2 \frac{d^2w(z)}{dz^2} + z(\alpha + 1 + 2z \cot z)\frac{dw(z)}{dz} + \\ \left[(\alpha+1) z(\cot z + 1) - (z \cot z)^2 - \left(\frac{z}{\sin z}\right)^2\right] w(z) = 0 \\ w(z) = \frac{z^{-\alpha/2}}{\sin z} C_\alpha(2\sqrt{z}) \end{gathered} \tag{2.13.305}$$

$$\begin{gathered} z^2 \frac{d^2w(z)}{dz^2} + z(\alpha + 1 + 2z \tan z)\frac{dw(z)}{dz} \\ + [(\alpha+1) z(\mathrm{tantz} + 1) - 2(z \tan z)^2 - z^2]\, w(z) = 0 \\ w(z) = \cos z\, C_\alpha(2\sqrt{z}) \end{gathered} \tag{2.13.306}$$

$$z^2\frac{d^2w(z)}{dz^2}+z[1+\beta\lambda+2\alpha z\tan(\alpha z)]\frac{dw(z)}{dz}+[\alpha^2z^2(1+2(\tan(\alpha z))^2$$
$$+\alpha z(1+\beta\lambda)\tan(\alpha z)+\beta^2\mu^2z^\lambda]w(z)=0 \tag{2.13.307}$$
$$w(z)=z^{-\beta\lambda/2}\cos(\alpha z)\,C_\beta(2\mu z^{\lambda/2})$$

$$z^2\frac{d^2w(z)}{dz^2}+z(\alpha+1-2z\cot z)\frac{dw(z)}{dz}$$
$$-[(\alpha+1)z(\cot z-1)-2(z\cot z)^2-z^2]w(z)=0 \tag{2.13.308}$$
$$w(z)=\sin z\,C_\alpha(2\sqrt{z})$$

$$z^2\frac{d^2w(z)}{dz^2}+z[1+\beta\lambda+2\alpha z\cot(\alpha z)]\frac{dw(z)}{dz}$$
$$+[\alpha^2z^2(1+2(\cot(\alpha z))^2-\alpha z(1+\beta\lambda)\cot(\alpha z) \tag{2.13.309}$$
$$+\beta^2\mu^2z^\lambda]w(z)=0\quad;\quad w(z)=z^{-\beta\lambda/2}\sin(\alpha z)\,C_\beta(2\mu z^{\lambda/2})$$

$$\frac{d^2w(z)}{dz^2}+\tan z\,\frac{dw(z)}{dz}+\alpha^2(\cos z)^2(\sin z)^{2n-2}\,w(z)=0 \tag{2.13.310}$$
$$w(z)=\sqrt{\sin z}\,C_{1/2n}\left[\frac{\alpha}{n}(\sin z)^n\right]$$

3 Differentiation and Integration with Respect to the Order of the Bessel and Related Functions

3.1 First Derivatives of the Bessel and Related Functions with Respect to the Order ν: Unrestricted Values of the Order ν

Two solutions of the Bessel differential equation (2.3.1)

$$z^2 \frac{d^2 w(z)}{dz^2} + z \frac{dw(z)}{dz} + (z^2 - \nu^2) w(z) = 0 \tag{3.1.1}$$

depend on continuously changing variable z (argument z) and a prescribed fixed value ν which is called the order of Bessel function or sometimes the index of it:

$$w(z) = A_1 J_\nu(z) + A_2 J_{-\nu}(z) \tag{3.1.2}$$

where A_1 and A_2 are the integration constants. These Bessel functions of the first kind in eq. (3.1.2) are linearly independent solutions of the Bessel differential equation for any real or complex value of ν. However, when the order ν is an integer n, it follows from (2.2.10) that

$$J_{-n}(z) = (-1)^n J_n(z) \tag{3.1.3}$$

and therefore in order to determine the second independent solution of differential equation, the Bessel functions of the second kind are introduced:

$$Y_n(z) = \frac{\cos(\pi n) J_n(z) - J_{-n}(z)}{\sin(\pi n)} \tag{3.1.4}$$

If $\nu = n$, the function $Y_n(z)$ is undefined, both the numerator and the denominator vanish and the resulting form of 0/0 is resolved by applying L'Hôpital rule to (3.1.4)

$$\begin{aligned} Y_n(z) &= \lim_{\nu \to n} Y_\nu(z) = \lim_{\nu \to n} \left[\frac{\cos(\pi n) J_n(z) - J_{-n}(z)}{\sin(\pi n)} \right] \\ Y_n(z) &= \frac{1}{\pi} \frac{\partial}{\partial \nu} [J_\nu(z) - (-1)^n J_{-\nu}(z)]_{\nu = n} \quad ; \quad n = 0, \pm 1, \pm 2, \pm 3, \ldots \\ Y_{-n}(z) &= (-1)^n Y_n(z) \end{aligned} \tag{3.1.5}$$

Thus, in the case of the Bessel functions of the second kind, the mathematical operation – differentiation of the Bessel functions of the first kind with regard to the order emerges. The explicit expression for (3.1.5) is received by differentiation of the ascending series of $J_\nu(z)$ from (2.9.1)

https://doi.org/10.1515/9783110681642-003

$$J_{\pm\nu}(z) = \sum_{k=0}^{\infty} \frac{(-1)^k \left(\frac{z}{2}\right)^{2k\pm\nu}}{k!\Gamma(k\pm\nu+1)} \tag{3.1.6}$$

which gives

$$\frac{\partial J_\nu(z)}{\partial\nu} = J_\nu(z)\ln\left(\frac{z}{2}\right) - \sum_{k=0}^{\infty}(-1)^k\left(\frac{z}{2}\right)^{\nu+2k}\frac{\psi(\nu+k+1)}{k!\Gamma(\nu+k+1)} \tag{3.1.7}$$

and

$$\frac{\partial J_{-\nu}(z)}{\partial\nu} = -J_{-\nu}(z)\ln\left(\frac{z}{2}\right) + \sum_{k=0}^{\infty}(-1)^k\left(\frac{z}{2}\right)^{-\nu+2k}\frac{\psi(-\nu+k+1)}{k!\Gamma(-\nu+k+1)} \tag{3.1.8}$$

where the logarithmic derivative of the gamma function, the psi function is defined by

$$\begin{aligned}
\psi(z) &= \frac{d}{dz}\ln\Gamma(z) = \frac{1}{\Gamma(z)}\frac{d\Gamma(z)}{dz} = -\gamma + \sum_{k=0}^{\infty}\left(\frac{1}{k+1} - \frac{1}{z+k}\right)\\
\psi\left(\frac{1}{2}\right) &= -\gamma - 2\ln 2\\
\psi(1) &= -\gamma = -0.5772157\ldots\\
\psi\left(\frac{1}{2}\right) &= -\gamma - 2\ln 2\\
\psi\left(\frac{3}{2}\right) &= -\gamma - 2\ln 2 + 2
\end{aligned} \tag{3.1.9}$$

and γ denotes the Euler constant.

In Figure 3.1, the first derivatives of the Bessel functions of the first kind with regard to the order are plotted. As can be observed, they change only for small values of the order, $\nu < 3$. For larger values of ν, irrespective of the argument t, these derivatives tend to be zero. Three-dimensional plot of the first derivatives and tables of their values are presented in Part 2 of this book.

The expressions in (3.1.7) and (3.1.8) serve to determine the first derivative of the Bessel function of the second kind with regard to the order ν

$$\frac{\partial Y_\nu(z)}{\partial\nu} = \cot(\pi\nu)\frac{\partial J_\nu(z)}{\partial\nu} - \operatorname{cosec}(\pi\nu)\frac{\partial J_{-\nu}(z)}{\partial\nu} - \pi\operatorname{cosec}(\pi\nu)Y_{-\nu}(z) \tag{3.1.10}$$

The modified Bessel functions of the second kind $I_\nu(z)$ and $K_\nu(z)$ are linearly independent solutions of the following differential equation

$$z^2\frac{d^2w(z)}{dz^2} + z\frac{dw(z)}{dz} \pm (z^2+\nu^2)w(z) = 0 \tag{3.1.11}$$

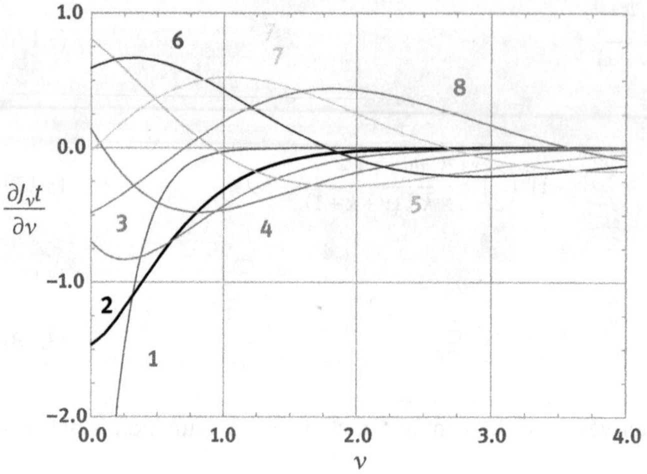

Figure 3.1: The first derivatives of the Bessel function of the first kind with regard to the order v as a function of v, at constant values of argument t.
1 – $t = 0.05$; 2 – $t = 0.25$; 3 – $t = 0.50$; 4 – $t = 1.0$; 5 – $t = 2.0$; 6 – $t = 3.0$; 7 – $t = 4.0$; 8 – $t = 5.0$

and similarly we have

$$w(z) = A_1 I_v(z) + A_2 I_{-v}(z) \quad ; \quad v \neq 0, \pm 1, \pm 2, \pm 3, \ldots$$
$$K_v(z) = \frac{\pi}{2}\left[\frac{I_{-v}(z) - I_v(z)}{\sin(\pi v)}\right] \tag{3.1.12}$$

and therefore for v being an integer n

$$K_n(z) = \frac{\pi}{2}\left[\frac{\frac{\partial}{\partial v}(I_{-v}(z) - I_v(z))}{\frac{\partial}{\partial v}(\sin(\pi v))}\right]_{v=n} \quad ; \quad n = 0, 1, 2, 3, \ldots$$
$$I_{-n}(z) = I_n(z) \tag{3.1.13}$$

Using the series expansion from (2.9.5)

$$I_v(z) = \sum_{k=0}^{\infty} \frac{\left(\frac{z}{2}\right)^{2k+v}}{k!\Gamma(k+v+1)} \tag{3.1.14}$$

direct differentiation gives

$$\frac{\partial I_v(z)}{\partial v} = I_v(z)\ln\left(\frac{z}{2}\right) - \sum_{k=0}^{\infty}\left(\frac{z}{2}\right)^{v+2k}\frac{\psi(v+k+1)}{k!\Gamma(v+k+1)} \tag{3.1.15}$$

and

$$\frac{\partial K_\nu(z)}{\partial \nu} = -\pi \cot(\pi\nu) K_\nu(z) + \frac{\pi}{2}\operatorname{cosec}(\pi\nu)\left[\frac{\partial I_{-\nu}(z)}{\partial \nu} - \frac{\partial I_\nu(z)}{\partial \nu}\right] \tag{3.1.16}$$

As can be observed, in expressions (3.1.7), (3.1.8) and (3.1.15), the term-by-term differentiation is performed in two places, the ν – powers of variable z, and the orders ν included in the gamma functions.

Introducing the series expansion of the Struve functions from (2.9.7)

$$H_\nu(z) = \sum_{k=0}^{\infty} (-1)^k \frac{\left(\frac{z}{2}\right)^{2k+\nu+1}}{\Gamma\left(k+\frac{3}{2}\right)\Gamma\left(k+\nu+\frac{3}{2}\right)} \tag{3.1.17}$$

it follows that

$$\frac{\partial H_\nu(z)}{\partial \nu} = H_\nu(z)\ln\left(\frac{z}{2}\right) - \sum_{k=0}^{\infty} (-1)^k \frac{\left(\frac{z}{2}\right)^{\nu+2k+1}\psi\left(\nu+k+\frac{3}{2}\right)}{\Gamma\left(k+\frac{3}{2}\right)\Gamma\left(\nu+k+\frac{3}{2}\right)} \tag{3.1.18}$$

and similarly from (2.9.10)

$$L_\nu(z) = \sum_{k=0}^{\infty} \frac{\left(\frac{z}{2}\right)^{2k+\nu+1}}{\Gamma\left(k+\frac{3}{2}\right)\Gamma\left(k+\nu+\frac{3}{2}\right)} \tag{3.1.19}$$

we have

$$\frac{\partial L_\nu(z)}{\partial \nu} = L_\nu(z)\ln\left(\frac{z}{2}\right) - \sum_{k=0}^{\infty} \frac{\left(\frac{z}{2}\right)^{\nu+2k+1}\psi\left(\nu+k+\frac{3}{2}\right)}{\Gamma\left(k+\frac{3}{2}\right)\Gamma\left(\nu+k+\frac{3}{2}\right)} \tag{3.1.20}$$

Considering that (2.9.19)

$$\operatorname{ber}_\nu(z) = \left(\frac{z}{2}\right)^\nu \sum_{k=0}^{\infty} \frac{\cos\left[\pi\left(\frac{3\nu+4k}{4}\right)\right]}{k!\Gamma(\nu+k+1)}\left(\frac{z^2}{2}\right)^k \tag{3.1.21}$$

and (2.9.21)

$$\operatorname{bei}_\nu(z) = \left(\frac{z}{2}\right)^\nu \sum_{k=0}^{\infty} \frac{\sin\left[\pi\left(\frac{3\nu+4k}{4}\right)\right]}{k!\Gamma(\nu+k+1)}\left(\frac{z^2}{2}\right)^k \tag{3.1.22}$$

the derivatives of the Kelvin functions are

$$\begin{aligned}\frac{\partial \operatorname{ber}_\nu(z)}{\partial \nu} &= \ln\left(\frac{z}{2}\right)\operatorname{ber}_\nu(z) - \frac{3\pi}{4}\operatorname{bei}_\nu(z)\\ &\quad - \left(\frac{z}{2}\right)^\nu \sum_{k=0}^{\infty} \frac{\cos\left[\pi\left(\frac{3\nu+4k}{4}\right)\right]\psi(\nu+k+1)}{k!\Gamma(\nu+k+1)}\left(\frac{z^2}{2}\right)^k\end{aligned} \tag{3.1.23}$$

and

$$\frac{\partial \operatorname{bei}_\nu(z)}{\partial \nu} = \ln\left(\frac{z}{2}\right)\operatorname{bei}_\nu(z) + \frac{3\pi}{4}\operatorname{ber}_\nu(z) - \left(\frac{z}{2}\right)^\nu \sum_{k=0}^{\infty} \frac{\sin\left[\pi\left(\frac{3\nu+4k}{4}\right)\right]\psi(\nu+k+1)}{k!\Gamma(\nu+k+1)}\left(\frac{z^2}{2}\right)^k \tag{3.1.24}$$

In the case of other related Bessel functions, the series expansions are less suitable to determine desired derivatives with regard to the order because differentiation should be performed at more than three places in the corresponding series, and therefore other approaches are more preferable, as will be discussed later.

The Bessel functions can also be expressed in terms of series of the Bessel functions having integer order

$$J_\nu(z) = \frac{1}{\Gamma(\nu)}\sum_{k=0}^{\infty}\frac{\left(\frac{z}{2}\right)^{k+\nu}}{k!(k+\nu)}J_k(z) \tag{3.1.25}$$

$$I_\nu(z) = \frac{1}{\Gamma(\nu)}\sum_{k=0}^{\infty}\frac{(-1)^k\left(\frac{z}{2}\right)^{k+\nu}}{k!(k+\nu)}I_k(z) \tag{3.1.26}$$

and therefore using

$$\psi(\nu+1) = \psi(\nu) + \frac{1}{\nu} \tag{3.1.27}$$

we have

$$\frac{\partial J_\nu(z)}{\partial \nu} = \left[\ln\left(\frac{z}{2}\right) - \psi(\nu+1)\right]J_\nu(z) + \frac{1}{\Gamma(\nu+1)}\sum_{k=1}^{\infty}\frac{k\left(\frac{z}{2}\right)^{k+\nu}}{k!(k+\nu)^2}J_k(z) \tag{3.1.28}$$

and

$$\frac{\partial I_\nu(z)}{\partial \nu} = \left[\ln\left(\frac{z}{2}\right) - \psi(\nu+1)\right]I_\nu(z) + \frac{1}{\Gamma(\nu+1)}\sum_{k=1}^{\infty}\frac{(-1)^k k\left(\frac{z}{2}\right)^{k+\nu}}{k!(k+\nu)^2}I_k(z) \tag{3.1.29}$$

Brychkov [45] presented the first derivatives of the Bessel and related functions with respect to the order in a more closed form by using the generalized hypergeometric functions:

$$\begin{aligned}\frac{\partial J_\nu(z)}{\partial\nu} &= \frac{\pi\left(\frac{z}{2}\right)^{2\nu}}{2\left[\Gamma(\nu+1)\right]^2}\left[Y_\nu(z)-\cot(\pi\nu)J_\nu(z)\right]f(\nu)\\ &\quad+J_\nu(z)\left[\frac{1}{2\nu}-\psi(\nu+1)+\ln\left(\frac{z}{2}\right)-\frac{z^2}{4(\nu^2-1)}g(\nu)\right]\\ f(\nu) &= {}_2F_3\left(\nu,\nu+\frac{1}{2};\nu+1,\nu+1,2\nu+1;-z^2\right)\\ g(\nu) &= {}_3F_4\left(1,1,\frac{3}{2};2,2,2-\nu,\nu+2;-z^2\right)\end{aligned} \tag{3.1.30}$$

and

$$\begin{aligned}\frac{\partial I_\nu(z)}{\partial\nu} &= \frac{\left(\frac{z}{2}\right)^{2\nu}}{\left[\Gamma(\nu+1)\right]^2}\left[K_\nu(z)+\frac{\pi}{2}\csc(\pi\nu)I_\nu(z)\right]f(\nu)\\ &\quad+I_\nu(z)\left[\frac{1}{2\nu}-\psi(\nu+1)+\ln\left(\frac{z}{2}\right)-\frac{z^2}{4(\nu^2-1)}g(\nu)\right]\\ f(\nu) &= {}_2F_3\left(\nu,\nu+\frac{1}{2};\nu+1,\nu+1,2\nu+1;z^2\right)\\ g(\nu) &= {}_3F_4\left(1,1,\frac{3}{2};2,2,2-\nu,\nu+2;z^2\right)\end{aligned} \tag{3.1.31}$$

Brychkov also derived more lengthy expressions for derivatives of $Y_\nu(z)$ and $K_\nu(z)$ by using the ${}_3F_4$ hypergeometric functions, and for derivatives of $\mathrm{ber}_\nu(z)$ and $\mathrm{bei}_\nu(z)$ the ${}_4F_7$ hypergeometric functions. Using these expressions, it is possible to obtain derivatives of $\mathrm{ker}_\nu(z)$ and $\mathrm{kei}_\nu(z)$ functions by applying interrelations between the Kelvin functions as given in (2.1.26) and (2.1.27). Recently, González-Santander [53, 54, 86, 87] investigated derivatives of the Bessel and the Kelvin functions and expressed them in terms of the hypergeometric and the Meijer-G functions.

Differentiation of the recurrence formulas between the Bessel functions $J_\nu(z)$ and $Y_\nu(z)$ gives interrelation between different orders, for example from (2.2.13)

$$C_{\nu+1}(z)=\frac{2\nu}{z}C_\nu(z)-C_{\nu-1}(z) \tag{3.1.32}$$

we have

$$\frac{\partial C_{\nu+1}(z)}{\partial\nu}=\frac{2\nu}{z}\frac{\partial C_\nu(z)}{\partial\nu}-\frac{\partial C_{\nu-1}(z)}{\partial\nu}+\frac{2}{z}C_\nu(z) \tag{3.1.33}$$

Similarly, from (2.3.6) and (2.3.21)

$$I_{\nu+1}(z)=I_{\nu-1}(z)-\frac{2\nu}{z}I_\nu(z) \tag{3.1.34}$$

$$K_{\nu+1}(z) = K_{\nu-1}(z) + \frac{2\nu}{z} K_\nu(z) \tag{3.1.35}$$

it follows that

$$\frac{\partial I_{\nu+1}(z)}{\partial \nu} = -\frac{2\nu}{z}\frac{\partial I_\nu(z)}{\partial \nu} + \frac{\partial I_{\nu-1}(z)}{\partial \nu} - \frac{2}{z} I_\nu(z) \tag{3.1.36}$$

and

$$\frac{\partial K_{\nu+1}(z)}{\partial \nu} = \frac{2\nu}{z}\frac{\partial K_\nu(z)}{\partial \nu} - \frac{\partial K_{\nu-1}(z)}{\partial \nu} + \frac{2}{z} K_\nu(z) \tag{3.1.37}$$

In the case of the Struve functions the recurrence formulas are (2.4.4) and (2.4.22)

$$H_{\nu+1}(z) = \frac{2\nu}{z} H_\nu(z) - H_{\nu-1}(z) + \frac{\left(\frac{z}{2}\right)^\nu}{\sqrt{\pi}\,\Gamma\left(\nu + \frac{3}{2}\right)} \tag{3.1.38}$$

$$L_{\nu+1}(z) = L_{\nu-1}(z) - \frac{2\nu}{z} L_\nu(z) - \frac{\left(\frac{z}{2}\right)^\nu}{\sqrt{\pi}\,\Gamma\left(\nu + \frac{3}{2}\right)} \tag{3.1.39}$$

and therefore

$$\begin{aligned} \frac{\partial H_{\nu+1}(z)}{\partial \nu} &= \frac{2\nu}{z}\frac{\partial H_\nu(z)}{\partial \nu} - \frac{\partial H_{\nu-1}(z)}{\partial \nu} + \frac{2}{z} H_\nu(z) \\ &\quad + \left[\ln\left(\frac{z}{2}\right) - \psi\left(\nu + \frac{3}{2}\right)\right] \frac{\left(\frac{z}{2}\right)^\nu}{\sqrt{\pi}\,\Gamma\left(\nu + \frac{3}{2}\right)} \end{aligned} \tag{3.1.40}$$

and

$$\begin{aligned} \frac{\partial L_{\nu+1}(z)}{\partial \nu} &= \frac{\partial L_{\nu-1}(z)}{\partial \nu} - \frac{2\nu}{z}\frac{\partial L_\nu(z)}{\partial \nu} - \frac{2}{z} L_\nu(z) \\ &\quad + \left[\psi\left(\nu + \frac{3}{2}\right) - \ln\left(\frac{z}{2}\right)\right] \frac{\left(\frac{z}{2}\right)^\nu}{\sqrt{\pi}\,\Gamma\left(\nu + \frac{3}{2}\right)} \end{aligned} \tag{3.1.41}$$

The Lommel differential equation

$$z^2 \frac{d^2 w(z)}{dz^2} + z\frac{dw(z)}{dz} + (z^2 - \nu^2) w(z) = z^\mu \tag{3.1.42}$$

includes two parameters (or variables) ν and μ, but here only ν is of importance because the homogenous equation is satisfied by the Bessel functions of the order ν and therefore differentiation with regard to it will only be performed. If two parameters ν and μ are identical, $\nu = \mu$, then the Lommel functions can be expressed in terms of the Neumann and Struve functions

$$s_{\nu,\nu}(z) = 2^{\nu-1}\sqrt{\pi}\,\Gamma\left(\nu+\frac{1}{2}\right) H_\nu(z) \tag{3.1.43}$$

$$S_{\nu,\nu}(z) = 2^{\nu-1}\sqrt{\pi}\,\Gamma\left(\nu+\frac{1}{2}\right) [H_\nu(z) - Y_\nu(z)] \tag{3.1.44}$$

which gives

$$\frac{\partial s_{\nu,\nu}(z)}{\partial \nu} = 2^{\nu-1}\sqrt{\pi}\,\Gamma\left(\nu+\frac{1}{2}\right)\left\{\frac{\partial H_\nu(z)}{\partial \nu} + \left[\ln 2 + \psi\left(\nu+\frac{1}{2}\right)\right] H_\nu(z)\right\} \tag{3.1.45}$$

and

$$\begin{aligned}\frac{\partial S_{\nu,\nu}(z)}{\partial \nu} &= \left\{\frac{\partial H_\nu(z)}{\partial \nu} - \frac{\partial Y_\nu(z)}{\partial \nu} + 2^{\nu-1}\sqrt{\pi}\,\Gamma\left(\nu+\frac{1}{2}\right)\right.\\ &\left.\left[\ln 2 + \psi\left(\nu+\frac{1}{2}\right)\right] [H_\nu(z) - Y_\nu(z)]\right\}\end{aligned} \tag{3.1.46}$$

In general case, from (2.9.11)

$$s_{\mu,\nu}(z) = \frac{z^{\mu+1}}{4}\sum_{k=0}^{\infty}\frac{(-1)^k\left(\frac{z}{2}\right)^{2k}\Gamma\left(\frac{\mu-\nu+1}{2}\right)\Gamma\left(\frac{\mu+\nu+1}{2}\right)}{\Gamma\left(\frac{\mu-\nu+2k+3}{2}\right)\Gamma\left(\frac{\mu+\nu+2k+3}{2}\right)} \tag{3.1.47}$$

differentiation yields

$$\begin{aligned}\frac{\partial s_{\mu,\nu}(z)}{\partial \nu} &= \frac{z^{\mu+1}}{4}\sum_{k=0}^{\infty}\frac{(-1)^k\left(\frac{z}{2}\right)^{2k}\Gamma\left(\frac{\mu-\nu+1}{2}\right)\Gamma\left(\frac{\mu+\nu+1}{2}\right)}{\Gamma\left(\frac{\mu-\nu+2k+3}{2}\right)\Gamma\left(\frac{\mu+\nu+2k+3}{2}\right)}\,a_k\\ a_k &= \left[\psi\left(\frac{\mu-\nu+1}{2}\right) + \psi\left(\frac{\mu+\nu+1}{2}\right) - \psi\left(\frac{\mu-\nu+2k+3}{2}\right) - \psi\left(\frac{\mu-\nu+2k+3}{2}\right)\right]\end{aligned} \tag{3.1.48}$$

Since

$$\begin{aligned}S_{\lambda,\nu}(z) &= s_{\mu,\nu}(z) + A_{\mu,\nu}B_{\mu,\nu}\\ A_{\mu,\nu} &= 2^{\mu-1}\Gamma\left(\frac{\mu-\nu+1}{2}\right)\Gamma\left(\frac{\mu+\nu+1}{2}\right)\\ B_{\mu,\nu} &= \left[\sin\left(\frac{\pi(\mu-\nu)}{2}\right) J_\nu(z) - \cos\left(\frac{\pi(\mu-\nu)}{2}\right) Y_\nu(z)\right]\end{aligned} \tag{3.1.49}$$

we have

$$\frac{\partial S_{\lambda,\nu}(z)}{\partial \nu} = \frac{\partial s_{\mu,\nu}(z)}{\partial \nu} + \frac{1}{2}A_{\mu,\nu}B_{\mu,\nu}\left[\psi\left(\frac{\mu+\nu+1}{2}\right) - \psi\left(\frac{\mu-\nu+1}{2}\right)\right]$$
$$+A_{\mu,\nu}\sin\left(\frac{\pi(\mu-\nu)}{2}\right)\left[\frac{\partial J_\nu(z)}{\partial \nu} - \frac{\pi}{2}Y_\nu(z)\right] \tag{3.1.50}$$
$$-A_{\mu,\nu}\cos\left(\frac{\pi(\mu-\nu)}{2}\right)\left[\frac{\partial Y_\nu(z)}{\partial \nu} + \frac{\pi}{2}J_\nu(z)\right]$$

3.2 First Derivatives of the Bessel and Related Functions with Respect to the Order *v*: Particular Values of the Order *v*

Since the Bessel functions of integer order can be expressed in the form of finite series, their derivatives with regard to the order are the finite sums of the corresponding Bessel and Hankel functions [18]

$$\left[\frac{\partial J_\nu(z)}{\partial \nu}\right]_{\nu=\pm n} = (\pm 1)^n\left[\frac{\pi}{2}Y_n(z) \pm \frac{n!}{2}\sum_{k=0}^{n-1}\left(\frac{z}{2}\right)^{k-n}\frac{J_k(z)}{k!(n-k)}\right] \tag{3.2.1}$$

$$\left[\frac{\partial Y_\nu(z)}{\partial \nu}\right]_{\nu=\pm n} = (\pm 1)^n\left[-\frac{\pi}{2}J_n(z) \pm \frac{n!}{2}\sum_{k=0}^{n-1}\left(\frac{z}{2}\right)^{k-n}\frac{Y_k(z)}{k!(n-k)}\right] \tag{3.2.2}$$

$$\left[\frac{\partial H_\nu^{(j)}(z)}{\partial \nu}\right]_{\nu=n} = (-1)^j\frac{\pi i}{2}H_n^{(j)}(z) + \left[\frac{n!}{2}\sum_{k=0}^{n-1}\left(\frac{z}{2}\right)^{k-n}\frac{H_k^{(j)}(z)}{k!(n-k)}\right] \tag{3.2.3}$$

$j=1,2$

$$\left[\frac{\partial H_\nu^{(j)}(z)}{\partial \nu}\right]_{\nu=-n} = (-1)^{j+n}\frac{\pi i}{2}H_n^{(j)}(z)$$
$$-(-1)^n\left[\frac{n!}{2}\sum_{k=0}^{n-1}\left(\frac{z}{2}\right)^{k-n}\frac{H_k^{(j)}(z)}{k!(n-k)}\right] \quad ; \quad j=1,2 \tag{3.2.4}$$

and

$$\left[\frac{\partial I_\nu(z)}{\partial \nu}\right]_{\nu=\pm n} = (-1)^n\left[-K_n(z) \pm \frac{n!}{2}\sum_{k=0}^{n-1}\left(\frac{z}{2}\right)^{k-n}\frac{(-1)^k I_k(z)}{k!(n-k)}\right] \tag{3.2.5}$$

$$\left[\frac{\partial K_\nu(z)}{\partial \nu}\right]_{\nu=\pm n} = \left[\pm\frac{n!}{2}\sum_{k=0}^{n-1}\left(\frac{z}{2}\right)^{k-n}\frac{K_k(z)}{k!(n-k)}\right] \tag{3.2.6}$$

Derivatives with respect to the integer order of the Lommel, Anger and Struve functions are

$$\left[\frac{\partial S_{\lambda,v}(z)}{\partial v}\right]_{v=n} = \frac{n!\Gamma\left(\frac{1-\lambda-n}{2}\right)}{2\Gamma\left(\frac{1-\lambda+n}{2}\right)} \sum_{k=0}^{n-1} \frac{z^{1-n}\left(\frac{1-\lambda-n}{2}\right)}{(n-k)\,k!} S_{\lambda+n-k,\,k}(z) - \frac{1}{2}\left[\psi\left(\frac{1-\lambda+n}{2}\right) - \psi\left(\frac{1-\lambda-n}{2}\right)\right] S_{\lambda,v}(z) \tag{3.2.7}$$

$$\left[\frac{\partial \mathbf{J}_v(z)}{\partial v}\right]_{v=n} = \frac{\pi}{2} H_n(z) + \frac{n!}{2}\sum_{k=0}^{n-1}\left(\frac{2}{z}\right)^{n-k} \frac{J_k(z)}{k!(n-k)} - \frac{1}{2}\sum_{k=0}^{n-1}\frac{\Gamma\left(k+\frac{1}{2}\right)}{\Gamma\left(n-k+\frac{1}{2}\right)}\left(\frac{z}{2}\right)^{n-2k-1} + \frac{(-1)^n}{z}\sum_{k=0}^{n-1}\frac{(-1)^k\Gamma\left(\frac{n+k+1}{2}\right)}{\Gamma\left(\frac{n-k+1}{2}\right)}\left(\frac{z}{2}\right)^k \tag{3.2.8}$$

and

$$\left[\frac{\partial \mathbf{J}_v(z)}{\partial v}\right]_{v=-n} = \frac{\pi}{2} H_{-n}(z) + \frac{(-1)^{n-1}n!}{2}\sum_{k=0}^{n-1}\left(\frac{2}{z}\right)^{n-k}\frac{J_k(z)}{k!(n-k)} + \frac{(-1)^n}{z}\sum_{k=0}^{n-1}\frac{\Gamma\left(\frac{n+k+1}{2}\right)}{\Gamma\left(\frac{n-k+1}{2}\right)}\left(\frac{z}{2}\right)^k \tag{3.2.9}$$

$$\left[\frac{\partial \mathbf{E}_v(z)}{\partial v}\right]_{v=n} = \frac{\pi}{2} J_n(z) + \frac{n!}{2}\sum_{k=0}^{n-1}\frac{\left(\frac{2}{z}\right)^{n-k}}{k!(n-k)}\left[-H_k(z) + \frac{1}{\pi}\sum_{j=0}^{k-1}\frac{\Gamma\left(j+\frac{1}{2}\right)}{\Gamma\left(k-j+\frac{1}{2}\right)}\left(\frac{z}{2}\right)^{k-2j-1}\right] + \frac{1}{2\pi}\sum_{k=0}^{n-1}\left[(-1)^k+(-1)^n\right]\frac{\Gamma\left(\frac{n+k+1}{2}\right)}{\Gamma\left(\frac{n-k+1}{2}\right)}\left(-\frac{2}{z}\right)^{k+1}\sum_{j=0}^{k-1}\frac{1}{n-k+2j+1} \tag{3.2.10}$$

$$\left[\frac{\partial \mathbf{E}_v(z)}{\partial v}\right]_{v=-n} = (-1)^n\frac{\pi}{2} J_n(z) + \frac{n!}{2}\sum_{k=0}^{n-1}\frac{\left(-\frac{2}{z}\right)^{n-k}}{k!(n-k)} H_{-k}(z) \quad \frac{1}{2\pi}\sum_{k=0}^{n-1}\left[(-1)^k+(-1)^n\right]\frac{\Gamma\left(\frac{n+k+1}{2}\right)}{\Gamma\left(\frac{n-k+1}{2}\right)}\left(\frac{2}{z}\right)^{k+1}\sum_{j=0}^{k-1}\frac{1}{n-k+2j+1} \tag{3.2.11}$$

The analogous expressions for derivatives with respect to the integer order of the Kelvin functions are

$$\left[\frac{\partial\, \mathrm{ber}_v(z)}{\partial v}\right]_{v=n} = -\frac{\pi}{2}\mathrm{bei}_n(z) - \mathrm{ker}_n(z) + \frac{n!}{2}\sum_{k=0}^{n-1}\frac{\left(\frac{z}{2}\right)^{k-n}}{k!(n-k)}\cos\left(\frac{5\pi(k-n)}{4}\right)\mathrm{ber}_k(z) + \frac{n!}{2}\sum_{k=0}^{n-1}\frac{\left(\frac{z}{2}\right)^{k-n}}{k!(n-k)}\sin\left(\frac{5\pi(k-n)}{4}\right)\mathrm{bei}_k(z) \tag{3.2.12}$$

$$\left[\frac{\partial \operatorname{bei}_\nu(z)}{\partial \nu}\right]_{\nu=n} = \frac{\pi}{2}\operatorname{ber}_n(z) - \operatorname{kei}_n(z) + \frac{n!}{2}\sum_{k=0}^{n-1}\frac{\left(\frac{z}{2}\right)^{k-n}}{k!(n-k)}\cos\left(\frac{5\pi(k-n)}{4}\right)\operatorname{bei}_k(z) - \frac{n!}{2}\sum_{k=0}^{n-1}\frac{\left(\frac{z}{2}\right)^{k-n}}{k!(n-k)}\sin\left(\frac{5\pi(k-n)}{4}\right)\operatorname{ber}_k(z) \tag{3.2.13}$$

and

$$\left[\frac{\partial \operatorname{ker}_\nu(z)}{\partial \nu}\right]_{\nu=n} = \frac{\pi}{2}\operatorname{kei}_n(z) + \frac{n!}{2}\sum_{k=0}^{n-1}\frac{\left(\frac{z}{2}\right)^{k-n}}{k!(n-k)}\cos\left(\frac{3\pi(k-n)}{4}\right)\operatorname{ker}_k(z) - \frac{n!}{2}\sum_{k=0}^{n-1}\frac{\left(\frac{z}{2}\right)^{k-n}}{k!(n-k)}\sin\left(\frac{3\pi(k-n)}{4}\right)\operatorname{kei}_k(z) \tag{3.2.14}$$

$$\left[\frac{\partial \operatorname{kei}_\nu(z)}{\partial \nu}\right]_{\nu=n} = -\frac{\pi}{2}\operatorname{ker}_n(z) + \frac{n!}{2}\sum_{k=0}^{n-1}\frac{\left(\frac{z}{2}\right)^{k-n}}{k!(n-k)}\sin\left(\frac{3\pi(k-n)}{4}\right)\operatorname{ker}_k(z) + \frac{n!}{2}\sum_{k=0}^{n-1}\frac{\left(\frac{z}{2}\right)^{k-n}}{k!(n-k)}\cos\left(\frac{3\pi(k-n)}{4}\right)\operatorname{kei}_k(z) \tag{3.2.15}$$

Derivatives of the Struve functions can be expressed for $\nu = \pm n$ only by using the Meijer-G functions

$$\left[\frac{\partial H_\nu(z)}{\partial \nu}\right]_{\nu=n} = -\frac{\pi}{2}J_n(z) + \frac{1}{2\pi}\left(\frac{2}{z}\right)^n G_{24}^{32}\left(\frac{z^2}{4}\middle|\begin{matrix}1/2 & 1/2\\ 1/2,\ 1/2, & n,\ 0\end{matrix}\right) + \frac{1}{\pi}\sum_{k=0}^{n-1}\frac{\Gamma\left(k+\frac{1}{2}\right)\left(\frac{z}{2}\right)^{n-2k-1}}{\Gamma\left(n-k+\frac{1}{2}\right)}\left[\ln\left(\frac{z}{2}\right) - \psi\left(n-k+\frac{1}{2}\right)\right] + \frac{n!}{2}\sum_{k=0}^{n-1}\frac{(-1)^k\left(\frac{z}{2}\right)^{k-n}}{k!(n-k)}H_{-k}(z) \quad ; \quad \operatorname{Re} z \ge 0 \tag{3.2.16}$$

$$\left[\frac{\partial H_\nu(z)}{\partial \nu}\right]_{\nu=-n} = (-1)^{n+1}\frac{\pi}{2}J_n(z) + (-1)^n\frac{1}{2\pi}\left(\frac{2}{z}\right)^n G_{24}^{32}\left(\frac{z^2}{4}\middle|\begin{matrix}1/2 & 1/2\\ 1/2,\ 1/2, & n,0\end{matrix}\right) - \frac{n!}{2}\sum_{k=0}^{n-1}\frac{\left(-\frac{z}{2}\right)^{k-n}}{k!(n-k)}H_{-k}(z) \quad ; \quad \operatorname{Re} z \ge 0 \tag{3.2.17}$$

and

$$\left[\frac{\partial L_\nu(z)}{\partial \nu}\right]_{\nu=n} = (-1)^n K_n(z) + \frac{(-1)^{n-1}}{2\pi^2}\left(\frac{2}{z}\right)^n G_{24}^{42}\left(\frac{z^2}{4}\middle|\begin{matrix}1/2 & 1/2\\ 0,1/2, & 1/2,n\end{matrix}\right)$$
$$-\frac{1}{\pi}\sum_{k=0}^{n-1}\frac{(-1)^k\Gamma\left(k+\frac{1}{2}\right)\left(\frac{z}{2}\right)^{n-2k-1}}{\Gamma\left(n-k+\frac{1}{2}\right)}\left[\ln\left(\frac{z}{2}\right)-\psi\left(n-k+\frac{1}{2}\right)\right] \tag{3.2.18}$$
$$+\frac{n!}{2}\sum_{k=0}^{n-1}\frac{\left(-\frac{z}{2}\right)^{k-n}}{k!(n-k)}L_{-k}(z) \quad ; \quad \operatorname{Re} z \geq 0$$

$$\left[\frac{\partial L_\nu(z)}{\partial \nu}\right]_{\nu=-n} = (-1)^n K_n(z) + \frac{(-1)^{n-1}}{2\pi^2}\left(\frac{2}{z}\right)^n G_{24}^{42}\left(\frac{z^2}{4}\middle|\begin{matrix}1/2 & 1/2\\ 0,1/2 & 1/2,n\end{matrix}\right)$$
$$-\frac{n!}{2}\sum_{k=0}^{n-1}\frac{\left(-\frac{z}{2}\right)^{k-n}}{k!(n-k)}L_{-k}(z) \tag{3.2.19}$$
$$\operatorname{Re} z \geq 0$$

Derivatives with respect to the order for positive and negative integer values of ν can be determined using general formulas given above, however the preferable technique is to obtain them consecutively, by applying the corresponding recurrence formulas that exist between the Bessel or related functions. The starting point is evident at $\nu = 0$ and $\nu = \pm 1$, and therefore a significant effort was directed to obtain them. Functional expressions for derivatives with respect to the order, especially for its zero value are well known and they are presented in the mathematical literature [9, 18]. There are a number of techniques to derive such expressions as is illustrated below. The consecutive derivation of derivatives in the simplest case of the Bessel functions of the first and second kind follows from (2.2.13):

$$\left[\frac{\partial J_\nu(z)}{\partial \nu}\right]_{\nu=0} = \frac{\pi}{2} Y_0(z) \tag{3.2.20}$$

$$\left[\frac{\partial J_\nu(z)}{\partial \nu}\right]_{\nu=-1} + \left[\frac{\partial J_\nu(z)}{\partial \nu}\right]_{\nu=1} = \frac{2}{z} J_0(z) \tag{3.2.21}$$

and

$$\left[\frac{\partial J_\nu(z)}{\partial \nu}\right]_{\nu=1} - \left[\frac{\partial J_\nu(z)}{\partial \nu}\right]_{\nu=-1} = \pi Y_1(z) \tag{3.2.22}$$

Combining these equations gives

$$\left[\frac{\partial J_\nu(z)}{\partial \nu}\right]_{\nu=-1} = \frac{1}{z} J_0(z) - \frac{\pi}{2} Y_1(z) \tag{3.2.23}$$

$$\left[\frac{\partial J_\nu(z)}{\partial \nu}\right]_{\nu=1} = \frac{1}{z}J_0(z) + \frac{\pi}{2}Y_1(z) \tag{3.2.24}$$

Similarly, we have

$$\left[\frac{\partial Y_\nu(z)}{\partial \nu}\right]_{\nu=-1} = \frac{1}{z}Y_0(z) + \frac{\pi}{2}J_1(z) \tag{3.2.25}$$

$$\left[\frac{\partial Y_\nu(z)}{\partial \nu}\right]_{\nu=0} = -\frac{\pi}{2}J_0(z) \tag{3.2.26}$$

$$\left[\frac{\partial Y_\nu(z)}{\partial \nu}\right]_{\nu=1} = \frac{1}{z}Y_0(z) - \frac{\pi}{2}J_1(z) \tag{3.2.27}$$

Since the Hankel functions are expressed simply by

$$H_\nu^{(1)}(z) = J_\nu(z) + i\,Y_\nu(z) \tag{3.2.28}$$

$$H_\nu^{(2)}(z) = J_\nu(z) - i\,Y_\nu(z) \tag{3.2.29}$$

Their first derivatives with respect to the order are

$$\left[\frac{\partial H_\nu^{(1)}(z)}{\partial \nu}\right]_{\nu=-1} = \frac{H_0^{(1)}(z)}{z} + \frac{\pi i}{2}H_1^{(1)}(z) \tag{3.2.30}$$

$$\left[\frac{\partial H_\nu^{(1)}(z)}{\partial \nu}\right]_{\nu=0} = -\frac{\pi i}{2}H_0^{(1)}(z) \tag{3.2.31}$$

$$\left[\frac{\partial H_\nu^{(1)}(z)}{\partial \nu}\right]_{\nu=1} = \frac{H_0^{(1)}(z)}{z} - \frac{\pi i}{2}H_1^{(1)}(z) \tag{3.2.32}$$

and

$$\left[\frac{\partial H_\nu^{(2)}(z)}{\partial \nu}\right]_{\nu=-1} = \frac{H_0^{(2)}(z)}{z} - \frac{\pi i}{2}H_1^{(2)}(z) \tag{3.2.33}$$

$$\left[\frac{\partial H_\nu^{(2)}(z)}{\partial \nu}\right]_{\nu=0} = \frac{\pi i}{2}H_0^{(2)}(z) \tag{3.2.34}$$

$$\left[\frac{\partial H_\nu^{(2)}(z)}{\partial \nu}\right]_{\nu=1} = \frac{H_0^{(2)}(z)}{z} + \frac{\pi i}{2}H_1^{(2)}(z) \tag{3.2.35}$$

In the case of the modified Bessel function of the first kind $I_\nu(z)$ we have from (2.3.6)

$$\left[\frac{\partial I_\nu(z)}{\partial \nu}\right]_{\nu=-1} - \left[\frac{\partial I_\nu(z)}{\partial \nu}\right]_{\nu=1} = \frac{2}{z} I_0(z) \tag{3.2.36}$$

$$I_{-\nu}(z) = I_\nu(z) + \frac{2}{\pi}\sin(\pi\nu)\, K_\nu(z) \tag{3.2.37}$$

and differentiation of (3.2.27) gives

$$-\frac{\partial I_{-\nu}(z)}{\partial \nu} = \frac{\partial I_\nu(z)}{\partial \nu} + 2\cos(\pi\nu)\, K_\nu(z) + \frac{2}{\pi}\sin(\pi\nu)\, \frac{\partial K_\nu(z)}{\partial \nu} \tag{3.2.38}$$

By introducing $\nu = 1$, it follows that

$$\left[\frac{\partial I_\nu(z)}{\partial \nu}\right]_{\nu=-1} + \left[\frac{\partial I_\nu(z)}{\partial \nu}\right]_{\nu=1} = 2K_1(z) \tag{3.2.39}$$

and finally

$$\left[\frac{\partial I_\nu(z)}{\partial \nu}\right]_{\nu=-1} = \frac{1}{z} I_0(z) + K_1(z) \tag{3.2.40}$$

$$\left[\frac{\partial I_\nu(z)}{\partial \nu}\right]_{\nu=0} = -K_0(z) \tag{3.2.41}$$

$$\left[\frac{\partial I_\nu(z)}{\partial \nu}\right]_{\nu=1} = K_1(z) - \frac{1}{z} I_0(z) \tag{3.2.42}$$

Considering that, for the modified Bessel function of the second kind we have

$$K_\nu(z) = K_{-\nu}(z) \tag{3.2.43}$$

$$K_{\nu-1}(z) - K_{\nu+1}(z) = -\frac{2\nu}{z} K_\nu(z) \tag{3.2.44}$$

and therefore from (3.2.43) and (3.2.44) it follows that

$$\frac{\partial K_\nu(z)}{\partial \nu} = -\frac{\partial K_{\nu-1}(z)}{\partial \nu} \tag{3.2.45}$$

$$\frac{\partial K_{\nu-1}(z)}{\partial \nu} + \frac{\partial K_{\nu+1}(z)}{\partial \nu} = \frac{2}{z} K_\nu(z) + \frac{2\nu}{z}\frac{\partial K_\nu(z)}{\partial \nu} \tag{3.2.46}$$

Introducing $\nu = 0$ in these equations gives

$$\left[\frac{\partial K_\nu(z)}{\partial \nu}\right]_{\nu=1} = \frac{1}{z} K_0(z) = -\left[\frac{\partial K_\nu(z)}{\partial \nu}\right]_{\nu=-1} \tag{3.2.47}$$

$$\left[\frac{\partial K_\nu(z)}{\partial \nu}\right]_{\nu=0} = 0 \tag{3.2.48}$$

Similarly, by using series expansions given in Section 3.1 and by differentiating the recurrence expressions of other related Bessel functions, it is possible to obtain their derivatives with respect to the order, initially for $\nu = 0$ and $\nu = \pm 1$ and after this for negative and positive integers. Final results for these values of ν, in the case of the integral Bessel functions $Ji_\nu(z)$, $Yi_\nu(z)$ and $Ki_\nu(z)$ are known from the Brychkov and Geddes paper [44]

$$\left[\frac{\partial \mathrm{Ji}_\nu(z)}{\partial \nu}\right]_{\nu=-1} = -\frac{1}{z}J_0(z) + \frac{\pi}{2}\mathrm{Yi}_1(z) - \mathrm{Ji}_1(z) \tag{3.2.49}$$

$$\left[\frac{\partial \mathrm{Ji}_\nu(z)}{\partial \nu}\right]_{\nu=0} = -\frac{\pi}{2}\mathrm{Yi}_0(z) \tag{3.2.50}$$

$$\left[\frac{\partial \mathrm{Ji}_\nu(z)}{\partial \nu}\right]_{\nu=1} = -\frac{1}{z}J_0(z) - \frac{\pi}{2}\mathrm{Yi}_1(z) - \mathrm{Ji}_1(z) \tag{3.2.51}$$

and

$$\left[\frac{\partial \mathrm{Yi}_\nu(z)}{\partial \nu}\right]_{\nu=-1} = -\frac{1}{z}Y_0(z) - \frac{\pi}{2}\mathrm{Ji}_1(z) - \mathrm{Yi}_1(z) \tag{3.2.52}$$

$$\left[\frac{\partial \mathrm{Yi}_\nu(z)}{\partial \nu}\right]_{\nu=0} = \frac{\pi}{2}\mathrm{Ji}_0(z) \tag{3.2.53}$$

$$\left[\frac{\partial \mathrm{Yi}_\nu(z)}{\partial \nu}\right]_{\nu=1} = -\frac{1}{z}Y_0(z) + \frac{\pi}{2}\mathrm{Ji}_1(z) - \mathrm{Yi}_1(z) \tag{3.2.54}$$

and finally for the integral modified Bessel function of the second kind

$$\left[\frac{\partial \mathrm{Ki}_\nu(z)}{\partial \nu}\right]_{\nu=-1} = \frac{1}{z}K_0(z) - \frac{\pi}{2}\mathrm{Ki}_1(z) \tag{3.2.55}$$

$$\left[\frac{\partial \mathrm{Ki}_\nu(z)}{\partial \nu}\right]_{\nu=0} = 0 \tag{3.2.56}$$

$$\left[\frac{\partial \mathrm{Ki}_\nu(z)}{\partial \nu}\right]_{\nu=1} = -\frac{1}{z}K_0(z) + \frac{\pi}{2}\mathrm{Ki}_1(z) \tag{3.2.57}$$

A number of additional expressions for derivatives with respect to the order of the Kelvin, Struve, Anger, Weber and Lommel functions at $\nu = 0$ or for other values of the order are also known and they are presented here

$$\left[\frac{\partial \mathrm{ber}_\nu(z)}{\partial \nu}\right]_{\nu=0} = -\frac{\pi}{2}\,\mathrm{bei}(z) - \mathrm{ker}(z) \tag{3.2.58}$$

$$\left[\frac{\partial \mathrm{bei}_\nu(z)}{\partial \nu}\right]_{\nu=0} = \frac{\pi}{2}\,\mathrm{ber}(z) - \mathrm{kei}(z) \tag{3.2.59}$$

$$\left[\frac{\partial \mathbf{H}_v(z)}{\partial v}\right]_{v=0} = -\frac{\pi}{2} J_0(z) + \frac{1}{\pi^2 z} G_{24}^{32}\left(\frac{z^2}{4}\middle| \begin{matrix} 1 & 1 \\ 1/2, 1, & 1, 1/2 \end{matrix}\right) \tag{3.2.60}$$

$$\operatorname{Re} z \geq 0$$

$$\left[\frac{\partial \mathbf{L}_v(z)}{\partial v}\right]_{v=0} = K_0(z) - \frac{1}{\pi^2 z} G_{24}^{42}\left(\frac{z^2}{4}\middle| \begin{matrix} 1 & 1 \\ 1/2, 1, & 1, 1/2 \end{matrix}\right) \tag{3.2.61}$$

$$\operatorname{Re} z \geq 0$$

$$\left[\frac{\partial \mathbf{J}_v(z)}{\partial v}\right]_{v=0} = \frac{\pi}{2} \mathbf{H}_0(z) \tag{3.2.62}$$

$$\left[\frac{\partial \mathbf{E}_v(z)}{\partial v}\right]_{v=0} = \frac{\pi}{2} J_0(z) \tag{3.2.63}$$

$$\left[\frac{\partial S_{0,v}(z)}{\partial v}\right]_{v=1} = 2S_{-2,1}(z) \tag{3.2.64}$$

Formulas for half odd integer order, $v = \pm n \pm 1/2$, $n = 0, 1, 2, 3, \ldots$ are much more longer than derivatives with the integer orders, and they can be found in the Brychkov publications [11, 44, 45]. In the mathematical literature, special attention was paid to the case $v = \pm 1/2$. Derivatives with respect to the order for half odd orders were derived by various techniques and they can be expressed in terms of trigonometric and exponential integrals. The applied methods will be discussed later, here the final results are only listed for the Bessel functions of the first and the second kind

$$\left[\frac{\partial J_v(z)}{\partial v}\right]_{v=-1/2} = \sqrt{\frac{2}{\pi z}}[\cos z \operatorname{Ci}(2z) + \sin z \operatorname{Si}(2z)] \tag{3.2.65}$$

$$\left[\frac{\partial J_v(z)}{\partial v}\right]_{v=1/2} = \sqrt{\frac{2}{\pi z}}[\sin z \operatorname{Ci}(2z) - \cos z \operatorname{Si}(2z)] \tag{3.2.66}$$

and

$$\left[\frac{\partial Y_v(z)}{\partial v}\right]_{v=-1/2} = -\sqrt{\frac{2}{\pi z}}\{\sin z \operatorname{Ci}(2z) + \cos z[\pi - \operatorname{Si}(2z)]\} \tag{3.2.67}$$

$$\left[\frac{\partial Y_v(z)}{\partial v}\right]_{v=1/2} = \sqrt{\frac{2}{\pi z}}\{\cos z \operatorname{Ci}(2z) - \sin z[\pi - \operatorname{Si}(2z)]\} \tag{3.2.68}$$

For the modified Bessel functions of the first and the second kind we have

$$\left[\frac{\partial I_v(z)}{\partial v}\right]_{v=-1/2} = \frac{1}{\sqrt{2\pi z}}[e^z \operatorname{Ei}(-2z) + e^{-z} \operatorname{Ei}(2z)] \tag{3.2.69}$$

$$\left[\frac{\partial I_\nu(z)}{\partial \nu}\right]_{\nu=1/2} = \frac{1}{\sqrt{2\pi z}}[e^z \mathrm{Ei}(-2z) - e^{-z}\mathrm{Ei}(2z)] \tag{3.2.70}$$

and

$$\left[\frac{\partial K_\nu(z)}{\partial \nu}\right]_{\nu=-1/2} = \sqrt{\frac{\pi}{2z}}\, e^z \mathrm{Ei}(-2z) \tag{3.2.71}$$

$$\left[\frac{\partial K_\nu(z)}{\partial \nu}\right]_{\nu=1/2} = -\sqrt{\frac{\pi}{2z}}\, e^z \mathrm{Ei}(-2z) \tag{3.2.72}$$

Derivatives with respect to the order of the Struve functions of the first and the second kind are

$$\left[\frac{\partial H_\nu(z)}{\partial \nu}\right]_{\nu=-1/2} = \sqrt{\frac{2}{\pi z}}\left\{\cos z\,[\mathrm{Si}(2z) - 2\mathrm{Si}(z)] - \sin z\,[\mathrm{Ci}(2z) - 2\mathrm{Ci}(z)]\right\} \tag{3.2.73}$$

$$\left[\frac{\partial H_\nu(z)}{\partial \nu}\right]_{\nu=1/2} = \sqrt{\frac{2}{\pi z}}\left\{\gamma + \ln\left(\frac{z}{2}\right) + \cos z\,[\mathrm{Ci}(2z) - 2\mathrm{Ci}(z)] + \sin z\,[\mathrm{Si}(2z) - 2\mathrm{Si}(z)]\right\} \tag{3.2.74}$$

$$\left[\frac{\partial L_\nu(z)}{\partial \nu}\right]_{\nu=-1/2} = \frac{1}{\sqrt{2\pi z}}\{e^{-z}[\mathrm{Ei}(2z) - 2\mathrm{Ei}(z)] - e^{z}[\mathrm{Ei}(-2z) - 2\mathrm{Ei}(-z)]\} \tag{3.2.75}$$

$$\left[\frac{\partial L_\nu(z)}{\partial \nu}\right]_{\nu=1/2} = -\frac{1}{\sqrt{2\pi z}}\left\{2\left[\gamma + \ln\left(\frac{z}{2}\right)\right] + e^{-z}[\mathrm{Ei}(2z) - 2\mathrm{Ei}(z)] + e^{z}[\mathrm{Ei}(-2z) - 2\mathrm{Ei}(-z)]\right\} \tag{3.2.76}$$

where the sine, cosine and exponential integrals are defined by

$$\mathrm{Si}(z) = \int_0^z \frac{\sin t}{t}\,dt \tag{3.2.77}$$

$$\mathrm{Ci}(z) = -\int_z^\infty \frac{\cos t}{t}\,dt = \gamma + \ln z + \int_0^z \frac{\cos t - 1}{t}\,dt \tag{3.2.78}$$

$$\mathrm{Ei}(z) = \gamma + \ln z + \int_0^z \frac{e^t - 1}{t}\,dt \tag{3.2.79}$$

$$-\mathrm{Ei}(-z) = -\gamma - \ln z + \int_0^z \frac{1 - e^{-t}}{t}\,dt \tag{3.2.80}$$

As pointed out above, the derivation of derivatives with respect to the order of the Bessel functions in the special case, $v = \pm 1/2$, was performed in a number of different ways. The results presented here were taken mainly from the Oberhettinger paper [56]. In his derivations, the integral representation of the modified Bessel function of the second kind was used as a starting point. Alternative procedures that are based on operational calculus, interrelations between Bessel functions and recurrence relations are illustrated below.

The substitution formula of the Laplace transformation is [32]

$$F(s) = L\{f(t)\} = \int_0^\infty e^{-st} f(t)\, dt$$
$$L^{-1}\left\{\frac{1}{(s^2+1)} F[\ln(s^2+1)]\right\} = \sqrt{\pi} \int_0^\infty \left(\frac{t}{2}\right)^{x+1/2} \frac{J_{x+1/2}(t)}{\Gamma(x+1)} f(x)\, dx \tag{3.2.81}$$

If the transform-inverse pair includes the derivative of the delta Dirac function $\delta'(t)$, then (3.2.81) can be written as

$$L\{\delta'(t)\} = s$$
$$L^{-1}\left\{\frac{1}{(s^2+1)} \ln(s^2+1)\right\} = \sqrt{\pi} \int_0^\infty \left\{\left(\frac{t}{2}\right)^{x+1/2} \frac{J_{x+1/2}(t)}{\Gamma(x+1)}\right\} \delta'(x)\, dx$$
$$= -\sqrt{\pi} \frac{d}{dx} \left\{\left(\frac{t}{2}\right)^{x+1/2} \frac{J_{x+1/2}(t)}{\Gamma(x+1)}\right\}_{x=0} \tag{3.2.82}$$

The inverse on the left-hand side of (3.2.82) is known [41]

$$L^{-1}\left\{\frac{\ln(s^2+1)}{(s^2+1)}\right\} = \left\{\cos t\, \mathrm{Si}(2t) - \sin t \left[\gamma + \ln\left(\frac{t}{2}\right) + \mathrm{Ci}(2t)\right]\right\} \tag{3.2.83}$$

and differentiation of the integrand in (3.2.82) gives

$$\frac{d}{dx}\left\{\left(\frac{t}{2}\right)^{x+1/2} \frac{J_{x+1/2}(t)}{\Gamma(x+1)}\right\} = \frac{\left(\frac{t}{2}\right)^{x+1/2}}{\Gamma(x+1)} \left\{\left[\ln\left(\frac{t}{2}\right) - \psi(x+1)\right] J_{x+1/2}(t) + \frac{\partial J_{x+1/2}(t)}{\partial x}\right\}$$
$$\psi(x+1) = \frac{\Gamma'(x+1)}{\Gamma(x+1)} \tag{3.2.84}$$

Considering that

$$J_{1/2}(t) = \sqrt{\frac{2}{\pi t}} \sin t$$
$$\psi(1) = -\gamma = -0.57721566... \tag{3.2.85}$$

and introducing $x = 0$ in (3.2.84) the expected result is reached.

$$\left\{\frac{\partial J_\nu(t)}{\partial \nu}\right\}_{\nu=1/2} = \sqrt{\frac{2}{\pi t}}\{\sin t\,\mathrm{Ci}(2t) - \cos t\,\mathrm{Si}(2t)\} \tag{3.2.86}$$

Following the Müller approach [73], it is possible to obtain the derivative with respect to the order for $\nu = -1/2$, by using the recurrence relation of the Bessel functions of the first kind

$$\frac{dJ_\nu(t)}{dt} = J_{\nu-1}(t) - \frac{\nu}{t}J_\nu(t) \tag{3.2.87}$$

Differentiation of (3.2.87) when ν is a variable yields

$$\frac{d}{dt}\left(\frac{\partial J_\nu(t)}{\partial \nu}\right) = \frac{\partial J_{\nu-1}(t)}{\partial \nu} - \frac{\nu}{t}\frac{\partial J_\nu(t)}{\partial \nu} - \frac{J_\nu(t)}{t} \tag{3.2.88}$$

and by introducing $\nu = 1/2$ into (3.2.88) we have

$$\left(\frac{\partial J_\nu(t)}{\partial \nu}\right)_{\nu=-1/2} = \left\{\frac{d}{dt}\left(\frac{\partial J_\nu(t)}{\partial \nu}\right)_{\nu=1/2} + \frac{1}{t}\left[\left(\frac{\partial J_\nu(t)}{\partial \nu}\right)_{\nu=1/2} + \frac{J_{1/2}(t)}{2}\right]\right\} \tag{3.2.89}$$

The first term in the parenthesis is

$$\begin{aligned}\frac{d}{dt}\left(\frac{\partial J_\nu}{\partial \nu}\right)_{\nu=1/2} &= \frac{d}{dt}\left\{\sqrt{\frac{2}{\pi t}}\left[-\sin t\int_{2t}^{\infty}\frac{\cos x}{x}dx - \cos t\int_{0}^{2t}\frac{\sin x}{x}dx\right]\right\}\\ &= \left\{\left(\cos t - \frac{\sin t}{t}\right)\mathrm{Ci}(2t) + \left(\sin t + \frac{\cos}{t}\right)\mathrm{Si}(2t)\right\}\end{aligned} \tag{3.2.90}$$

and two additional terms are known and therefore once again we have the expected result

$$\left(\frac{\partial J_\nu(t)}{\partial \nu}\right)_{\nu=-1/2} = \sqrt{\frac{2}{\pi t}}\{\cos t\,\mathrm{Ci}(2t) + \sin t\,\mathrm{Si}(2t)\} \tag{3.2.91}$$

As has been demonstrated by Petiau [23], derivatives with respect to the order of the Bessel function of the second kind are available from the interrelations between $J_\nu(t)$ and $Y_\nu(t)$ functions

$$J_\nu(t) = \frac{\cos(\pi\nu)J_\nu(t) - J_{-\nu}(t)}{\sin(\pi\nu)} = \frac{1}{\sin(\pi\nu)}\left[\frac{\partial Y_{-\nu}(t)}{\partial\nu} - \cos(\pi\nu)Y_\nu(t)\right]$$
$$\frac{\partial Y_\nu(t)}{\partial\nu} = \cos(\pi\nu)\frac{\partial J_\nu(t)}{\partial\nu} - \csc(\pi\nu)\frac{\partial J_{-\nu}(t)}{\partial\nu} - \pi\csc(\pi\nu)\,Y_{-\nu}(t) \tag{3.2.92}$$

Introducing $\nu = 1/2$ into (3.2.92) gives the already known result

$$\left(\frac{\partial Y_\nu(t)}{\partial\nu}\right)_{\nu=1/2} = \left(\frac{\partial J_\nu(t)}{\partial\nu}\right)_{\nu=-1/2} - \pi\,Y_{-1/2}(t)$$
$$Y_{-1/2}(t) = \sqrt{\frac{2}{\pi t}}\sin t \tag{3.2.93}$$
$$\left(\frac{\partial Y_\nu(t)}{\partial\nu}\right)_{\nu=1/2} = \sqrt{\frac{2}{\pi t}}\{\cos t\,\mathrm{Ci}(2t) + \sin t\,[\mathrm{Si}(2t) - \pi]\}$$

Similarly, if $\nu = -1/2$ is introduced into (3.2.92), we have

$$\left(\frac{\partial Y_\nu(t)}{\partial\nu}\right)_{\nu=-1/2} = -\left(\frac{\partial J_\nu(t)}{\partial\nu}\right)_{\nu=-1/2} + \pi\,Y_{1/2}(t)$$
$$Y_{1/2}(t) = \sqrt{\frac{2}{\pi t}}\cos t \tag{3.2.94}$$
$$\left(\frac{\partial Y_\nu(t)}{\partial\nu}\right)_{\nu=1/2} = \sqrt{\frac{2}{\pi t}}\{\cos t\,[\mathrm{Si}(2t) - \pi] - \sin t\,\mathrm{Ci}(2t)\}$$

The interrelation between the Bessel function $J_\nu(t)$ and the modified Bessel function of the first kind $I_\nu(t)$

$$I_\nu(t) = e^{-i\pi\nu/2}J_\nu(it) \tag{3.2.95}$$

permits to obtain derivatives with respect to the order from (3.2.95)

$$\frac{\partial I_\nu(t)}{\partial\nu} = -\frac{\pi i}{2}I_\nu(t) + e^{-i\pi\nu/2}\frac{\partial J_\nu(it)}{\partial\nu} \tag{3.2.96}$$

Taking into account that for $\nu = 1/2$ we have

$$\sin(it) = i\sinh t \quad ; \quad \cos(it) = \cosh t$$
$$\mathrm{Si}(it) = \frac{i}{2}[\mathrm{Ei}(2t) + E_1(2t)]$$
$$\mathrm{Ci}(it) = \frac{1}{2}[\mathrm{Ei}(2t) - E_1(2t)] = \frac{1}{2}[\mathrm{Ei}(2t) + \mathrm{Ei}(-2t)] \tag{3.2.97}$$
$$I_{1/2}(t) = \sqrt{\frac{2}{\pi t}}\sinh t \quad ; \quad I_{-1/2}(t) = \sqrt{\frac{2}{\pi t}}\cosh t$$

and

$$e^{-i\pi/4}\left(\frac{\partial J_\nu(it)}{\partial \nu}\right)_{\nu=1/2} = -i\sqrt{\frac{2}{\pi t}}\{\sin(it)\,\mathrm{Ci}(2it) - \cos(it)\,\mathrm{Si}(2it)\}$$
$$= \sqrt{\frac{2}{\pi t}}\{\pi i \sinh t - e^t E_1(2t) - e^{-t}\mathrm{Ei}(2t)\} \tag{3.2.98}$$

the derivative with respect to the order for $\nu = 1/2$ is

$$\left(\frac{\partial I_\nu(t)}{\partial \nu}\right)_{\nu=1/2} = \frac{1}{\sqrt{2\pi t}}\{e^t\mathrm{Ei}(-2t) - e^{-t}\mathrm{Ei}(2t)\} \tag{3.2.99}$$

Similarly, it is possible to obtain for $\nu = -1/2$

$$\left(\frac{\partial I_\nu(t)}{\partial \nu}\right)_{\nu=-1/2} = \frac{1}{\sqrt{2\pi t}}\{e^t\mathrm{Ei}(-2t) + e^{-t}\mathrm{Ei}(2t)\} \tag{3.2.100}$$

The modified Bessel functions of the first and second kind, $I_\nu(t)$ and $K_\nu(t)$, are also interrelated

$$K_\nu(t) = \frac{\pi}{2}\frac{[I_{-\nu}(t) - I_\nu(t)]}{\sin(\pi\nu)} \tag{3.2.101}$$

and differentiation of (3.2.101) with respect to the order ν gives

$$\frac{\partial K_\nu(t)}{\partial \nu} = -\pi\cot(\pi\nu)K_\nu(t) - \frac{\pi}{2}\operatorname{cosec}(\pi\nu)\left\{\left(\frac{\partial I_{-\nu}(t)}{\partial \nu}\right) + \left(\frac{\partial I_\nu(t)}{\partial \nu}\right)\right\} \tag{3.2.102}$$

Introducing derivatives for $\nu = \pm 1/2$, from (3.2.99) and (3.2.100) we have

$$\left(\frac{\partial K_\nu(t)}{\partial \nu}\right)_{\nu=\pm 1/2} = \mp\sqrt{\frac{\pi}{2t}}e^t\mathrm{Ei}(-2t) \tag{3.2.103}$$

Struve function of the first kind $H_\nu(t)$ is interrelated with the Bessel function of the second kind $Y_\nu(t)$ in the following way

$$H_\nu(z) = Y_\nu(z) + \frac{2\left(\frac{z}{2}\right)^\nu}{\sqrt{\pi}\Gamma\left(\nu+\frac{1}{2}\right)}\int_0^\infty e^{-zt}(1+t^2)^{\nu-1/2}dt$$
$$\mathrm{Re}\,\nu > -\frac{1}{2} \tag{3.2.104}$$

Differentiation of (3.2.105) with respect to the order ν gives

$$\frac{\partial H_\nu(z)}{\partial \nu} = \frac{\partial Y_\nu(z)}{\partial \nu} + \left[\ln\left(\frac{z}{2}\right) - \psi(\nu + \frac{1}{2})\right] [H_\nu(z) - Y_\nu(z)] \\ + \frac{2\left(\frac{z}{2}\right)^\nu}{\sqrt{\pi}\Gamma(\nu+\frac{1}{2})} \int_0^\infty e^{-zt}(1+t^2)^{\nu-1/2} \ln(1+t^2)\, dt \tag{3.2.105}$$

For $\nu = 1/2$ from (3.2.105) it follows that

$$\left(\frac{\partial H_\nu(z)}{\partial \nu}\right)_{\nu=1/2} = \left(\frac{\partial Y_\nu(z)}{\partial \nu}\right)_{\nu=1/2} + \left[\ln\left(\frac{z}{2}\right) - \psi(1)\right] [H_{1/2}(z) - Y_{1/2}(z)] \\ + \sqrt{\frac{2z}{\pi}} \int_0^\infty e^{-zt} \ln(1+t^2)\, dt \tag{3.2.106}$$

$$H_{1/2}(z) = \sqrt{\frac{2}{\pi z}}(1 - \cos z)$$

The Laplace transform of the integral in (3.2.107) is known [36]

$$\int_0^\infty e^{-zt} \ln(1+t^2)\, dt = -\frac{2}{z}\left\{\cos z \operatorname{Ci}(z) + \sin z \left[\operatorname{Si}(z) - \frac{\pi}{2}\right]\right\} \tag{3.2.107}$$

and therefore the derivative with respect to the order ν is

$$\left(\frac{\partial H_\nu}{\partial \nu}\right)_{\nu=1/2} = \sqrt{\frac{2}{\pi z}} I \\ I = \left\{\gamma + \ln\left(\frac{z}{2}\right) + \cos z[\operatorname{Ci}(2z) - 2\operatorname{Ci}(z)] + \sin z\,[\operatorname{Si}(2z) - 2\operatorname{Si}(z)]\right\} \tag{3.2.108}$$

If recurrence relations of the Struve functions are differentiated with regard to the argument z and the order ν

$$\frac{d}{dz}\left(\frac{\partial H_\nu(z)}{\partial \nu}\right) = \left(\frac{\partial H_{\nu-1}(z)}{\partial \nu}\right) - \frac{\nu}{z}\left(\frac{\partial H_\nu(z)}{\partial \nu}\right) - \frac{1}{z} H_\nu(z) \tag{3.2.109}$$

and taking into account that

$$\left(\frac{\partial H_{\nu-1}(z)}{\partial \nu}\right) + \left(\frac{\partial H_{\nu+1}(z)}{\partial \nu}\right) = \frac{2\nu}{z}\left(\frac{\partial H_\nu(z)}{\partial \nu}\right) + \frac{2}{z} H_\nu(z) \\ + \frac{\left(\frac{z}{2}\right)^\nu}{\sqrt{\pi}\Gamma(\nu+\frac{3}{2})}\left[\ln\left(\frac{z}{2}\right) - \psi\left(\nu + \frac{3}{2}\right)\right] \tag{3.2.110}$$

it is possible to evaluate the derivative with respect to the order for $\nu = -1/2$

$$\left(\frac{\partial H_\nu}{\partial \nu}\right)_{\nu=-1/2} = \sqrt{\frac{2}{\pi z}} I$$
$$I = \{\cos z[\mathrm{Si}(2z) - 2\mathrm{Ci}(z)] - \sin z\,[\mathrm{Ci}(2z) - 2\mathrm{Ci}(z)]\} \tag{3.2.111}$$

and for $\nu = 3/2$

$$\left(\frac{\partial H_\nu}{\partial \nu}\right)_{\nu=3/2} = \sqrt{\frac{2}{\pi z}} \left\{ \left[\ln\left(\frac{z}{2}\right) + \gamma\right] \left(\frac{z}{2} + \frac{1}{z}\right) - \frac{z}{2} + \frac{2(1-\cos z)}{z} \right.$$
$$+ \left(\sin z + \frac{\cos z}{z}\right) [\mathrm{Ci}(2z) - 2\mathrm{Ci}(z)] \tag{3.2.112}$$
$$\left. + \left(\frac{\sin z}{z} - \cos z\right) \left[\mathrm{Si}(2z) - 2\mathrm{Si}(z)\right] \right\}$$

Thus, by using (3.2.108), (3.2.111) and (3.2.112) it is possible, consecutively to determine derivatives with respect to the order for $\nu = n + 1/2$, $n = 1,2,3, \ldots$

Similarly, as the Bessel function in (3.2.95), the Struve functions $H_\nu(t)$ and $L_\nu(t)$ are interrelated

$$L_\nu(t) = -ie^{-i\pi\nu/2} H_\nu(it) \tag{3.2.113}$$

Differentiation of (3.2.113) with respect to the order ν gives

$$\frac{\partial L_\nu(t)}{\partial \nu} = -\frac{\pi i}{2} L_\nu(t) - ie^{-i\pi\nu/2} \frac{\partial H_\nu(it)}{\partial \nu} \tag{3.2.114}$$

and the recurrence relation is

$$\left(\frac{\partial L_{\nu-1}(z)}{\partial \nu}\right) - \left(\frac{\partial L_{\nu+1}(z)}{\partial \nu}\right) = \frac{2\nu}{z} \left(\frac{\partial L_\nu(z)}{\partial \nu}\right) + \frac{2}{z} L_\nu(z)$$
$$+ \frac{\left(\frac{z}{2}\right)^\nu}{\sqrt{\pi}\,\Gamma\left(\nu + \frac{3}{2}\right)} \left[\ln\left(\frac{z}{2}\right) - \psi\left(\nu + \frac{3}{2}\right)\right] \tag{3.2.115}$$

Using (3.2.114) and (3.2.115) we have

$$\left(\frac{\partial L_\nu}{\partial \nu}\right)_{\nu=-1/2} = \frac{1}{\sqrt{2\pi z}} \{ e^{-z}[\mathrm{Ei}(2z) - 2\mathrm{Ei}(z)] + e^{z}[E_1(2z) - 2E_1(z)]\}$$
$$\mathrm{Ei}(z) = \gamma + \ln z + \int_0^z (e^x - 1)\,\frac{dx}{x} \tag{3.2.116}$$
$$E_1(z) = \int_z^\infty e^{-x} \frac{dx}{x} = -\mathrm{Ei}(-z) = \Gamma(0, z)$$

and

$$\left(\frac{\partial L_\nu}{\partial \nu}\right)_{\nu=1/2} = -\sqrt{\frac{2}{\pi z}}\,I$$

$$I = \left\{\left[\ln\left(\frac{z}{2}\right) + \gamma\right] - \frac{e^{-z}}{2}\left[\mathrm{Ei}(2z) + 2\mathrm{Ei}(z)\right] + \frac{e^{z}}{2}\left[E_1(2z) - 2E_1(z)\right]\right\} \tag{3.2.117}$$

Using the general formula for derivatives of the modified Struve functions

$$\left(\frac{\partial L_\nu(z)}{\partial \nu}\right) = \left[\ln\left(\frac{z}{2}\right) - \psi\left(\nu + \frac{1}{2}\right)\right] L_\nu(z) + \frac{2\left(\frac{z}{2}\right)^\nu}{\sqrt{\pi}\,\Gamma\left(\nu+\frac{1}{2}\right)} \int_0^1 (1-t^2)^{\nu-1/2} \ln(1-t^2)\,\sinh(zt)\,dt \quad ; \quad \mathrm{Re}\,\nu > -\frac{1}{2} \tag{3.2.118}$$

and (3.2.117), it possible, as a by-product, to evaluate the following integral

$$\int_0^1 \ln(1-t^2)\,\sinh(zt)\,dt = -\frac{1}{z}\left[\ln\left(\frac{z}{2}\right) + \gamma\right] \cosh z + \frac{e^{-z}}{2}\left[\mathrm{Ei}(2z) - 2\mathrm{Ei}(z)\right] - \frac{e^{z}}{2}\left[E_1(2z) - 2E_1(z)\right] \tag{3.2.119}$$

3.3 Derivatives with Respect to the Order ν of the Bessel and Related Functions Based on Integral Representations

Using the Leibniz rule of differentiating integrals, almost any integral with integrand containing Bessel or related functions is a starting point in derivation of derivatives with respect to the order. Evidently, the same happens with the integral representations of the Bessel and related functions. One of well-known examples is

$$J_\nu(z)\frac{\partial Y_\nu(z)}{\partial \nu} - Y_\nu(z)\frac{\partial J_\nu(z)}{\partial \nu} = -\frac{4}{\pi}\int_0^\infty e^{-2\nu t} K_0(2z\sinh t)\,dt \tag{3.3.1}$$

$\mathrm{Re}\,z > 0$

The expression (3.3.1) was derived similar to the Nicholson integral [7]

$$[J_\nu(z)]^2 + [Y_\nu(z)]^2 = \frac{8}{\pi^2}\int_0^\infty K_0(2z\sinh t)\,\cosh(2\nu t)\,dt \quad ; \quad \mathrm{Re}\,z > 0 \tag{3.3.2}$$

Dunster [48], by solving the Bessel differential equation with respect to the order by the method of variation of parameters showed that for $\nu > 0$ and Re $z > 0$ it is possible to derive similar expressions

$$\frac{\partial J_\nu(z)}{\partial \nu} = \pi\nu Y_\nu(z) \int_0^z \frac{[J_\nu(t)]^2}{t} dt + \pi\nu J_\nu(z) \int_z^\infty \frac{J_\nu(t)\, Y_\nu(t)}{t} dt \tag{3.3.3}$$

$$\frac{\partial Y_\nu(z)}{\partial \nu} = \pi\nu J_\nu(z) \int_0^z \frac{[Y_\nu(t)]^2}{t} dt - \pi\nu Y_\nu(z) \int_z^\infty \frac{J_\nu(t)\, Y_\nu(t)}{t} dt - \frac{\pi}{2} J_\nu(z) \tag{3.3.4}$$

Thus, derivatives with respect to the order can be expressed in a closed form by using integral representations of Bessel and related functions. Usually, differentiations should be performed in three places in integral representations, one place is the differentiation under integral sign (differentiation of integrand). In some cases, the differentiated integral can be evaluated for particular values of ν and the final result can be expressed in terms of elementary or special functions.

Oberhettinger [56] applied the following integral representation of the modified Bessel function of the second kind to obtain derivatives for $\nu = \pm 1/2$

$$K_\nu(z) = \frac{\sqrt{\pi}\left(\frac{z}{2}\right)^\nu}{\Gamma\left(\nu+\frac{1}{2}\right)} \int_1^\infty e^{-zt}(t^2-1)^{\nu-1/2} dt \qquad \mathrm{Re}\, z > 0 \;;\; \mathrm{Re}\,\nu > -\frac{1}{2} \tag{3.3.5}$$

Differentiation of (3.3.5) gives

$$\begin{aligned}\frac{\partial K_\nu(z)}{\partial \nu} &= \left[\ln\left(\frac{z}{2}\right) - \psi\left(\nu+\frac{1}{2}\right)\right] \frac{\sqrt{\pi}\left(\frac{z}{2}\right)^\nu}{\Gamma\left(\nu+\frac{1}{2}\right)} \int_1^\infty e^{-zt}(t^2-1)^{\nu-1/2} dt \\ &+ \frac{\sqrt{\pi}\left(\frac{z}{2}\right)^\nu}{\Gamma\left(\nu+\frac{1}{2}\right)} \int_1^\infty e^{-zt} \ln(t^2-1)\,(t^2-1)^{\nu-1/2} dt\end{aligned} \tag{3.3.6}$$

or in the equivalent form

$$\frac{\partial K_\nu(z)}{\partial \nu} = \left[\ln\left(\frac{z}{2}\right) - \psi\left(\nu+\frac{1}{2}\right)\right] K_\nu(z) + \frac{\sqrt{\pi}\left(\frac{z}{2}\right)^\nu}{\Gamma\left(\nu+\frac{1}{2}\right)} \int_1^\infty e^{-zt} \ln(t^2-1)\,(t^2-1)^{\nu-1/2} dt \tag{3.3.7}$$

Introducing $\nu = 1/2$ into (3.3.7) we have

$$\left[\frac{\partial K_v(z)}{\partial v}\right]_{v=1/2} = \sqrt{\frac{\pi}{2z}}e^{-z}\left[\gamma + \ln\left(\frac{z}{2}\right)\right] + \sqrt{\frac{\pi z}{2}}I\left(z, \frac{1}{2}\right)$$

$$I\left(z, \frac{1}{2}\right) = \int_1^\infty e^{-zt}\ln(t^2 - 1)\,dt \tag{3.3.8}$$

$$K_{1/2}(z) = \sqrt{\frac{\pi}{2z}}e^{-z} \quad ; \quad \psi(1) = -\gamma$$

Oberhettinger showed that this integral can be integrated by parts and the final result of integration is

$$\begin{aligned} I\left(z, \frac{1}{2}\right) &= \int_1^\infty e^{-zt}\ln(t^2 - 1)\,dt = \int_1^\infty \ln(t^2-1)\,d\left(-\frac{e^{-zt}}{z}\right)dt \\ &= -\frac{1}{z}e^{z}\mathrm{Ei}(-2z) - \frac{e^{-z}}{z}\left[\gamma + \ln\left(\frac{z}{2}\right)\right] \end{aligned} \tag{3.3.9}$$

Combining (3.3.8) with (3.3.9) gives the same expression as in (3.2.103)

$$\left[\frac{\partial K_v(z)}{\partial v}\right]_{v=1/2} = -\sqrt{\frac{\pi}{2z}}e^{z}\mathrm{Ei}(-2z) \tag{3.3.10}$$

and using

$$\begin{aligned} &K_v(z) = K_{-v}(z) \\ &\frac{\partial K_{-v}(z)}{\partial v} = -\frac{\partial K_v(z)}{\partial v} \end{aligned} \tag{3.3.11}$$

the derivative for $v = -1/2$ is determined as

$$\left[\frac{\partial K_v(z)}{\partial v}\right]_{v=-1/2} = \sqrt{\frac{\pi}{2z}}e^{z}\mathrm{Ei}(-2z) \tag{3.3.12}$$

Derivatives with respect to the order for other Bessel functions with $v = \pm 1/2$ (see, for example, expressions (3.2.106) or (3.2.117)) were derived by Oberhettinger. He differentiated the Bessel functions' relations when argument z is reduced to the principal branch $|z| < \pi$ (for details see [56]).

The integral representation of the Bessel and related functions (denoted here as $F_v(z)$) can be formally written in the following form

$$F_v(z) = A\frac{\left(\frac{z}{2}\right)^v}{\Gamma\left(v + \frac{1}{2}\right)}\int_a^b G(t, v)\,dt \tag{3.3.13}$$

where A is a numerical constant and the limits of integration a and b are constants or infinity. The general result for derivatives with respect to the order ν based on integral representations is available from a direct differentiation of (3.3.13)

$$\frac{\partial F_\nu(z)}{\partial \nu} = \left[\ln\left(\frac{z}{2}\right) - \psi\left(\nu + \frac{1}{2}\right)\right] F_\nu(z) + A\frac{\left(\frac{z}{2}\right)^\nu}{\Gamma\left(\nu + \frac{1}{2}\right)} I(t,\nu)$$
$$I(t,\nu) = \int_a^b \frac{\partial G(t,\nu)}{\partial \nu}\, dt \tag{3.3.14}$$

Thus, derivatives of different Bessel functions have the same form, but they contain different expressions for integrals $I(t,\nu)$. In mathematical manipulations, the formulas for derivatives given in (3.3.14) are superior than those expressed by infinite or finite series. They include only the function under consideration and one integral that should be evaluated. In some cases, these integrals can be presented in closed form, as demonstrated by Oberhettinger for $\nu = \pm 1/2$ [56]. Usually, these integrals can be evaluated by rather simple numerical methods.

For the first time, in a systematic way, the expression (3.3.14) is applied here to obtain derivatives with respect to the order by using different integral representations of the Bessel and related functions given in Section 2.12. Let us start with the Bessel function of the first kind which has the following integral representation (2.12.13)

$$J_\nu(z) = \frac{2\left(\frac{z}{2}\right)^\nu}{\sqrt{\pi}\,\Gamma\left(\nu + \frac{1}{2}\right)} \int_0^1 (1-t^2)^{\nu-1/2} \cos(zt)\, dt$$
$$\operatorname{Re} z > 0 \quad ; \quad \operatorname{Re}\nu > -\frac{1}{2} \tag{3.3.15}$$

From (3.3.14) we have

$$\frac{\partial J_\nu(z)}{\partial \nu} = \left[\ln\left(\frac{z}{2}\right) - \psi\left(\nu + \frac{1}{2}\right)\right] J_\nu(z) + \frac{2\left(\frac{z}{2}\right)^\nu}{\sqrt{\pi}\,\Gamma\left(\nu + \frac{1}{2}\right)} I(t,\nu)$$
$$I(t,\nu) = \int_0^1 (1-t^2)^{\nu-1/2} \ln(1-t^2)\, \cos(zt)\, dt \tag{3.3.16}$$

For some values of ν the above integral can be expressed in terms of special functions. For $\nu = 0$, we have

$$\left[\frac{\partial J_\nu(z)}{\partial \nu}\right]_{\nu=0} = \left[\ln\left(\frac{z}{2}\right) - \psi\left(\frac{1}{2}\right)\right] J_0(z) + \frac{2}{\pi} I(t,0)$$

$$I(t,0) = \int_0^1 \frac{\ln(1-t^2)\,\cos(zt)}{\sqrt{1-t^2}}\,dt \tag{3.3.17}$$

$$\psi\left(\frac{1}{2}\right) = -\gamma - 2\ln 2$$

This derivative with respect to the order is given in (3.2.20) and therefore by comparing it with (3.3.17) as a by-product the integral in (3.3.17) can be evaluated

$$\int_0^1 \frac{\ln(1-t^2)\,\cos(zt)}{\sqrt{1-t^2}}\,dt = \frac{\pi}{4}\{\pi\, Y_0(z) - 2[\ln(2z) + \gamma]\, J_0(z)\} \tag{3.3.18}$$

Introducing $\nu = 1$ into (3.3.16) gives

$$\left[\frac{\partial J_\nu(z)}{\partial \nu}\right]_{\nu=1} = \left[\ln\left(\frac{z}{2}\right) - \psi\left(\frac{3}{2}\right)\right] J_1(z) + \frac{2z}{\pi} I(t,1)$$

$$I(t,1) = \int_0^1 \sqrt{1-t^2}\,\ln(1-t^2)\,\cos(zt)\,dt \tag{3.3.19}$$

and by comparing (3.2.22) with (3.3.319) the logarithmic integral is

$$\int_0^1 \sqrt{1-t^2}\,\ln(1-t^2)\,\cos(zt)\,dt = \frac{\pi}{2z}\left[\frac{1}{z} J_0(z) + \frac{\pi}{2} Y_1(z) - (\ln(2z) - 2 + \gamma)\, J_1(z)\right] \tag{3.3.20}$$

Similarly, for $\nu = 1/2$

$$\left[\frac{\partial J_\nu(z)}{\partial \nu}\right]_{\nu=1/2} = \sqrt{\frac{2}{\pi z}}\left[\ln\left(\frac{z}{2}\right) + \gamma\right]\ \sin z + \sqrt{\frac{2z}{\pi}}\, I\left(t, \frac{1}{2}\right)$$

$$I\left(t, \frac{1}{2}\right) = \int_0^1 \ln(1-t^2)\,\cos(zt)\,dt \tag{3.3.21}$$

with help of (3.2.66) we have

$$\int_0^1 \ln(1-t^2)\,\cos(zt)\,dt = \left\{\frac{1}{z}[\cos z\,\mathrm{Ci}(2z) + \sin z\,\mathrm{Si}(2z)] - \sin z\left[\ln\left(\frac{z}{2}\right) + \gamma\right]\right\} \tag{3.3.22}$$

If the integrand in (3.3.14) includes trigonometric functions

$$J_\nu(z) = \frac{\left(\frac{z}{2}\right)^\nu}{\sqrt{\pi}\,\Gamma\left(\nu+\frac{1}{2}\right)} \int_0^\pi \cos(z\cos t)(\sin t)^{2\nu} dt$$
$$\operatorname{Re}\nu > -\frac{1}{2} \tag{3.3.23}$$

then the differentiation of this expression with respect to ν under the integral sign gives

$$\frac{\partial J_\nu(z)}{\partial \nu} = \left[\ln\left(\frac{z}{2}\right) - \psi\left(\nu+\frac{1}{2}\right)\right] J_\nu(z) + \frac{2\left(\frac{z}{2}\right)^\nu}{\sqrt{\pi}\,\Gamma\left(\nu+\frac{1}{2}\right)} I(t,\nu)$$
$$I(t,\nu) = \int_0^\pi \ln(\sin t)\, \cos(z\cos t)\, (\sin t)^{2\nu} dt \tag{3.3.24}$$

and from (3.3.16) and (3.3.24) we have the equality of integrals

$$\int_0^1 (1-t^2)^{\nu-1/2} \ln(1-t^2)\, \cos(zt)\, dx = \int_0^\pi \ln(\sin t)\, \cos(z\cos t)(\sin t)^{2\nu} dt \tag{3.3.25}$$

If integral representations are in the form given in (3.3.14), but they have different integrands and limits of integration, then the equality of integrals in general case can be written as

$$A_1 \int_a^b \frac{\partial G_1(t,\nu)}{\partial \nu} dt = A_2 \int_c^d \frac{\partial G_2(t,\nu)}{\partial \nu} dt$$

The first integral representations of the Bessel functions was given by Schlaefli [7]

$$J_\nu(z) = \frac{1}{\pi} \int_0^\pi \cos(z\sin t - \nu t)\, dt - \frac{\sin(\pi\nu)}{\pi} \int_0^\infty e^{-z\sinh t - \nu t} dt$$
$$\operatorname{Re} z > 0 \tag{3.3.26}$$

and by a direct differentiation of it we have

$$\frac{\partial J_\nu(z)}{\partial \nu} = \frac{1}{\pi} \int_0^\pi t\sin(z\sin t - \nu t)\, dt + \frac{1}{\pi} \int_0^\infty e^{-z\sinh t - \nu t}[t\, \sin(\pi\nu) - \pi\cos(\pi\nu)]\, dt \tag{3.3.27}$$

For $\nu = 0$ it reduces to

$$\left[\frac{\partial J_\nu(z)}{\partial \nu}\right]_{\nu=0} = \frac{1}{\pi} \int_0^\pi t\sin(z\sin t)\, dt - \int_0^\infty e^{-z\sinh t}\, dt \tag{3.3.28}$$

but

$$\frac{\pi}{2} Y_0(z) = \frac{1}{2} \int_0^{\pi} \sin(z \sin t)\, dt - \int_0^{\infty} e^{-z \sinh t}\, dt \tag{3.3.29}$$

and

$$\frac{1}{2} \int_0^{\pi} \sin(z \sin t)\, dt = \frac{1}{\pi} \int_0^{\pi} t \sin(z \sin t)\, dt \tag{3.3.30}$$

and therefore the expected result is derived

$$\left[\frac{\partial J_\nu(z)}{\partial \nu}\right]_{\nu=0} = \frac{\pi}{2} Y_0(z) \tag{3.3.31}$$

Differentiation of the Schlaefli integral representation for the Bessel function of the second kind

$$Y_\nu(z) = \frac{1}{\pi} \int_0^{\pi} \sin(z \sin t - \nu t)\, dt - \frac{1}{\pi} \int_0^{\infty} e^{-z \sinh t} [e^{\nu t} + \cos(\pi\nu)\, e^{-\nu t}]\, dt$$
$$\operatorname{Re} z > 0 \tag{3.3.32}$$

gives

$$\frac{\partial Y_\nu(z)}{\partial \nu} = -\frac{1}{\pi} \int_0^{\pi} t \cos(z \sin t - \nu t)\, dt$$
$$-\frac{1}{\pi} \int_0^{\infty} e^{-z \sinh t} \{t\, [e^{\nu t} - \cos(\pi\nu)\, e^{-\nu t}] - \pi \sin(\pi\nu)\, e^{-\nu t}\}\, dt \tag{3.3.33}$$

When introducing $\nu = 0$, the second integral in (3.3.33) vanishes, combines with (3.3.26) and we have

$$\left[\frac{\partial Y_\nu(z)}{\partial \nu}\right]_{\nu=0} = -\frac{1}{\pi} \int_0^{\pi} t \cos(z \sin t)\, dt$$
$$J_0(z) = \frac{2}{\pi^2} \int_0^{\pi} t \cos(z \sin t)\, dt \tag{3.3.34}$$
$$\left[\frac{\partial Y_\nu(z)}{\partial \nu}\right]_{\nu=0} = -\frac{\pi}{2} J_0(z)$$

and the equivalent to (3.3.30) expression for the equality of integrals

$$\frac{1}{2}\int_0^{\pi}\cos(z\sin t)\,dt = \frac{1}{\pi}\int_0^{\pi} t\cos(z\sin t)\,dt \tag{3.3.35}$$

Integral representation of the modified Bessel function of the first kind also includes two integrals

$$I_\nu(z) = \frac{1}{\pi}\int_0^{\pi} e^{z\cos t}\cos(\nu t)\,dt - \frac{\sin(\pi\nu)}{\pi}\int_0^{\infty} e^{-z\cosh t-\nu t}dt \tag{3.3.36}$$
$$\operatorname{Re} z > 0 \quad ; \quad \operatorname{Re}\nu > 0$$

and therefore

$$\frac{\partial I_\nu(z)}{\partial\nu} = -\frac{1}{\pi}\int_0^{\pi} t e^{z\cos t}\sin(\nu t)\,dt + \frac{1}{\pi}\int_0^{\infty} e^{-z\cosh t-\nu t}[t\sin(\pi\nu) - \pi\cos(\pi\nu)]\,dt \tag{3.3.37}$$

From the integral representation of the modified Bessel function of the second kind

$$K_\nu(z) = \int_0^{\infty} e^{-z\cosh t}\cosh(\nu t)\,dt \quad ; \quad \operatorname{Re} z > 0 \tag{3.3.38}$$

it is possible to obtain

$$\frac{\partial K_\nu(z)}{\partial\nu} = \int_0^{\infty} t e^{-z\cosh t}\sinh(\nu t)\,dt \tag{3.3.39}$$

The derivatives of the Bessel functions of the first and second kind derived above are useful in the case of the Anger and Weber functions

$$\mathbf{J}_\nu(z) = J_\nu(z) + \frac{\sin(\pi\nu)}{\pi}\int_0^{\infty} e^{-z\sinh t-\nu t}dt \tag{3.3.40}$$

$$\mathbf{E}_\nu(z) = -Y_\nu(z) - \frac{1}{\pi}\int_0^{\infty} e^{-z\sinh t}[e^{\nu t} + e^{-\nu t}\cos(\pi\nu)]\,dt \tag{3.3.41}$$

and therefore

$$\frac{\partial \mathbf{J}_\nu(z)}{\partial\nu} = \frac{\partial J_\nu(z)}{\partial\nu} + \frac{1}{\pi}\int_0^{\infty} e^{-z\sinh t-\nu t}[\pi\cos(\pi\nu) - t\sin(\pi\nu)]\,dt \tag{3.3.42}$$

$$\frac{\partial \mathbf{E}_\nu(z)}{\partial \nu} = -\frac{\partial Y_\nu(z)}{\partial \nu} - \frac{1}{\pi}\int_0^\infty e^{-z\sinh t}\{t\,[e^{\nu t} - \cos(\pi\nu)\,e^{-\nu t}] - \pi\,\sin(\pi\nu)\,e^{-\nu t}\}dt \tag{3.3.43}$$

In particular case of $\nu = 0$, we have

$$\left(\frac{\partial \mathbf{J}_\nu(z)}{\partial \nu}\right)_{\nu=0} = \frac{\pi}{2}Y_0(z) + \int_0^\infty e^{-z\sinh t}\,dt \tag{3.3.44}$$

but the Anger function is defined by the integral

$$\mathbf{J}_\nu(z) = \frac{1}{\pi}\int_0^\pi \cos(\nu t - z\sin t)\,dt \tag{3.3.45}$$

and its differentiation yields

$$\frac{\partial \mathbf{J}_\nu(z)}{\partial \nu} = -\frac{1}{\pi}\int_0^\pi t\,\sin(\nu t - z\sin t)\,dt \tag{3.3.46}$$

therefore for $\nu = 0$, the expression in (3.3.46) becomes

$$\left(\frac{\partial \mathbf{J}_\nu(z)}{\partial \nu}\right)_{\nu=0} = \frac{1}{\pi}\int_0^\pi t\,\sin(z\sin t)\,dt \tag{3.3.47}$$

Comparing (3.3.47) with (3.3.32.) gives

$$Y_0(z) = -\frac{2}{\pi}\int_0^\infty e^{-z\sinh t}\,dt - \frac{2}{\pi^2}\int_0^\pi t\,\sin(z\sin t)\,dt \tag{3.3.48}$$

and since [18]

$$\left(\frac{\partial \mathbf{J}_\nu(z)}{\partial \nu}\right)_{\nu=0} = \frac{\pi}{2}\mathbf{H}_0(z) \tag{3.3.49}$$

we have

$$\mathbf{H}_0(z) = Y_0(z) + \frac{2}{\pi}\int_0^\infty e^{-z\sinh t}\,dt \tag{3.3.50}$$

and

$$H_0(z) = \frac{2}{\pi^2}\int_0^{\pi} t\,\sin(z\,\sin t)\,dt \tag{3.3.51}$$

Similarly, it is possible to obtain

$$\left(\frac{\partial \mathbf{E}_\nu(z)}{\partial \nu}\right)_{\nu=0} = -\left(\frac{\partial Y_\nu(z)}{\partial \nu}\right)_{\nu=0} = \frac{\pi}{2}J_0(z) \tag{3.3.52}$$

From

$$\mathbf{E}_\nu(z) = \frac{1}{\pi}\int_0^{\pi} \sin(\nu t - z\,\sin t)\,dt \tag{3.3.53}$$

by differentiation of (3.3.53) we have

$$\frac{\partial \mathbf{E}_\nu(z)}{\partial \nu} = \frac{1}{\pi}\int_0^{\pi} t\,\cos(\nu t - z\,\sin t)\,dt \tag{3.3.54}$$

and for $\nu = 0$, once again the result given in (3.3.52) is reached.

If a number of evaluated integrals is considered, the Schlaefli integral representations are less convenient to obtain derivatives of the Bessel and related functions with respect to the order than those based on (3.3.14) expression. Using integral representations given in Section 2.12, it is possible to present a short list of first derivatives. In all cases of derived formulas, conditions $\operatorname{Re} z > 0$ and $\operatorname{Re} \nu > -1/2$ are imposed.

The first expression is a different form of derivative with respect to the order of the Bessel function of the first kind

$$\begin{aligned}\frac{\partial J_\nu(z)}{\partial \nu} &= \left[\ln\left(\frac{z}{2}\right) - \psi\left(\nu + \frac{1}{2}\right)\right] J_\nu(z) \\ &\quad + \frac{\left(\frac{z}{2}\right)^\nu}{\sqrt{\pi}\,\Gamma\left(\nu+\frac{1}{2}\right)}\int_0^1 \frac{t^\nu \ln t}{\sqrt{t(1-t)}}\cos(z\sqrt{1-t})\,dt\end{aligned} \tag{3.3.55}$$

For $\nu = 0$ and $\nu = 1$ it becomes that

$$\begin{aligned}\left[\frac{\partial J_\nu(z)}{\partial \nu}\right]_{\nu=0} &= [\ln z + \gamma + \ln 2]\,J_0(z) \\ &\quad + \frac{1}{\pi}\int_0^1 \frac{\ln t}{\sqrt{t(1-t)}}\cos(z\sqrt{1-t})\,dt\end{aligned} \tag{3.3.56}$$

$$\left[\frac{\partial J_\nu(z)}{\partial \nu}\right]_{\nu=1} = [\ln z + \gamma + \ln 2 - 2]\, J_1(z) + \frac{z}{\pi}\int_0^1 \sqrt{\frac{t}{1-t}}\ln t\, \cos(z\sqrt{1-t})\, dt \tag{3.3.57}$$

These integrals can be evaluated by using (3.3.31) and (3.2.24).

In case of the modified Bessel functions, we have

$$\begin{aligned}\frac{\partial I_\nu(z)}{\partial \nu} &= \left[\ln\left(\frac{z}{2}\right) - \psi\left(\nu + \frac{1}{2}\right)\right] I_\nu(z) \\ &+ \frac{2\left(\frac{z}{2}\right)^\nu}{\sqrt{\pi}\,\Gamma\left(\nu+\frac{1}{2}\right)}\int_0^\pi \ln(\sin t)(\sin t)^{2\nu}\cosh(zt)\, dt\end{aligned} \tag{3.3.58}$$

$$\begin{aligned}\frac{\partial K_\nu(z)}{\partial \nu} &= \left[\ln\left(\frac{z}{2}\right) - \psi\left(\nu + \frac{1}{2}\right)\right] K_\nu(z) \\ &+ \frac{2\sqrt{\pi}\left(\frac{z}{2}\right)^\nu}{\Gamma\left(\nu+\frac{1}{2}\right)}\int_0^\infty e^{-z\cosh t}\ln(\sinh t)(\sinh t)^{2\nu}\, dt\end{aligned} \tag{3.3.59}$$

Also in this case, integrals in (3.3.58) and (3.3.59) can be expressed in a closed form for $\nu = 0$ and $\nu = 1$.

Derivatives with respect to the order of the Struve functions are

$$\begin{aligned}\frac{\partial H_\nu(z)}{\partial \nu} &= \left[\ln\left(\frac{z}{2}\right) - \psi\left(\nu + \frac{1}{2}\right)\right] H_\nu(z) \\ &+ \frac{2\left(\frac{z}{2}\right)^\nu}{\sqrt{\pi}\,\Gamma\left(\nu+\frac{1}{2}\right)}\int_0^1 (1-t^2)^{\nu-1/2}\ln(1-t^2)\sin(zt)\, dt\end{aligned} \tag{3.3.60}$$

$$\begin{aligned}\frac{\partial H_\nu(z)}{\partial \nu} &= \left[\ln\left(\frac{z}{2}\right) - \psi\left(\nu + \frac{1}{2}\right)\right] H_\nu(z) \\ &+ \frac{4\left(\frac{z}{2}\right)^\nu}{\sqrt{\pi}\,\Gamma\left(\nu+\frac{1}{2}\right)}\int_0^{\pi/2} \sin(z\cos t)\, \ln(\sin t)\, (\sin t)^{2\nu}\, dt\end{aligned} \tag{3.3.61}$$

$$\begin{aligned}\frac{\partial L_\nu(z)}{\partial \nu} &= \left[\ln\left(\frac{z}{2}\right) - \psi\left(\nu + \frac{1}{2}\right)\right] L_\nu(z) \\ &+ \frac{4\left(\frac{z}{2}\right)^\nu}{\sqrt{\pi}\,\Gamma\left(\nu+\frac{1}{2}\right)}\int_0^{\pi/2} \sinh(z\cos t)\, \ln(\sin t)\, (\sin t)^{2\nu}\, dt\end{aligned} \tag{3.3.62}$$

For $\nu = 0$ and $\nu = 1$, by using different integral representations of the Struve functions, it is possible to obtain [5]:

$$\left(\frac{\partial H_\nu(z)}{\partial \nu}\right)_{\nu=0} = -\frac{\pi}{2} J_0(z) + \left[\ln\left(\frac{z}{2}\right) + \gamma + 2\ln 2\right] [H_0(z) - Y_0(z)] + \frac{2}{\pi}\int_0^\infty e^{-zt} \frac{\ln(1+t^2)}{\sqrt{1+t^2}}\, dt \tag{3.3.63}$$

$$\left(\frac{\partial H_\nu(z)}{\partial \nu}\right)_{\nu=1} = \left[\ln\left(\frac{z}{2}\right) + \gamma + 2\ln 2 - 2\right] [H_1(z) - Y_1(z)] + \frac{2z}{\pi}\int_0^\infty e^{-zt} \sqrt{1+t^2}\, \ln(1+t^2)\, dt - \frac{1}{z} Y_0(z) - \frac{\pi}{2} J_1(z) \tag{3.3.64}$$

$$\left(\frac{\partial L_\nu(z)}{\partial \nu}\right)_{\nu=0} = -\frac{\pi}{2} J_0(z) + \left[\ln\left(\frac{z}{2}\right) + \gamma + 2\ln 2\right] L_0(z) + \frac{2}{\pi}\int_0^1 \frac{\ln(1-t^2)}{\sqrt{1-t^2}} \sinh(zt)\, dt \tag{3.3.65}$$

$$\left(\frac{\partial L_\nu(z)}{\partial \nu}\right)_{\nu=1} = \left[\ln\left(\frac{z}{2}\right) + \gamma + 2\ln 2 - 2\right] L_1(z) + \frac{2z}{\pi}\int_0^1 e^{-zt} \sqrt{1-t^2}\, \ln(1-t^2)\, \sinh(zt)\, dt \tag{3.3.66}$$

The asymptotic behaviour of integrals given in the above expressions is known for small and large values of argument z [5].

The Struve functions can be expressed also with the help of the Bessel functions and therefore it is possible to obtain from (2.12.60) and (2.12.64)

$$\frac{\partial H_\nu(z)}{\partial \nu} = \frac{\partial Y_\nu(z)}{\partial \nu} + \left[\ln\left(\frac{z}{2}\right) - \psi\left(\nu + \frac{1}{2}\right)\right] H_\nu(z) + \frac{2\left(\frac{z}{2}\right)^\nu}{\sqrt{\pi}\,\Gamma\left(\nu+\frac{1}{2}\right)} \int_0^\infty e^{-zt}(1+t^2)^{\nu-1/2} \ln(1+t^2)\, dt \tag{3.3.67}$$

$$\frac{\partial L_\nu(z)}{\partial \nu} = \frac{\partial I_\nu(z)}{\partial \nu} - \left[\ln\left(\frac{z}{2}\right) - \psi\left(\nu + \frac{1}{2}\right)\right] L_\nu(z) - \frac{2\left(\frac{z}{2}\right)^\nu}{\sqrt{\pi}\,\Gamma\left(\nu+\frac{1}{2}\right)} \int_0^1 e^{-zt}(1-t^2)^{\nu-1/2} \ln(1-t^2)\, dt \tag{3.3.68}$$

These integrals for $\nu = 0$ and $\nu = 1/2$, can be determined because

$$\left[\frac{\partial \mathbf{H}_\nu(z)}{\partial \nu}\right]_{\nu=0} = -\frac{\pi}{2} J_0(z) + \frac{1}{\pi^2 z}\, G_{24}^{32}\left(\frac{z^2}{4}\left|\begin{matrix} 1 & 1 \\ 1/2,\ 1, & 1,\ 1/2 \end{matrix}\right.\right) \quad ; \quad \operatorname{Re} z \geq 0 \tag{3.3.69}$$

and

$$\left[\frac{\partial \mathbf{H}_\nu(z)}{\partial \nu}\right]_{\nu=1/2} = \sqrt{\frac{2}{\pi z}}\left\{\gamma + \ln\left(\frac{z}{2}\right) + \cos z\,[\operatorname{Ci}(2z) - 2\operatorname{Ci}(z)] + \sin z\,[\operatorname{Si}(2z) - 2\operatorname{Si}(z)]\right\} \tag{3.3.70}$$

Thus, we have

$$\int_0^\infty \frac{e^{-zt}\ln(1+t^2)}{\sqrt{1+t^2}}\, dt = -\frac{\pi}{2}(\ln z + \gamma + \ln 2)\,\mathbf{H}_0(z) + \frac{1}{2\pi z}\, G_{24}^{32}\left(\frac{z^2}{4}\left|\begin{matrix} 1 & 1 \\ 1/2,\ 1, & 1,\ 1/2 \end{matrix}\right.\right) \quad ; \quad \operatorname{Re} z \geq 0 \tag{3.3.71}$$

and

$$\int_0^\infty e^{-zt}\ln(1+t^2)\, dt = [\pi - 2\operatorname{Si}(z)]\sin z + [\ln z - \ln 2 + \gamma - 2\operatorname{Ci}(z)]\cos z \tag{3.3.72}$$

In the case of the modified Struve function for $\nu = 0$ and $\nu = 1/2$, from

$$\begin{aligned}\left[\frac{\partial \mathbf{L}_\nu(z)}{\partial \nu}\right]_{\nu=0} &= K_0(z) - \frac{1}{\pi^2 z}\, G_{24}^{42}\left(\frac{z^2}{4}\left|\begin{matrix} 1 & 1 \\ 1/2,\ 1, & 1,\ 1/2 \end{matrix}\right.\right) \\ &= -K_0(z) - \frac{2}{z}\, G_{46}^{42}\left(\frac{z^2}{4}\left|\begin{matrix} 1,\ 1,\ 1/4,\ 3/4 \\ 1/2,\ 1/2,\ 1,\ 1,\ 1/4,\ 3/4 \end{matrix}\right.\right) \quad ; \quad \operatorname{Re} z \geq 0\end{aligned} \tag{3.3.73}$$

$$\left[\frac{\partial I_\nu(z)}{\partial \nu}\right]_{\nu=0} = -K_0(z) \tag{3.3.74}$$

it is possible to derive that

$$\left[\frac{\partial \mathbf{L}_\nu(z)}{\partial \nu}\right]_{\nu=1/2} = -\frac{1}{\sqrt{2\pi z}}\left\{2\left[\gamma + \ln\left(\frac{z}{2}\right)\right] + e^{-z}[\operatorname{Ei}(2z) - 2\operatorname{Ei}(z)] + e^{z}[\operatorname{Ei}(-2z) - 2\operatorname{Ei}(-z)]\right\} \tag{3.3.75}$$

$$\left[\frac{\partial I_\nu(z)}{\partial \nu}\right]_{\nu=1/2} = \frac{1}{\sqrt{2\pi z}}[e^{z}\operatorname{Ei}(-2z) - e^{-z}\operatorname{Ei}(2z)] \tag{3.3.76}$$

and

$$\int_0^1 \frac{e^{-zt}\ln(1-t^2)}{\sqrt{1-t^2}} = \frac{\pi}{z} G_{46}^{42}\left(\frac{z^2}{4}\middle| \begin{matrix} 1, 1, 1/4, 3/4 \\ 1/2, 1/2, 1, 1, 1/4, 3/4 \end{matrix}\right) - \frac{\pi}{2}[\ln z + \gamma + \ln 2]\, L_0(z) \tag{3.3.77}$$

$$\int_0^1 e^{-zt}\ln(1-t^2)\,dt = -\frac{1}{z}\left[\ln\left(\frac{z}{2}\right) + \gamma + e^z \mathrm{Ei}(-z) - e^{-z}\mathrm{Ei}(z)\right] \tag{3.3.78}$$

In some cases, differentiation with respect to the order is easier to perform if in the integral representation of the Bessel and related functions the order ν is absent in the pre-integral factor. Such situation appears in the Mehler–Sonine formulas [7]

$$J_\nu(z) = \frac{2}{\pi}\int_0^\infty \sin\left(z\cosh t - \frac{\pi\nu}{2}\right)\cosh(\nu t)\,dt \quad ; \quad -1 < \mathrm{Re}\,\nu < 1 \tag{3.3.79}$$

$$Y_\nu(z) = -\frac{2}{\pi}\int_0^\infty \cos\left(z\cosh t - \frac{\pi\nu}{2}\right)\cosh(\nu t)\,dt \quad ; \quad -1 < \mathrm{Re}\,\nu < 1 \tag{3.3.80}$$

In this case, these integral representations are valid for a limited range of the order ν. Differentiation of (3.3.79) gives

$$\frac{\partial J_\nu(z)}{\partial \nu} = \int_0^\infty \left[\frac{2t}{\pi}\sin\left(z\cosh t - \frac{\pi\nu}{2}\right)\sinh(\nu t) - \cos\left(z\cosh t - \frac{\pi\nu}{2}\right)\cosh(\nu t)\right] dt \tag{3.3.81}$$

and for $\nu = 0$, the expected result is achieved

$$\left(\frac{\partial J_\nu(z)}{\partial \nu}\right)_{\nu=0} = \frac{\pi}{2} Y_0(z) = -\int_0^\infty \cos(z\cosh t)\,dt \tag{3.3.82}$$

Similarly from

$$\frac{\partial Y_\nu(z)}{\partial \nu} = -\int_0^\infty \left[\frac{2t}{\pi}\cos\left(z\cosh t - \frac{\pi\nu}{2}\right)\sinh(\nu t) + \sin\left(z\cosh t - \frac{\pi\nu}{2}\right)\cosh(\nu t)\right] dt \tag{3.3.83}$$

we have for $\nu = 0$

$$\left(\frac{\partial J_\nu(z)}{\partial \nu}\right)_{\nu=0} = -\frac{\pi}{2} J_0(z) = -\int_0^\infty \sin(z\cosh t)\, dt \tag{3.3.84}$$

Integral representations of the modified Bessel functions of the second kind which are given in (2.12.45) – (2.12.55), have the order ν in one or two places only. The simplest case is

$$K_\nu(z) = \int_0^\infty e^{-z\cosh t}\cosh(\nu t)\, dt \tag{3.3.85}$$

and therefore

$$\frac{\partial K_\nu(z)}{\partial \nu} = \int_0^\infty t\, e^{-z\cosh t}\sinh(\nu t)\, dt \tag{3.3.86}$$

For $\nu = 0$, this expression shows that the first derivative of $K_\nu(z)$ with the respect to the order ν is zero. In the case $\nu = 1$, using (3.3.85) and (3.3.86), we have

$$K_0(z) = z\int_0^\infty t\, e^{-z\cosh t}\sinh t\, dt \tag{3.3.87}$$

and for $\nu = 1/2$ it is possible to deduce that

$$-\operatorname{Ei}(-2z) = \sqrt{\frac{2z}{\pi}}\, e^{-z}\int_0^\infty t\, e^{-z\cosh t}\sinh\left(\frac{t}{2}\right) dt \tag{3.3.88}$$

The Kelvin functions of the first kind and the order ν have the following integral representations of the Schlaefli type [6]

$$\begin{aligned}\operatorname{ber}_\nu(\sqrt{2}z) = \frac{1}{\pi}\int_0^\pi &[\cos(\pi\nu)\cos(z\sin t - \nu t)\cosh(z\sin t)\\ &-\sin(\pi\nu)\sin(z\sin t - \nu t)\sinh(z\sin t)]\, dt\\ &-\frac{\sin(\pi\nu)}{\pi}\int_0^\infty e^{-\nu t - z\sinh t}\cos(z\sinh t + \pi\nu)\, dt\end{aligned} \tag{3.3.89}$$

and the Kelvin functions of the second kind and the order ν are

$$\operatorname{bei}_\nu(\sqrt{2}z) = \frac{1}{\pi}\int_0^\pi [\cos(\pi\nu)\,\sin(z\sin t - \nu t)\,\sinh(z\sin t)$$
$$+ \sin(\pi\nu)\,\cos(z\sin t - \nu t)\,\cosh(z\sin t)]\,dt \tag{3.3.90}$$
$$- \frac{\sin(\pi\nu)}{\pi}\int_0^\infty e^{-\nu t - z\sinh t}\sin(z\sinh t + \pi\nu)\,dt$$

Direct differentiation of these formulas gives

$$\frac{\partial \operatorname{ber}_\nu(\sqrt{2}z)}{\partial\nu} = -\pi\operatorname{bei}_\nu(\sqrt{2}z)$$
$$+ \frac{1}{\pi}\int_0^\pi t\,[\sin(\pi\nu)\,\cos(z\sin t - \nu t)\,\sinh(z\sin t)$$
$$- \cos(\pi\nu)\,\sin(z\sin t - \nu t)\,\cosh(z\sin t)]\,dt$$
$$+ \frac{1}{\pi}\int_0^\infty e^{-\nu t - z\sinh t}[t\,\sin(\pi\nu) - \pi\,\cos(\pi\nu)]\cos(z\sinh t + \pi\nu)\,dt \tag{3.3.91}$$

and

$$\frac{\partial \operatorname{bei}_\nu(\sqrt{2}z)}{\partial\nu} = -\pi\operatorname{ber}_\nu(\sqrt{2}z)$$
$$+ \frac{1}{\pi}\int_0^\pi t\,[\sin(\pi\nu)\,\cos(z\sin t - \nu t)\,\cosh(z\sin t)$$
$$- \cos(\pi\nu)\,\cos(z\sin t - \nu t)\,\sinh(z\sin t)]\,dt$$
$$+ \frac{1}{\pi}\int_0^\infty e^{-\nu t - z\sinh t}[t\,\sin(\pi\nu) - \pi\,\cos(\pi\nu)]\sin(z\sinh t + \pi\nu)\,dt \tag{3.3.92}$$

These formulas have a much simpler form if $\nu = 0$

$$\left(\frac{\partial \operatorname{ber}_\nu(\sqrt{2}z)}{\partial\nu}\right)_{\nu=0} = -\pi\operatorname{bei}(\sqrt{2}z) - \frac{1}{\pi}\int_0^\pi t\,\sin(z\sin t)\,\cosh(z\sin t)]\,dt \tag{3.3.93}$$
$$- \int_0^\infty e^{-z\sinh t}\cos(z\sinh t)\,dt$$

and

$$\left(\frac{\partial \operatorname{bei}_v(\sqrt{2}z)}{\partial v}\right)_{v=0} = -\pi \operatorname{ber}(\sqrt{2}z) - \frac{1}{\pi}\int_0^{\pi} t\,\cos(z\sin t)\,\sinh(z\sin t)]\,dt + \int_0^{\infty} e^{-z\sinh t}\sin(z\sinh t)\,dt \tag{3.3.94}$$

In the case of the modified Kelvin functions $\operatorname{ker}_v(z)$, $\operatorname{kei}_v(z)$ and $\operatorname{her}_v(z)$, $\operatorname{hei}_v(z)$, the first step is to derive the integral representations from their definitions

$$\begin{aligned} &\operatorname{ker}_v(z) + i\operatorname{kei}_v(z) = e^{-\pi i/2}K_v(ze^{\pi i/4}) \\ &\operatorname{ker}_v(z) - i\operatorname{kei}_v(z) = e^{\pi i/2}K_v(ze^{-\pi i/4}) \\ &e^{\pm\pi i/4} = \frac{1 \pm i}{\sqrt{2}} \end{aligned} \tag{3.3.95}$$

and from (2.12.45) we have

$$K_v(z) = \int_0^{\infty} e^{-z\cosh t}\cosh(vt)\,dt \quad ; \quad \operatorname{Re} z > 0 \tag{3.3.96}$$

Combining (3.3.95) with (3.3.96), after many but elementary steps, the integral representations of the modified Kelvin functions are

$$\operatorname{ker}_v(\sqrt{2}z) = \int_0^{\infty} e^{-z\cosh t}\cos\left(z\cosh t + \frac{\pi v}{2}\right)\cosh(vt)dt \tag{3.3.97}$$

and

$$\operatorname{kei}_v(\sqrt{2}z) = -\int_0^{\infty} e^{-z\cosh t}\sin\left(z\cosh t + \frac{\pi v}{2}\right)\cosh(vt)dt \tag{3.3.98}$$

Thus, the derivatives of the modified Kelvin functions are available by a direct differentiation of (3.3.97) and (3.3.98) with respect to the order v

$$\frac{\partial \operatorname{ker}_v(\sqrt{2}z)}{\partial v} = \int_0^{\infty} e^{-z\cosh t}\left[t\,\cos\left(z\cosh t + \frac{\pi v}{2}\right)\sinh(vt) - \left(\frac{\pi}{2}\right)\sin\left(z\cosh t + \frac{\pi v}{2}\right)\cosh(vt)\right]dt \tag{3.3.99}$$

and

$$\frac{\partial \operatorname{kei}_\nu(\sqrt{2}z)}{\partial \nu} = -\int_0^\infty e^{-z\cosh t}\left[t \sin\left(z\cosh t + \frac{\pi\nu}{2}\right)\sinh(\nu t) + \left(\frac{\pi}{2}\right)\cos\left(z\cosh t + \frac{\pi\nu}{2}\right)\cosh(\nu t)\right] dt \tag{3.3.100}$$

For $\nu = 0$, both the expressions take the form

$$\left(\frac{\partial \operatorname{ker}_\nu(\sqrt{2}z)}{\partial \nu}\right)_{\nu=0} = -\frac{\pi}{2}\int_0^\infty e^{-z\cosh t}\sin(z\cosh t)\, dt \tag{3.3.101}$$

and

$$\left(\frac{\partial \operatorname{kei}_\nu(\sqrt{2}z)}{\partial \nu}\right)_{\nu=0} = -\frac{\pi}{2}\int_0^\infty e^{-z\cosh t}\cos(z\cosh t)\, dt \tag{3.3.102}$$

Derivatives with respect to the order of the Lommel functions can be derived for the symmetrical case $\nu = \mu$ with the help of derivatives of the Bessel or the Struve functions. However, if $\mu = 0$ and $\mu = \pm 1$, the integral representations of the Lommel functions can be directly used. From Section 2.12. we have

$$s_{-1,\nu}(z) = -\frac{1}{\nu\sin(\pi\nu)}\int_0^\pi \cos(z\sin t)\cos(\nu t)\, dt \tag{3.3.103}$$

$$s_{1,\nu}(z) = 1 + \nu^2 s_{-1,\nu}(z) \tag{3.3.104}$$

$$s_{0,\nu}(z) = \frac{1}{\sin(\pi\nu)}\int_0^\pi \sin(z\sin t)\cos(\nu t)\, dt \tag{3.3.105}$$

$$S_{0,\nu}(z) = \int_0^\infty e^{-z\sinh t}\cosh(\nu t)\, dt \quad ; \quad \operatorname{Re} z > 0 \tag{3.3.106}$$

$$S_{1,\nu}(z) = z\int_0^\infty e^{-z\sinh t}\cosh(\nu t)\cosh t\, dt \quad ; \quad \operatorname{Re} z > 0 \tag{3.3.107}$$

$$S_{-1,\nu}(z) = \frac{1}{\nu^2}\left[S_{1,\nu}(z) - 1\right] \tag{3.3.108}$$

and therefore

$$\frac{\partial s_{-1,\nu}(z)}{\partial \nu} = \frac{1}{\nu}\left\{ [1-\pi\nu\cot(\pi\nu)]\, s_{-1,\nu}(z) - \csc(\pi\nu)\int_0^{\pi} t\cos(z\sin t)\sin(\nu t)\,dt \right\} \tag{3.3.109}$$

$$\frac{\partial s_{1,\nu}(z)}{\partial \nu} = \nu\left[2s_{-1,\nu}(z) + \nu\,\frac{\partial s_{-1,\nu}(z)}{\partial \nu}\right] \tag{3.3.110}$$

$$\frac{\partial s_{0,\nu}(z)}{\partial \nu} = -\frac{1}{\sin(\pi\nu)}\left\{ [\pi\cot(\pi\nu)]\, s_{0,\nu}(z) + \int_0^{\pi} t\sin(z\sin t)\sin(\nu t)\,dt \right\} \tag{3.3.111}$$

$$\frac{\partial S_{0,\nu}(z)}{\partial \nu} = \int_0^{\infty} t\,e^{-z\sinh t}\sinh(\nu t)\,dt \tag{3.3.112}$$

$$\frac{\partial S_{1,\nu}(z)}{\partial \nu} = \int_0^{\infty} t\,e^{-z\sinh t}\sinh(\nu t)\cosh t\,dt \tag{3.3.113}$$

$$\frac{\partial S_{-1,\nu}(z)}{\partial \nu} = \frac{1}{\nu^2}\frac{\partial S_{1,\nu}(z)}{\partial \nu} - \frac{2}{\nu}S_{-1,\nu}(z) \tag{3.3.114}$$

The above expressions can be related to the Anger and the Weber functions

$$\boldsymbol{J}_\nu(z) = \frac{1}{\pi}\sin(\pi\nu)\,[s_{0,\nu}(z) - \nu s_{-1,\nu}(z)] \tag{3.3.115}$$

$$\boldsymbol{E}_\nu(z) = -\frac{1+\cos(\pi\nu)}{\pi}s_{0,\nu}(z) - \frac{\nu\,[1-\cos(\pi\nu)]}{\pi}s_{-1,\nu}(z) \tag{3.3.116}$$

and therefore we have

$$\frac{\partial \boldsymbol{J}_\nu(z)}{\partial \nu} = \pi\cot(\pi\nu)\,\boldsymbol{J}_\nu(z) + \frac{1}{\pi}\sin(\pi\nu)\left[\frac{\partial s_{0,\nu}(z)}{\partial \nu} - \frac{\partial s_{-1,\nu}(z)}{\partial \nu} - s_{-1,\nu}(z)\right] \tag{3.3.117}$$

$$\begin{aligned}\frac{\partial \boldsymbol{E}_\nu(z)}{\partial \nu} = &-\sin(\pi\nu)\,s_{0,\nu}(z) + \left[\frac{\cos(\pi\nu) - (\pi\nu)\sin(\pi\nu) - 1}{\pi}\right] s_{-1,\nu}(z)] \\ &+ \left[\frac{1+\cos(\pi\nu)}{\pi}\right]\frac{\partial s_{0,\nu}(z)}{\partial \nu} + \left[\frac{\nu\cos(\pi\nu) - \nu}{\pi}\right]\frac{\partial s_{-1,\nu}(z)}{\partial \nu}\end{aligned} \tag{3.3.118}$$

3.4 Higher Order Derivatives of the Bessel and Related Functions with Respect to the Order *ν*

Our knowledge about higher derivatives of the Bessel functions with respect to the order is very limited. Their significance considering properties of the Bessel functions is still unclear. Derivations of them is not always easy, especially if performed with expressions having series. The first investigation devoted to numerical evaluation of the second derivative of the Bessel function $J_\nu(z)$ with respect to the order for a wide range of orders and arguments is that of Airey [43] from 1935. For particular case of $\nu = 0$, the second order derivatives were given by Luke [35] in his book dealing with integrals of the Bessel functions

$$\left(\frac{\partial^2 J_\nu(z)}{\partial \nu^2}\right)_{\nu=0} = \pi \left[\gamma + \ln\left(\frac{z}{2}\right)\right] Y_0(z) - \left\{\frac{\pi^2}{6} + \left[\gamma + \ln\left(\frac{z}{2}\right)\right]^2\right\} J_0(z) - 2\sum_{k=1}^{\infty} \frac{\left(\frac{z}{2}\right)^k}{k^2 k!} J_k(z) \tag{3.4.1}$$

and

$$\frac{\partial^2 I_\nu(z)}{\partial \nu^2} = -2\left[\gamma + \ln\left(\frac{z}{2}\right)\right] K_0(z) - \left\{\frac{\pi^2}{6} + \left[\gamma + \ln\left(\frac{z}{2}\right)\right]^2\right\} I_0(z) - 2\sum_{k=1}^{\infty} \frac{(-1)^k \left(\frac{z}{2}\right)^k}{k^2 k!} I_k(z) \tag{3.4.2}$$

In 1977, Wienke [57] proposed to replace the power series representation of the derivatives with respect to the order of the Bessel function $J_\nu(z)$ with the Neumann type expansion.

$$\begin{gathered}\frac{\partial J_\nu(z)}{\partial \nu} = \left[\gamma + \ln\left(\frac{z}{2}\right)\right] J_\nu(z) - \sum_{k=0}^{\infty} a_{2k} J_{\nu+2k}(z) \\ a_0 = \gamma + \frac{1}{\nu} + \psi(\nu) \quad ; \quad a_{2k} = \frac{(-1)^k}{k}\left(\frac{2k+\nu}{k+\nu}\right) \\ \nu \neq -1, -2, -3, \ldots\end{gathered} \tag{3.4.3}$$

and

$$\begin{gathered}\frac{\partial^2 J_\nu(z)}{\partial \nu^2} = \left[\gamma + \ln\left(\frac{z}{2}\right)\right]^2 J_\nu(z) - 2\left[\gamma + \ln\left(\frac{z}{2}\right)\right] \sum_{k=0}^{\infty} a_{2k} J_{\nu+2k}(z) \\ - \sum_{k=0}^{\infty} \frac{\partial a_{2k}}{\partial \nu} J_{\nu+2k}(z) + \sum_{l=0}^{\infty}\sum_{k=0}^{\infty} a_{2l} a_{2k} J_{\nu+2l+2k}(z) \\ \frac{\partial a_0}{\partial \nu} = -\frac{1}{\nu^2} + \psi'(\nu) \quad ; \quad \frac{\partial a_{2k}}{\partial \nu} = -\frac{(-1)^k}{(\nu+k)^2} \quad ; \quad k > 0\end{gathered} \tag{3.4.4}$$

Wienke also gave the first derivatives with respect to the order when ν takes positive and negative integer values, they are presented in the Hansen Tables [15].

Evidently, by the direct differentiation of the series expansion of the Bessel function $J_\nu(z)$ it is possible to obtain the following result

$$\frac{\partial^2 J_\nu(z)}{\partial \nu^2} = 2\ \ln\left(\frac{z}{2}\right)\frac{\partial J_\nu(z)}{\partial \nu} - \left[\ln\left(\frac{z}{2}\right)\right]^2 J_\nu(z)$$
$$+\sum_{k=0}^{\infty}(-1)^k\left(\frac{z}{2}\right)^{2k+\nu}\frac{k!\left\{[\psi(\nu+k+1)]^2-\psi'(\nu+k+1)\right\}}{\Gamma(\nu+k+1)} \tag{3.4.5}$$

A more systematic investigation of the second order and third order derivatives of the Bessel functions has been recently performed by Brychkov. Only one of his expressions for $\nu = 0$ is presented here

$$\left[\frac{\partial^2 J_\nu(z)}{\partial \nu^2}\right]_{\nu=0} = \pi\, Y_0(z)\left[\gamma + \ln\left(\frac{z}{2}\right)\right] - \frac{z^2}{4}\ {}_3F_4\left(1,1,\frac{3}{2};2,2,2,2,;-z^2\right)$$
$$-\frac{\pi^2}{4}J_0(z) + \frac{\sqrt{\pi}}{2}J_0(z)\,G^{31}_{35}\left(z^2\left|\begin{matrix}1/2, & -1/2 & & 1\\ 0, & 0, & 0, & -1/2, & 0\end{matrix}\right.\right)$$
$$\operatorname{Re} z \ge 0 \tag{3.4.6}$$

Other formulas for derivatives of the Bessel functions with integer orders are very long and complex.

Since most of integral representations of the Bessel and related functions $F_\nu(z)$ have the form given in (3.3.13)

$$F_\nu(z) = A\frac{\left(\frac{z}{2}\right)^\nu}{\Gamma\left(\nu+\frac{1}{2}\right)}\int_a^b G(t,\nu)\,dt \tag{3.4.7}$$

the second derivative with respect to the order becomes

$$\frac{\partial^2 F_\nu(z)}{\partial \nu^2} = -\psi'\left(\nu+\frac{1}{2}\right)F_\nu(z) + \left[\ln\left(\frac{z}{2}\right) - \psi\left(\nu+\frac{1}{2}\right)\right]^2 F_\nu(z)$$
$$+A\frac{\left(\frac{z}{2}\right)^\nu}{\Gamma\left(\nu+\frac{1}{2}\right)}\left\{2\left[\ln\left(\frac{z}{2}\right) - \psi\left(\nu+\frac{1}{2}\right)\right]\int_a^b \frac{\partial G(t,\nu)}{\partial \nu}dt + \int_a^b \frac{\partial^2 G(t,\nu)}{\partial \nu^2}dt\right\} \tag{3.4.8}$$

Thus, the formula for the second derivative with respect to the order includes only one additional mathematical operation, the differentiation of integrand of already known integral $I(t,\nu)$

$$\frac{\partial I(t,\nu)}{\partial \nu} = \frac{\partial}{\partial \nu}\int_a^b \frac{\partial G(t,\nu)}{\partial \nu}dt = \int_a^b \frac{\partial^2 G(t,\nu)}{\partial \nu^2}dt \tag{3.4.9}$$

If first derivative from (3.3.14) is preserved, (3.4.8) takes the form

$$\begin{aligned}\frac{\partial^2 F_\nu(z)}{\partial\nu^2} &= -\psi^{(1)}\left(\nu+\frac{1}{2}\right)F_\nu(z) + \left[\ln\left(\frac{z}{2}\right)-\psi\left(\nu+\frac{1}{2}\right)\right]\frac{\partial F_\nu(z)}{\partial\nu}\\ &+A\frac{\left(\frac{z}{2}\right)^\nu}{\Gamma\left(\nu+\frac{1}{2}\right)}\left[\ln\left(\frac{z}{2}\right)-\psi\left(\nu+\frac{1}{2}\right)\right]\int_a^b\frac{\partial G(t,\nu)}{\partial\nu}\,dt\\ &+A\frac{\left(\frac{z}{2}\right)^\nu}{\Gamma\left(\nu+\frac{1}{2}\right)}\int_a^b\frac{\partial^2 G(t,\nu)}{\partial\nu^2}\,dt\end{aligned} \tag{3.4.10}$$

For example in the case of the Bessel function $J_\nu(z)$ we have

$$\begin{aligned}\frac{\partial^2 J_\nu(z)}{\partial\nu^2} &= -\psi^{(1)}\left(\nu+\frac{1}{2}\right)J_\nu(z) + \left[\ln\left(\frac{z}{2}\right)-\psi\left(\nu+\frac{1}{2}\right)\right]^2 J_\nu(z)\\ &+\frac{4\left(\frac{z}{2}\right)^\nu\left[\ln\left(\frac{z}{2}\right)-\psi\left(\nu+\frac{1}{2}\right)\right]}{\sqrt{\pi}\,\Gamma\left(\nu+\frac{1}{2}\right)}\int_0^1(1-t^2)^{\nu-1/2}\ln(1-t^2)\,\cos(zt)\,dt\\ &+\frac{2\left(\frac{z}{2}\right)^\nu}{\sqrt{\pi}\,\Gamma\left(\nu+\frac{1}{2}\right)}\int_0^1(1-t^2)^{\nu-1/2}[\ln(1-t^2)]^2\cos(zt)\,dt\end{aligned} \tag{3.4.11}$$

For $\nu = 0$, the expression in (3.4.11) becomes

$$\begin{aligned}\left(\frac{\partial^2 J_\nu(z)}{\partial\nu^2}\right)_{\nu=0} &= \left[-\frac{\pi^2}{2}+(\ln z+\gamma+\ln 2)\right]^2 J_0(z)\\ &+\frac{4\,(\ln z+\gamma+\ln 2)}{\pi}\int_0^1\frac{\ln(1-t^2)}{\sqrt{1-t^2}}\cos(zt)\,dt\\ &+\frac{2}{\pi}\int_0^1\frac{[\ln(1-t^2)]^2}{\sqrt{1-t^2}}\cos(zt)\,dt\end{aligned} \tag{3.4.12}$$

where the first integral is known

$$\int_0^1\frac{\ln(1-t^2)\,\cos(zt)}{\sqrt{1-t^2}}\,dt = \frac{\pi}{4}\{\pi Y_0(z)-2\,[\ln(2z)+\gamma]\,J_0(z)\} \tag{3.4.13}$$

For $\nu = 1/2$, it follows from (3.4.11) that

$$\left(\frac{\partial^2 J_\nu(z)}{\partial \nu^2}\right)_{\nu=1/2} = \sqrt{\frac{2}{\pi z}}\left[-\frac{\pi^2}{6} + (\ln z + \gamma - \ln 2)\right]^2 \sin z$$
$$+ \frac{2^{5/2}\sqrt{z}\,(\ln z + \gamma - \ln 2)}{\pi}\int_0^1 \ln(1-t^2)\cos(zt)\,dt \qquad (3.4.14)$$
$$+ \frac{2^{3/2}\sqrt{z}}{\pi}\int_0^1 [\ln(1-t^2)]^2 \cos(zt)\,dt$$

where the first integral is also known

$$\int_0^1 \ln(1-t^2)\cos(zt)\,dt = \left\{\frac{1}{z}[\cos z\,\mathrm{Ci}(2z) + \sin z\,\mathrm{Si}(2z)] - \sin z\left[\ln\left(\frac{z}{2}\right) + \gamma\right]\right\} \qquad (3.4.15)$$

In the same way, using (3.4.10), it is possible to determine the second order derivatives for other Bessel functions when the first order derivatives are given as in Section 3.3. Third or higher order derivatives with respect to the order of the Bessel functions include repeated differentiation of the integral $I(t,\nu)$ and evidently they become longer and less convenient. As observed in (3.4.11), the only change in this expression or in those for higher derivatives will be powers of the logarithmic function in integrands of integrals.

Integral representations of Bessel functions which have in their expressions less then three places of ν are also suitable for the determination of higher order derivatives with respect to the order. This occurs in the Schlaefli integral representations of the Bessel functions

$$J_\nu(z) = \frac{1}{\pi}\int_0^\pi \cos(z\sin t - \nu t)\,dt - \frac{\sin(\pi\nu)}{\pi}\int_0^\infty e^{-z\sinh t - \nu t}\,dt \qquad (3.4.16)$$

$$\operatorname{Re} z > 0$$

and nth – derivative with respect to the order is

$$\frac{\partial^n J_\nu(z)}{\partial \nu^n} = \frac{1}{\pi}\int_0^\pi \frac{\partial^n}{\partial \nu^n}[\cos(z\sin t - \nu t)]\,dt$$
$$-\frac{1}{\pi}\int_0^\infty e^{-z\sinh t}\frac{\partial^n}{\partial \nu^n}[\sin(\pi\nu)\,e^{-\nu t}]\,dt \quad ; \quad n = 1, 2, 3, \ldots \qquad (3.4.17)$$

Before performing the differentiation, it is convenient to present the product of sine and exponential function by using the Euler complex form

$$\frac{\partial^n J_\nu(z)}{\partial \nu^n} = \frac{1}{\pi}\int_0^\pi \frac{\partial^n}{\partial \nu^n}[\cos(z\sin t - \nu t)]\,dt - \frac{1}{2\pi i}\int_0^\infty e^{-z\sinh t}\frac{\partial^n}{\partial \nu^n}[e^{(\pi i - t)\nu} - e^{-(\pi i + t)\nu}]\,dt \tag{3.4.18}$$

and therefore

$$\begin{aligned}\frac{\partial^n J_\nu(z)}{\partial \nu^n} &= \frac{1}{\pi}\int_0^\pi t^n\left\{\frac{(-1)^{n/2}[1+(-1)^n]}{2}\cos(z\sin t - \nu t)\right.\\ &\quad \left.+\frac{(-1)^{(n-1)/2}[1-(-1)^n]}{2}\sin(z\sin t - \nu t)\right\}dt\\ &\quad -\frac{1}{2\pi i}\int_0^\infty e^{-z\sinh t - \nu t}\{(\pi i - t)^n e^{i\pi\nu} - (-1)^n(\pi i + t)^n e^{-i\pi\nu}\}dt\\ &\quad n = 0, 1, 2, 3, \ldots\end{aligned} \tag{3.4.19}$$

In the first integral, the sine function disappears for even order derivatives and the cosine function for odd order derivatives. Thus, from (3.4.19) first three order derivatives with respect to the order are

$$\frac{\partial J_\nu(z)}{\partial \nu} = \frac{1}{\pi}\int_0^\pi t\sin(z\sin t - \nu t)\,dt + \frac{1}{\pi}\int_0^\infty e^{-z\sinh t - \nu t}[t\sin(\pi\nu) - \pi\cos(\pi\nu)]\,dt \tag{3.4.20}$$

$$\frac{\partial^2 J_\nu(z)}{\partial \nu^2} = -\frac{1}{\pi}\int_0^\pi t^2\cos(z\sin t - \nu t)\,dt + \frac{1}{\pi}\int_0^\infty e^{-z\sinh t - \nu t}[(\pi^2 - t^2)\sin(\pi\nu) + 2\pi t\cos(\pi\nu)]\,dt \tag{3.4.21}$$

and

$$\frac{\partial^3 J_\nu(z)}{\partial \nu^3} = -\frac{1}{\pi}\int_0^\pi t^3\sin(z\sin t - \nu t)\,dt + \frac{1}{\pi}\int_0^\infty e^{-z\sinh t - \nu t}[(t^3 - 3\pi^2 t)\sin(\pi\nu) + (\pi^3 - 3\pi t^2)\cos(\pi\nu)]\,dt \tag{3.4.22}$$

In the case of $v = 0$, we will encounter three types of integrals

$$\begin{aligned} I_1(z) &= \int_0^{\pi} t^n \sin(z \sin t)\, dt \quad ; \quad n = 0, 1, 2, 3, \ldots \\ I_2(z) &= \int_0^{\pi} t^n \cos(z \sin t)\, dt \\ I_3(z) &= \int_0^{\pi} t^n e^{-z \sinh t}\, dt \end{aligned} \tag{3.4.23}$$

From the Schlaefli integral representation of the Bessel function of second kind in (3.3.32) we have

$$\begin{aligned} \frac{\partial^n Y_v(z)}{\partial v^n} &= \frac{1}{\pi} \int_0^{\pi} t^n \left\{ \frac{(-1)^{n/2}[1+(-1)^n]}{2} \sin(z \sin t - v t) \right. \\ &\quad \left. + \frac{(-1)^{(n+1)/2}[1-(-1)^n]}{2} \cos(z \sin t - v t) \right\} dt \\ &\quad - \frac{1}{2\pi} \int_0^{\infty} e^{-z \sinh t} \{2t^n e^{vt} + e^{-vt}[(\pi i - t)^n e^{i\pi v} + (-1)^n (\pi i + t)^n e^{-i\pi v}]\} dt \\ &\quad n = 0, 1, 2, 3, \ldots \quad ; \quad \operatorname{Re} z > 0 \end{aligned} \tag{3.4.24}$$

which for $v = 0$ it reduces to

$$\begin{aligned} \left(\frac{\partial^n Y_v(z)}{\partial v^n} \right)_{v=0} &= \frac{1}{\pi} \int_0^{\pi} t^n \left\{ \frac{(-1)^{n/2}[1+(-1)^n]}{2} \sin(z \sin t) \right. \\ &\quad \left. + \frac{(-1)^{(n+1)/2}[1-(-1)^n]}{2} \cos(z \sin t) \right\} dt \\ &\quad - \frac{1}{2\pi} \int_0^{\infty} e^{-z \sinh t} \{2t^n + [(\pi i - t)^n + (-1)^n (\pi i + t)^n]\}\, dt \end{aligned} \tag{3.4.25}$$

The first three derivatives with respect to the order are

$$\begin{aligned} \frac{\partial Y_v(z)}{\partial v} &= -\frac{1}{\pi} \int_0^{\pi} t \cos(z \sin t - v t)\, dt \\ &\quad - \frac{1}{\pi} \int_0^{\infty} e^{-z \sinh t} \left\{ t e^{vt} - [t \cos(\pi v) + \pi \sin(\pi v)]\, e^{-vt} \right\} dt \end{aligned} \tag{3.4.26}$$

$$\frac{\partial^2 Y_\nu(z)}{\partial \nu^2} = -\frac{1}{\pi}\int_0^\pi t^2 \sin(z\sin t - \nu t)\,dt - \frac{1}{\pi}\int_0^\infty e^{-z\sinh t}\{t^2 e^{\nu t} + [(t^2-\pi^2)\cos(\pi\nu) - 2t\pi\sin(\pi\nu)]\,e^{-\nu t}\}\,dt \tag{3.4.27}$$

and

$$\frac{\partial^3 Y_\nu(z)}{\partial \nu^3} = \frac{1}{\pi}\int_0^\pi t^3 \cos(z\sin t - \nu t)\,dt - \frac{1}{\pi}\int_0^\infty e^{-z\sinh t}\{t^3 e^{\nu t} + [(\pi^3 - 3\pi t^2)\sin(\pi\nu) + (3\pi^2 t - t^3)\cos(\pi\nu)]\,e^{-\nu t}\}dt \tag{3.4.28}$$

Derivatives with respect to the order of the modified Bessel function of the first kind are derived using (3.3.36)

$$\frac{\partial^n I_\nu(z)}{\partial \nu^n} = \frac{1}{\pi}\int_0^\pi t^n e^{z\cos t}\left\{\frac{(-1)^{n/2}[1+(-1)^n]}{2}\cos(\nu t) + \frac{(-1)^{(n+1)/2}[1-(-1)^n]}{2}\sin(\nu t)\right\}dt - \frac{(-1)^n}{2\pi i}\int_0^\infty e^{-z\cosh t - \nu t}[(t-\pi i)^n e^{i\pi\nu} - (t+\pi i)^n e^{-i\pi\nu}]\,dt$$

$$n = 0, 1, 2, 3, \ldots \quad ; \quad \operatorname{Re} z > 0 \tag{3.4.29}$$

and they are for $\nu = 0$

$$\left(\frac{\partial^n I_\nu(z)}{\partial \nu^n}\right)_{\nu=0} = \frac{(-1)^{n/2}[1+(-1)^n]}{2\pi}\int_0^\pi t^n e^{z\cos t}\,dt - \frac{(-1)^n}{2\pi i}\int_0^\infty e^{-z\cosh t}[(t-\pi i)^n - (t+\pi i)^n]\,dt \tag{3.4.30}$$

where first integrals in (3.4.30) disappear for odd n. The first derivative with respect to the order of the modified Bessel function is given in (3.3.37), the second and third derivatives are presented below

$$\frac{\partial^2 I_\nu(z)}{\partial \nu^2} = -\frac{1}{\pi}\int_0^\pi t^2 e^{z\cos t}\cos(\nu t)\,dt + \frac{1}{\pi}\int_0^\infty e^{-z\cosh t - \nu t}[(\pi^2 - t^2)\sin(\pi\nu) + 2\pi t\cos(\pi\nu)]\,dt \tag{3.4.31}$$

$$\frac{\partial^3 I_\nu(z)}{\partial \nu^3} = \frac{1}{\pi}\int_0^\pi t^3 e^{z\cos t}\sin(\nu t)\,dt$$
$$-\frac{1}{\pi}\int_0^\infty e^{-z\cosh t-\nu t}[(t^3-3\pi^2 t)\sin(\nu t)+(\pi^3-3\pi t^2)\cos(\nu t)]\,dt \tag{3.4.32}$$

From the integral representations of $I_\nu(z)$ given in (3.3.37) and that of $K_\nu(z)$ in (3.3.38)

$$K_\nu(z) = \int_0^\infty e^{-z\cosh t}\cosh(\nu t)\,dt \quad ; \quad \mathrm{Re}\, z > 0 \tag{3.4.33}$$

we have the expected result by using (3.4.30)

$$\left[\frac{\partial I_\nu(z)}{\partial \nu}\right]_{\nu=0} = -\int_0^\infty e^{-z\cosh t}dt = -K_0(z) \tag{3.4.34}$$

Differentiation of (3.4.33) with respect to ν gives

$$\frac{\partial^n K_\nu(z)}{\partial \nu^n} = \int_0^\infty t^n e^{-z\cosh t}\cosh(\nu t)\,dt \quad ; \quad n = 0, 2, 4, 6, \ldots \tag{3.4.35}$$

$$\frac{\partial^n K_\nu(z)}{\partial \nu^n} = \int_0^\infty t^n e^{-z\cosh t}\sinh(\nu t)\,dt \quad ; \quad n = 1, 3, 5, 7, \ldots \tag{3.4.36}$$

and

$$\left(\frac{\partial^n K_\nu(z)}{\partial \nu^n}\right)_{\nu=0} = \int_0^\infty t^n e^{-z\cosh t}\,dt \quad ; \quad n = 0, 2, 4, 6, \ldots \tag{3.4.37}$$

$$\left(\frac{\partial^n K_\nu(z)}{\partial \nu^n}\right)_{\nu=0} = 0 \quad ; \quad n = 1, 3, 5, 7 \ldots \tag{3.4.38}$$

Similar integrals as in (3.4.33), define particular cases of modified Lommel functions

$$S_{0,\nu}(z) = \int_0^\infty e^{-z\sinh t}\cosh(\nu t)\,dt \quad ; \quad \mathrm{Re}\, z > 0 \tag{3.4.39}$$

and therefore we have

$$\frac{\partial^n S_{0,\nu}(z)}{\partial \nu^n} = \int_0^\infty t^n e^{-z\sinh t}\cosh(\nu t)\,dt \quad ; \quad n = 0, 2, 4, 6, \ldots \tag{3.4.40}$$

$$\frac{\partial^n S_{0,\nu}(z)}{\partial \nu^n} = \int_0^\infty t^n e^{-z\sinh t} \sinh(\nu t)\, dt \quad ; \quad n=1,3,5,7,\ldots \tag{3.4.41}$$

with

$$\left(\frac{\partial^n S_{0,\nu}(z)}{\partial \nu^n}\right)_{\nu=0} = \int_0^\infty t^n e^{-z\sinh t}\, dt \quad ; \quad n=0,2,4,6,\ldots \tag{3.4.42}$$

$$\left(\frac{\partial^n S_{0,\nu}(z)}{\partial \nu^n}\right)_{\nu=0} = 0 \quad ; \quad n=1,3,5,7\ldots \tag{3.4.43}$$

Since

$$S_{1,\nu}(z) = z\int_0^\infty e^{-z\sinh t} \cosh(\nu t)\, \cosh t\, dt \quad ; \quad \mathrm{Re}\, z>0 \tag{3.4.44}$$

similar expressions, as (3.4.40)–(3.4.33), can be written for the function $S_{1,\nu}(z)$.

In case of the Lommel functions which have the same orders we have

$$S_{\nu,\nu}(z) = \int_0^\infty e^{-z\sinh t}\left[z\,(\cosh t)^2\right]^\nu dt = \int_0^\infty e^{-z\sinh t+\nu\,[\ln z+2\ln(\cosh t)]}\, dt \tag{3.4.45}$$

$\mathrm{Re}\, z>0$

and higher derivatives with respect to the order are

$$\frac{\partial^n S_{\nu,\nu}(z)}{\partial \nu^n} = z^\nu \int_0^\infty e^{-z\sinh t}(\cosh t)^{2\nu}[\ln z + 2\ln(\cosh t)]^n\, dt \tag{3.4.46}$$

$n=1,2,3,\ldots$

For $\nu = 0$ it becomes

$$\left(\frac{\partial^n S_{\nu,\nu}(z)}{\partial \nu^n}\right)_{\nu=0} = \int_0^\infty e^{-z\sinh t}[\ln z + 2\,\ln(\cosh t)]^n\, dt \tag{3.4.47}$$

and in this case the first derivative with respect to the order is

$$\begin{aligned}\left(\frac{\partial S_{\nu,\nu}(z)}{\partial \nu}\right)_{\nu=0} &= \ln z \int_0^\infty e^{-z\sinh t}\, dt + 2\int_0^\infty e^{-z\sinh t}\ln(\cosh t)\, dt \\ &= \ln z\, S_{0,0}(z) + 2\int_0^\infty \frac{e^{-zx}}{\sqrt{1+x^2}} \ln(\sqrt{1+x^2})\, dx\end{aligned} \tag{3.4.48}$$

The integral representation of the Lommel function with the same orders $s_{\nu,\nu}(z)$ is

$$s_{\nu,\nu}(z) = \sqrt{\pi}\, z^{\nu} \int_0^{\pi/2} (\cos t)^{2\nu} \sin(z \sin t)\, dt \quad ; \quad \operatorname{Re} \nu > -\frac{1}{2} \tag{3.4.49}$$

and therefore

$$\frac{\partial^n s_{\nu,\nu}(z)}{\partial \nu^n} = z^{\nu} \int_0^{\pi/2} (\cos t)^{2\nu} [\ln z + 2 \ln(\cos t)]^n \sin(z \sin t)\, dt \qquad n = 1, 2, 3, \ldots \tag{3.4.50}$$

with

$$\left(\frac{\partial^n s_{\nu,\nu}(z)}{\partial \nu^n}\right)_{\nu=0} = \int_0^{\pi/2} [\ln z + 2 \ln(\cos t)]^n \sin(z \sin t)\, dt \tag{3.4.51}$$

For $\nu = 0$, the first derivative with respect to the order is

$$\begin{aligned} \left(\frac{\partial s_{\nu,\nu}(z)}{\partial \nu}\right)_{\nu=0} &= \ln z \int_0^{\pi/2} \sin(z \sin t)\, dt + 2 \int_0^{\pi/2} \ln(\cos t) \sin(z \sin t)\, dt \\ &= \ln z\, s_{0,0}(z) + 2 \int_0^1 \frac{\ln(\sqrt{1-x^2}}{(\sqrt{1-x^2})} \sin(zx)\, dx \end{aligned} \tag{3.4.52}$$

Since the Lommel functions with the same orders are interrelated with the Struve functions

$$s_{\nu,\nu}(z) = 2^{\nu-1} \sqrt{\pi}\, \Gamma\left(\nu + \frac{1}{2}\right) H_{\nu}(z) \tag{3.4.53}$$

$$S_{\nu,\nu}(z) = 2^{\nu-1} \sqrt{\pi}\, \Gamma\left(\nu + \frac{1}{2}\right) [H_{\nu}(z) - Y_{\nu}(z)] \tag{3.4.54}$$

it is possible to derive higher derivatives with respect to the order of the Struve functions. However, differentiation starts to be arduous with increasing value of n, and (3.4.53) and (3.4.54) are actually convenient only for the first derivatives of the Struve functions.

The higher order derivatives with respect to the order of the Anger and Weber functions can directly be determined by taking into account their definitions

$$\mathbf{J}_{\nu}(z) = \frac{1}{\pi} \int_0^{\pi} \cos(\nu t - z \sin t)\, dt \tag{3.4.55}$$

and therefore,

$$\frac{\partial^n \mathbf{J}_\nu(z)}{\partial \nu^n} = \frac{(-1)^{n/2}}{\pi} \int_0^\pi t^n \cos(\nu t - z \sin t)\, dt \quad ; \quad 0, 2, 4, 6, \ldots \tag{3.4.56}$$

$$\frac{\partial^n \mathbf{J}_\nu(z)}{\partial \nu^n} = \frac{(-1)^{(n+1)/2}}{\pi} \int_0^\pi t^n \sin(\nu t - z \sin t)\, dt \quad ; \quad 1, 3, 5, 7, \ldots \tag{3.4.57}$$

For $\nu = 0$ we have

$$\left(\frac{\partial^n \mathbf{J}_\nu(z)}{\partial \nu^n}\right)_{\nu=0} = \frac{(-1)^{n/2}}{\pi} \int_0^\pi t^n \cos(z \sin t)\, dt \quad ; \quad 0, 2, 4, 6, \ldots \tag{3.4.58}$$

$$\left(\frac{\partial^n \mathbf{J}_\nu(z)}{\partial \nu^n}\right)_{\nu=0} = -\frac{(-1)^{(n+1)/2}}{\pi} \int_0^\pi t^n \sin(z \sin t)\, dt \quad ; \quad 1, 3, 5, 7, \ldots \tag{3.4.59}$$

Using the integral representation of the Weber function

$$\mathbf{E}_\nu(z) = \frac{1}{\pi} \int_0^\pi \sin(\nu t - z \sin t)\, dt \tag{3.4.60}$$

Higher derivatives with respect to the order are

$$\frac{\partial^n \mathbf{E}_\nu(z)}{\partial \nu^n} = \frac{(-1)^{n/2}}{\pi} \int_0^\pi t^n \sin(\nu t - z \sin t)\, dt \quad ; \quad 0, 2, 4, 6, \ldots \tag{3.4.61}$$

$$\frac{\partial^n \mathbf{E}_\nu(z)}{\partial \nu^n} = \frac{(-1)^{(n+1)/2}}{\pi} \int_0^\pi t^n \cos(\nu t - z \sin t)\, dt \quad ; \quad 1, 3, 5, 7, \ldots \tag{3.4.62}$$

and for $\nu = 0$ we have

$$\left(\frac{\partial^n \mathbf{E}_\nu(z)}{\partial \nu^n}\right)_{\nu=0} = \frac{(-1)^{n/2}}{\pi} \int_0^\pi t^n \sin(z \sin t)\, dt \quad ; \quad 0, 2, 4, 6, \ldots \tag{3.4.63}$$

$$\left(\frac{\partial^n \mathbf{E}_\nu(z)}{\partial \nu^n}\right)_{\nu=0} = -\frac{(-1)^{(n+1)/2}}{\pi} \int_0^\pi t^n \cos(z \sin t)\, dt \quad ; \quad 1, 3, 5, 7, \ldots \tag{3.4.64}$$

As expected, the first derivatives with respect to the order determined above by using the Schlaefli integral representations of the Anger and Weber functions are identical to those expressions derived in (3.4.61)–(3.4.64).

Finally, it is possible to obtain the higher order derivatives with respect to the order of the integral Bessel functions $\mathrm{Ji}_\nu(z)$ and $\mathrm{Ki}_\nu(z)$. From the integral representation [4]

$$\mathrm{Ji}_\nu(z) = \frac{1}{2\nu} + \frac{1}{\pi\nu}\int_0^\pi \cot t \sin(z\sin t - \nu t)\,dt - \frac{\sin(\pi\nu)}{\pi\nu}\int_0^\infty e^{-z\sinh t - \nu t}\coth t\,dt \tag{3.4.65}$$

$$\mathrm{Re}\,\nu > 0$$

by direct differentiation of (3.4.65) we have

$$\begin{aligned}\frac{\partial^n \mathrm{Ji}_\nu(z)}{\partial \nu^n} &= \frac{(-1)^n n!}{2\nu^{n+1}} + \frac{1}{\pi\nu}\int_0^\pi t^n\left\{\frac{(-1)^{n/2}[1+(-1)^n]}{2}\sin(z\sin t - \nu t)\right.\\ &\quad \left. + \frac{(-1)^{(n-1)/2}[1-(-1)^n]}{2}\cos(z\sin t - \nu t)\right\}dt\\ &\quad - \int_0^\infty \left\{\frac{F_n(\nu)}{\pi} + (-1)^n\frac{\sin(\pi\nu)\,t^n}{\pi\nu}\right\}e^{-z\sinh t - \nu t}\coth t\,dt\\ n &= 0, 1, 2, 3, \ldots\end{aligned} \tag{3.4.66}$$

where $F_n(\nu)$ are products of two functions $f(\nu)$ and $g(\nu)$ defined in (3.4.67). It follows from the Leibniz theorem for differentiation of a products that

$$\begin{aligned} f(\nu) &= \frac{1}{\nu} \quad ; \quad g(\nu) = \sin(\pi\nu)\\ F_n(\nu) &= \frac{d^n[f(\nu)g(\nu)]}{d\nu^n} = \sum_{k=0}^n \binom{n}{k}\frac{d^{n-k}f(\nu)}{d\nu^{n-k}}\frac{d^k g(\nu)}{d\nu^k}\\ &= \sum_{k=0}^n \binom{n}{k} a_{n-k}(\nu)\,b_k(\nu)\\ a_{n-k}(\nu) &= (-1)^{n-k}\frac{(n-k)!}{\nu^{n-k+1}} \quad ; \quad b_k(\nu) = \pi^k \sin\left(\pi\nu + \frac{\pi k}{2}\right)\end{aligned} \tag{3.4.67}$$

Glasser suggested to express $F_n(\nu)$ in an alternative way, in terms of elementary trigonometric integrals

$$F_n(\nu) = \frac{d^n}{d\nu^n}\left(\frac{\sin(\pi\nu)}{\nu}\right) = \frac{\pi^{n+1}}{2}\int_{-1}^{1} t^n \cos(\pi\nu t)\,dt \tag{3.4.68}$$

$$n = 0, 2, 4, 6, \ldots$$

and

$$F_n(\nu) = \frac{d^n}{d\nu^n}\left(\frac{\sin(\pi\nu)}{\nu}\right) = \frac{\pi^{n+1}}{2}\sin\left(\frac{\pi n}{2}\right)\int_{-1}^{1} t^n \sin(\pi\nu t)\,dt$$

$$n = 0, 1, 3, 5, \ldots$$

The first three $F_n(\nu)$ functions are

$$F_1(\nu) = \frac{\pi\cos(\pi\nu)}{\nu} - \frac{\sin(\pi\nu)}{\nu^2}$$

$$F_2(\nu) = -\frac{\pi^2\sin(\pi\nu)}{\nu} - \frac{2\pi\cos(\pi\nu)}{\nu^2} - \frac{2\sin(\pi\nu)}{\nu^3} \tag{3.4.69}$$

$$F_3(\nu) = -\frac{\pi^3\cos(\pi\nu)}{\nu} + \frac{3\pi^2\sin(\pi\nu)}{\nu^2} + \frac{6\pi\cos(\pi\nu)}{\nu^3} - \frac{6\sin(\pi\nu)}{\nu^4}$$

Since the integral representation of the modified Bessel function of the second kind has a simple form

$$K_\nu(z) = \int_0^\infty e^{-z\cosh t}\cosh(\nu t)\,dt \tag{3.4.70}$$

and the integral function $Ki_\nu(z)$ is defined by

$$\mathrm{Ki}_\nu(z) = \int_z^\infty \frac{K_\nu(x)}{x}\,dx \tag{3.4.71}$$

it is possible to derive higher derivatives with respect to the order by introducing (3.4.70) into (3.4.71)

$$\mathrm{Ki}_\nu(z) = \int_z^\infty \frac{1}{x}\left[\int_0^\infty e^{-x\cosh t}\cosh(\nu t)\,dt\right]dx \tag{3.4.72}$$

Changing order of integration in (3.4.72) we have

$$\mathrm{Ki}_\nu(z) = \int_0^\infty \cosh(\nu t)\left[\int_z^\infty \frac{e^{-x\cosh t}}{x}\,dx\right]dt \tag{3.4.73}$$

The inner integral in (3.4.73) is the exponential integral [9]

$$E_1(az) = \int_z^\infty \frac{e^{-at}}{t} dt = -\mathrm{Ei}(-az) \tag{3.4.74}$$

and therefore the integral representation of $\mathrm{Ki}_\nu(z)$ is given by

$$\mathrm{Ki}_\nu(z) = \int_0^\infty E_1(z \cosh t) \cosh(\nu t)\, dt \tag{3.4.75}$$

Direct differentiation of (3.4.75) with respect to the order gives the higher derivatives

$$\frac{\partial^n \mathrm{Ki}_\nu(z)}{\partial \nu^n} = \int_0^\infty t^n E_1(z \cosh t) \cosh(\nu t)\, dt \qquad n = 0, 2, 4, 6, \ldots \tag{3.4.76}$$

and

$$\frac{\partial^n \mathrm{Ki}_\nu(z)}{\partial \nu^n} = \int_0^\infty t^n E_1(z \cosh t) \sinh(\nu t)\, dt \qquad n = 1, 3, 5, 7, \ldots \tag{3.4.77}$$

For $\nu = 0$, we have analogous expressions to those in (3.4.37) and (3.4.38)

$$\begin{aligned} \left(\frac{\partial^n \mathrm{Ki}_\nu(z)}{\partial \nu^n}\right)_{\nu=0} &= \int_0^\infty t^n E_1(z \cosh t)\, dt \quad ; \quad n = 0, 2, 4, 6, \ldots \\ \left(\frac{\partial^n \mathrm{Ki}_\nu(z)}{\partial \nu^n}\right)_{\nu=0} &= 0 \quad ; \quad n = 1, 3, 5, 7, \ldots \end{aligned} \tag{3.4.78}$$

4 Mathematical Operations with Respect to the Order of the Bessel and Related Functions – Integration, Differentiation, Series and Limits

The Application of Laplace Transform Methods and Other Related Topics

4.1 Integration with Respect to the Order of the Bessel and Related Functions

Contrary to derivatives with respect to the order which were sporadically investigated in mathematical literature, at least with an enduring interest, the integration with respect to the order appeared only on very special occasions. Usually, this occurred as a by-product of solving of mathematical problems or illustrating some mathematical techniques. Only a few such cases can be found, and they all will be presented here. Besides, the next section (Section 4.2) will discuss the systematic application of the Laplace transformation to the integration with respect to the order.

As pointed out above, a number of known integrals with respect to the order of Bessel functions is small, but exists as a large group of integrals when the order is an imaginary number, they are connected with the Kontorovich-Lebedev or with other integral transforms [18]. These integrals are entirely omitted here.

The first example of integration with respect to the order is quoted by Watson in his book [7] He considered one of the extraordinary integrals derived by Ramanujan in 1920

$$I(t) = \int_{-\infty}^{\infty} \frac{J_{\mu+\xi}(x)}{x^{\mu+\xi}} \frac{J_{\nu-\xi}(x)}{x^{\nu-\xi}} e^{i\,t\xi} d\xi \qquad x, y > 0 \quad ; \quad \operatorname{Re}(\mu+\nu) > 1 \tag{4.1.1}$$

and for real values of t, this integral is

$$I(t) = \begin{cases} \left(\frac{2\cos(t/2)}{x^2 e^{-it/2} + y^2 e^{it/2}}\right)^{(\mu+\nu)/2} e^{it(\nu-\mu)/2} J_{\mu+\nu}(\phi) & ; \quad |t| < \pi \\ 0 \quad : \quad |t| < \pi \end{cases} \qquad \phi = \sqrt{2\cos\left(\frac{t}{2}\right)[x^2 e^{-it/2} + y^2 e^{it/2}]} \tag{4.1.2}$$

For $t = 0$ and $y = x$ it reduces to

https://doi.org/10.1515/9783110681642-004

$$I(0) = \int_{-\infty}^{\infty} J_{\mu+\xi}(x) J_{\nu-\xi}(x)\, d\xi = J_{\mu+\nu}(2x) \tag{4.1.3}$$

In 1954, Cooke [46] considered only cosine function in (4.1.1) and derived it in a way which was different from the Ramanujan result by starting from the integral which represents the product of two Bessel functions [7]

$$J_\mu(z) J_\nu(z) = \int_0^{\infty} J_{\mu+\nu}(2z\cos\theta) \cos[(\mu-\nu)\theta]\, d\theta \qquad \mathrm{Re}(\mu+\nu) > -1 \tag{4.1.4}$$

He replaced the independent variable and orders of the Bessel functions in the following way $t = 2\theta$, $\mu = \alpha + \xi$ and $\nu = \beta - \xi$. In the next step, Cooke multiplied both sides of (4.1.4) by $\cos(\phi\xi)$ and integrated from $-\infty$ to $+\infty$ with respect to ξ. The result of integration is

$$\int_{-\infty}^{\infty} J_{\alpha+\xi}(z) J_{\beta-\xi}(z) \cos(\phi\xi)\, d\xi = \begin{cases} \cos\left[\left(\frac{\alpha-\beta}{z}\right)\phi\right] J_{\alpha+\beta}\left[2z\cos\left(\frac{\phi}{2}\right)\right] & ; \quad |\phi| < \pi \\ 0 & ; \quad |\phi| > \pi \end{cases} \tag{4.1.5}$$

where for $\phi = 0$ and $\beta = \alpha$ (4.1.5) is identical with (4.1.2).

Using the same procedure Cooke derived few more integrals with respect to the order for the modified Bessel and the Anger functions, and they are presented below:

$$I(\theta) = \int_0^{\infty} I_\nu(z) \cos(\theta\nu)\, d\nu = \frac{e^{z\cos\theta}}{2} - \frac{1}{2\pi}\int_0^{\infty} e^{-z\cosh t}\left\{\frac{\pi+\theta}{(\pi+\theta)^2+t^2} + \frac{\pi-\theta}{(\pi-\theta)^2+t^2}\right\} dt \tag{4.1.6}$$

$$|\mathrm{Re}\nu| > 0 \quad ; \quad |\arg z| \le \frac{1}{2} \quad ; \quad |\theta| < \pi$$

$$I(0) = \int_0^{\infty} I_\nu(z)\, d\nu = \frac{e^z}{2} - \int_0^{\infty} \frac{e^{-z\cosh t}}{\pi^2+t^2}\, dt \tag{4.1.7}$$

$$\frac{1}{2\pi i}\int_{\alpha-i\infty}^{\alpha+i\infty} K_\nu(z) \cosh[(\nu-\alpha)\theta]\, d\nu = e^{-z\cosh\theta}\cosh(\alpha\theta) \qquad |\arg z| < \frac{\pi}{2} \tag{4.1.8}$$

$$\frac{1}{2\pi i}\int_{\alpha-i\infty}^{\alpha+i\infty} K_v(z)\sinh[(v-\alpha)\theta]\,dv = -\,e^{-z\cosh\theta}\sinh(\alpha\theta) \tag{4.1.9}$$

$$\frac{1}{2\pi i}\int_{\alpha-i\infty}^{\alpha+i\infty} K_v(z)\,dv = e^{-z} \tag{4.1.10}$$

$$\int_{-\infty}^{\infty} J_v(z)\cos(\theta v)\,dv = \begin{cases} \cos(z\sin\theta) & ; \ |\theta|<\pi \\ 1/z & ; \ |\theta|=\pi \\ 0 & ; \ |\theta|>\pi \end{cases} \tag{4.1.11}$$

$$\int_{-\infty}^{\infty} J_v(z)\sin(\theta v)\,dv = \begin{cases} \sin(z\sin\theta) & ; \ |\theta|<\pi \\ 0 & ; \ |\theta|>\pi \end{cases} \tag{4.1.12}$$

and

$$\int_{-\infty}^{\infty} J_v(z)\,dv = 1 \tag{4.1.13}$$

where the last integral is valid for any value of argument z.

The Cooke results given in (4.1.6) were expressed in terms of an infinite series by Fényes [50, 51] in 1993. He stated that such integrals are important in the heat-conduction problems, when solutions of an integral equation of convolution type are required.

$$\begin{aligned}\int_0^{\infty} I_v(z)\sin(\theta v)\,dv = {} & \frac{e^{z\cos\theta}-e^{-\theta}}{2\pi}\ln\left|\frac{\pi+\theta}{\pi-\theta}\right| \\ & -\frac{z}{4\pi}\int_0^{\infty} e^{-z\cosh t}\ln\left[\frac{t^2+(\theta-\pi)^2}{t^2+(\theta+\pi)^2}\right]\sinh t\,dt \\ & +\frac{1}{\pi}\sum_{k=1}^{\infty} I_k(z)\cos(k\theta)\{c[(\pi-\theta)k]-c[(\pi+\theta)k]\} \\ & +\frac{1}{\pi}\sum_{k=1}^{\infty} I_k(z)\sin(k\theta)\{Si[(\pi-\theta)k]+Si[(\pi+\theta)k]\}\end{aligned} \tag{4.1.14}$$

$$Si(z)=\int_0^{z}\frac{\sin t}{t}dt \quad ; \quad c(z)=\int_0^{z}\frac{1-\cos t}{t}dt$$

and

$$\begin{aligned}\int_0^\infty \nu I_\nu(z)\,d\nu &= \frac{e^{-z}}{\pi^2} - 2\int_0^\infty \frac{te^{-z\cosh t}}{(\pi^2+t^2)^2} + dt + \frac{2}{\pi}\sum_{k=1}^\infty k\,Si(\pi k) I_k(z) \\ &= \frac{e^{-z}}{\pi^2} - 2\int_0^\infty \frac{te^{-z\cosh t}}{(\pi^2+t^2)^2} + \frac{t}{\pi}\int_0^\pi e^{z\cos t}\sin t\,\frac{dt}{t}\end{aligned} \tag{4.1.15}$$

Considering heat transfer in cylindrical coordinates, Fényes solved the following integral equation

$$\int_0^\tau f(t)\,K_0(\tau - t)\,dt = 1 \quad ; \quad \tau > 0 \tag{4.1.16}$$

and found that its solution contains the convolution operation

$$f(\tau) = \frac{1}{\tau}\int_0^\infty \nu I_\nu(\tau)\,d\nu - \frac{1}{\tau}\int_0^\infty \nu I_\nu(\tau)\,d\nu * \int_0^\tau \frac{I_1(x)}{x}\,dx \tag{4.1.17}$$

Introducing (4.1.15) into (4.1.17), the solution of (4.1.16) can be written as

$$\begin{aligned} f(\tau) &= \frac{2}{\pi\tau}\sum_{k=1}^\infty k\,Si(\pi k)\,I_k(\tau) \\ &\quad - \frac{2}{\pi}\sum_{k=1}^\infty (k+1)\,Si(\pi k)\,I_k(\tau)\int_0^\tau \frac{I_{k+1}(t)}{t}\,dt \\ &\quad + \frac{e^{-\tau}}{\pi^2\tau} - \frac{2}{\tau}\int_0^\infty \frac{te^{-\tau\cosh t}}{(\pi^2+t^2)^2}\,dt - \left\{\frac{e^{-\tau}}{\pi^2\tau} - \frac{2}{\tau}\int_0^\infty \frac{te^{-\tau\cosh t}}{(\pi^2+t^2)^2}\,dt\right\} * \int_0^\tau \frac{I_1(t)}{t}\,dt \end{aligned} \tag{4.1.18}$$

If the modified Bessel function of the first kind $I_\nu(z)$ is replaced by the Bessel functions $J_\nu(z)$, Fényes found that

$$\begin{aligned}\int_0^\infty J_\nu(z)\cos(\theta\nu)\,d\nu &= \frac{\cos(\sin\theta)}{2} + \frac{\sin(\sin\theta)}{2\pi}\ln\left|\frac{\pi-\theta}{\pi+\theta}\right| \\ &\quad - \frac{1}{2\pi}\int_0^\infty e^{-z\sinh t}\left\{\frac{\pi+\theta}{t^2+(\theta+\pi)^2} + \frac{\pi-\theta}{t^2+(\pi-\theta)^2}\right\}dt \\ &\quad + \frac{1}{\pi}\sum_{k=0}^\infty J_{2k+1}(z)\sin[(2k+1)\theta]\,A_k(\theta) \\ &\quad + \frac{1}{\pi}\sum_{k=0}^\infty J_{2k+1}(z)\cos[(2k+1)\theta]\,B_k(\theta)\end{aligned} \tag{4.1.19}$$

where

$$\begin{aligned} A_k(\theta) &= c[(\pi+\theta)(2k+1)] - c[(\pi-\theta)(2k+1)] \\ B_k(\theta) &= Si[(\pi+\theta)(2k+1)] + Si[(\pi-\theta)(2k+1)] \end{aligned} \tag{4.1.20}$$

The expression (4.1.19) is valid for $|\theta| < \pi$ and $z \geq 0$, for $|\theta| > \pi$, the first term is replaced by zero, and for $\theta = \pi$, the value of first term is 1/4 and the second term is zero. If $\theta = 0$, the integral in (4.1.19) becomes

$$\int_0^\infty J_\nu(z)\,d\nu = \frac{1}{2} - \int_0^\infty \frac{e^{-z\sinh t}}{\pi^2+t^2}\,dt + \frac{2}{\pi}\sum_{k=0}^{\infty} Si[(2k+1)\pi]\,J_{2k+1}(z) \tag{4.1.21}$$

and similar integral in (4.1.15) is

$$\int_0^\infty \nu J_\nu(z)\,d\nu = \frac{1}{\pi^2} + \frac{z}{2} - 2\int_0^\infty \frac{te^{-z\sinh t}}{(\pi^2+t^2)^2}\,dt + \frac{4}{\pi}\sum_{k=1}^{\infty} k\,Si(2\pi k) J_{2k}(z) \tag{4.1.22}$$

Fényes [52, 76] also derived integral representations of moments of the Bessel functions which he defined by

$$M_n(t) = \int_0^\infty \nu^n J_\nu(t)\,dt \quad ; \quad t>0 \quad ; \quad n = 0,1,2,3,\ldots \tag{4.1.23}$$

$$m_n(t) = \int_0^\infty \nu^n I_\nu(t)\,dt \quad ; \quad t>0 \quad ; \quad n = 0,1,2,3,\ldots \tag{4.1.24}$$

and the Laplace transforms of these moments by

$$\begin{aligned} J_\nu^{(n)}(t,s) &= \frac{1}{\pi}\int_0^\pi \cos(t\sin x)\left[\int_0^\infty e^{-\nu s}\nu^n\cos(\nu x)\,d\nu\right]dx \\ &+ \frac{1}{\pi}\int_0^\pi \sin(t\sin x)\left[\int_0^\infty e^{-\nu s}\nu^n\sin(\nu x)\,d\nu\right]dx \\ &- \frac{1}{\pi}\int_0^\infty e^{-t\sinh x}\left[\int_0^\infty e^{-\nu s}\sin(\nu x)\,d\nu\right]dx \end{aligned} \tag{4.1.25}$$

These Laplace transforms are based on the Schlaefli integral representations of the Bessel functions. Performing the change of order of integration in (4.1.25) we have

$$\begin{aligned} I_\nu^{(n)}(t,s) &= \frac{1}{\pi}\int_0^\pi e^{t\cos x}\left[\int_0^\infty e^{-\nu s}\nu^n\cos(\nu x)\,d\nu\right]dx \\ &- \frac{1}{\pi}\int_0^\infty e^{-t\cosh x}\left[\int_0^\infty e^{-\nu(x+s)}\nu^n\,d\nu\right]dx \end{aligned} \tag{4.1.26}$$

where the inner integrals in (4.1.25) and (4.1.26) are known [36]

$$\begin{aligned} \int_0^\infty e^{-\nu s}\nu^n\sin(\nu x)\,d\nu &= (-1)^n\frac{\partial^n}{\partial s^n}\left(\frac{x}{s^2+x^2}\right) \\ &= n!\left(\frac{s}{s^2+x^2}\right)^{n+1}\sum_{0\le 2k\le n}(-1)^k\binom{n+1}{2k+1}\left(\frac{x}{s}\right)^{2k+1} \end{aligned} \tag{4.1.27}$$

$$\begin{aligned} \int_0^\infty e^{-\nu s}\nu^n\cos(\nu x)\,d\nu &= (-1)^n\frac{\partial^n}{\partial s^n}\left(\frac{s}{s^2+x^2}\right) \\ &= n!\left(\frac{s}{s^2+x^2}\right)^{n+1}\sum_{0\le 2k\le n+1}(-1)^k\binom{n+1}{2k}\left(\frac{x}{s}\right)^{2k} \end{aligned} \tag{4.1.28}$$

$$\int_0^\infty e^{-\nu(s+1)}\nu^n\,d\nu = \frac{n!}{(s+x)^{n+1}} \tag{4.1.29}$$

In such way, Fényes derived general expressions for the Laplace transforms of the moments in (4.1.23) and (4.1.24). However, his interest was limited to behaviour of the moments for large times, $t\to\infty$, not to the inversion of the Laplace transforms. This is equivalent for behaviour of the Laplace transforms for $s\to 0$

$$\begin{aligned} M_n(t) &= \int_0^\infty \nu^n J_\nu(t)\,d\nu = Q_n(t) \\ &+ \frac{2^{n+2}n!}{\sqrt{2\pi t}\,\pi^{n+1}}\left\{\sin\left(t-\frac{\pi n}{2}-\frac{\pi}{4}\right)+o(1)\right\} \quad ; \quad t\to\infty \end{aligned} \tag{4.1.30}$$

Using the polynomials $Q_n(t)$, Fényes derived the first 11 moments in an explicit form

$$
\begin{aligned}
&Q_0(t)=1 \quad ; \quad Q_1(t)=t \quad ; \quad Q_2(t)=t^2 \\
&Q_3(t)=t^3+t \quad ; \quad Q_4(t)=t^4+4t^2 \quad ; \quad Q_5(t)=t^5+10t^3+t \\
&Q_6(t)=t^6+20t^4+16t^2 \quad ; \quad Q_7(t)=t^7+35t^5+191t^3+t \\
&Q_8(t)=t^8+56t^6+336t^4+64t^2 \\
&Q_9(t)=t^9+84t^7+996t^5+820t^3+t \\
&Q_{10}(t)=t^{10}+120t^8+2352t^6+544t^4+256t^2
\end{aligned}
\tag{4.1.31}
$$

In case of moments of the modified Bessel function of the first kind, the final results of his calculations are

$$
\begin{aligned}
&m_0(t)\sim\frac{e^t}{2} \quad ; \quad m_1(t)\sim\frac{e^t t^{1/2}}{\sqrt{2\pi}}\left(1-\frac{1}{24t}\right) \quad ; \quad t\to\infty \\
&m_2(t)\sim\frac{e^t t}{2} \quad ; \quad m_3(t)\sim\frac{e^t t^{3/2}}{\sqrt{2\pi}}\left(2+\frac{1}{4t}+\frac{7}{960t^2}\right) \\
&m_4(t)\sim\frac{e^t(t+3t^2)}{2} \quad ; \quad m_5(t)\sim\frac{e^t t^{5/2}}{\sqrt{2\pi}}\left(8+\frac{5}{t}+\frac{16}{t^2}-\frac{31}{8064t^3}\right) \\
&m_6(t)\sim\frac{e^t(t+15t^2+15t^3)}{2} \\
&m_7(t)\sim\frac{e^t t^{7/2}}{\sqrt{2\pi}}\left(48+\frac{50}{t}+\frac{91}{8t^2}+\frac{64}{t^3}+\frac{127}{30720t^4}\right) \\
&m_8(t)\sim\frac{e^t(t+63t^2+210t^3+105t^4)}{2}
\end{aligned}
\tag{4.1.32}
$$

A number of integrals of the modified Bessel functions of the second kind and the imaginary order is presented in [18], the simplest two are

$$
\int_{-\infty}^{\infty} K_{\alpha-i\nu}(a)K_{i\nu-\beta}(b)\,d\nu=\pi K_{\alpha+\beta}(a+b) \qquad \operatorname{Re} a,b>0
\tag{4.1.33}
$$

and

$$
\int_{0}^{\infty} K_{i\nu}(a)K_{i\nu}(b)\cosh[(\pi-|\theta|)\nu]d\nu=\frac{\pi}{2}K_0\left(\sqrt{a^2+b^2-2ab\cos\theta}\right) \qquad 0\le\theta\le 2\pi
\tag{4.1.34}
$$

Recently, Becker [88] in 2009 demonstrated that

$$\int_{-\infty}^{\infty} \frac{J_{i\nu}(iz)\, e^{\pi\nu/2}}{\sinh(\pi\nu)} d\nu = -ie^{-z} \tag{4.1.35}$$

$$\int_{-\infty}^{\infty} \frac{I_{i\nu}(z)}{\sinh(\pi\nu)} d\nu = -ie^{-z} \tag{4.1.36}$$

$$\int_{0}^{\infty} K_{i\nu}\left(\frac{z}{2}\right) d\nu = \frac{\pi}{2} e^{-z/2} \tag{4.1.37}$$

$$\int_{-\infty}^{\infty} \nu\,\Gamma(\lambda + i\nu)\,\Gamma(\lambda + i\nu)\, I_{i\nu}(z)\, d\nu$$
$$= \begin{cases} 2^{\lambda}\sqrt{\pi} i \Gamma\left(\lambda + \frac{1}{2}\right)\left[\frac{2^{\lambda+1}}{\Gamma(-\lambda)} K_{\lambda}(z) - z^{\lambda} e^{-z}\right] & ; \quad -1 < \operatorname{Re} \lambda \le 0 \\ -2^{\lambda}\sqrt{\pi} i \Gamma\left(\lambda + \frac{1}{2}\right) z^{\lambda} e^{-z} & ; \quad \operatorname{Re} \lambda > 0 \end{cases} \tag{4.1.38}$$

$$\int_{0}^{\infty} \nu \sinh(\pi\nu)\Gamma(\lambda + i\nu)\,\Gamma(\lambda + i\nu)\, K_{i\nu}(z)\, d\nu$$
$$= \begin{cases} -2^{\lambda-1}\pi^{3/2}\,\Gamma\left(\lambda + \frac{1}{2}\right)\left[\frac{2^{\lambda+1}}{\Gamma(-\lambda)} K_{\lambda}(z) - z^{\lambda} e^{-z}\right] & ; \quad -1 < \operatorname{Re} \lambda \le 0 \\ 2^{\lambda-1}\pi^{3/2}\,\Gamma\left(\lambda + \frac{1}{2}\right) z^{\lambda} e^{-z} & ; \quad \operatorname{Re} \lambda > 0 \end{cases} \tag{4.1.39}$$

These integrals have some application in a time-dependent radiation transport theory.

Finally, it is worthwhile to mention an equality of integrals proved by van der Pol [82].

$$\int_{-\infty}^{\infty} K_{\nu}(a)\, e^{-b\cosh\nu} d\nu = \int_{-\infty}^{\infty} K_{\nu}(b)\, e^{-a\cosh\nu} d\nu \qquad \operatorname{Re} a, b > 0 \tag{4.1.40}$$

All integrals presented in Section 4.1 were evaluated in a rather long and messy way, and this shows that the operation of the integration is considerably more complex than differentiation with respect to the order of Bessel functions.

4.2 Integration and Differentiation with Respect to the Order of the Bessel Function of the First Kind by Applying the Laplace Transform Approach

The use of the convolution (product) theorem of the Laplace transformation to determine derivatives with respect to the order, in a compact and elegant form, was initiated by van der Pol [81] in 1929. The finding of inverse transforms in the Laplace transformation is facilitated if in the Laplace transformation

$$L\{f(t)\} = \int_0^\infty e^{-st} f(t)\,dt = F(s) \tag{4.2.1}$$

$$f(t) = L^{-1}\{F(s)\} = \frac{1}{2\pi i}\int_{c-i\infty}^{c+i\infty} e^{st}F(s)\,ds \qquad \operatorname{Re} s > c \tag{4.2.2}$$

the transforms $F(s)$ can be written as a product $F(s) = F_1(s)\cdot F_2(s)$ of two functions of s and inverses $L^{-1}\{F_1(s)\} = f_1(t)$ and $L^{-1}\{F_2(s)\} = f_2(t)$ are known. Then, the inverse $f(t)$, the original function, is given by the convolution integral

$$\begin{aligned} f(t) &= L^{-1}\{F(s)\} = f_1(t) * f_2(t) \\ &= \int_0^t f_1(u) f_2(t-u)\,du = \int_0^t f_1(t-u) f_2(u)\,du \\ & t > 0 \end{aligned} \tag{4.2.3}$$

where the convolution of two functions $f_1(t)$ and $f_2(t)$ is denoted by symbol$*$. The convolution of two or more functions obeys the commutative law and the associative law and is distributive with respect to the addition [31].

Van der Pol [82] started with the Laplace transforms of the Bessel function of the first kind of the order ν and the integral Bessel function of the order zero

$$L\{J_\nu(t)\} = \frac{1}{\sqrt{s^2+1}\,\left(s+\sqrt{s^2+1}\right)^\nu} \qquad \operatorname{Re} s > 0 \;;\; \operatorname{Re}\nu > -1 \tag{4.2.4}$$

$$L\{Ji_0(t)\} = -\frac{\ln\left(s+\sqrt{s^2+1}\right)}{s} \qquad \operatorname{Re} s > 0 \tag{4.2.5}$$

and he differentiated the Laplace transform in (4.2.4) with respect to ν

$$L\left\{\frac{\partial J_\nu(t)}{\partial \nu}\right\} = -\frac{\ln\left(s+\sqrt{s^2+1}\right)}{\sqrt{s^2+1}\left(s+\sqrt{s^2+1}\right)^\nu} \tag{4.2.6}$$

$$\operatorname{Re} s > 0 \quad ; \quad \operatorname{Re}\nu > -1$$

If both Laplace transforms from (4.2.5) and (4.2.6) are written as a product in the following way

$$L\left\{\frac{\partial J_\nu(t)}{\partial \nu}\right\} = -\frac{\ln\left[s+\sqrt{s^2+1}\right]}{s} \cdot \frac{s}{\sqrt{s^2+1}\left(s+\sqrt{s^2+1}\right)^\nu} \tag{4.2.7}$$

the inverse according to (4.2.3) is

$$\frac{\partial J_\nu(t)}{\partial \nu} = \int_0^t Ji_0(t-x)\, J'_\nu(x)\, dx \tag{4.2.8}$$

$$\operatorname{Re}\nu > 0$$

where the derivative of the Bessel function with respect to argument t is

$$\frac{\partial J_\nu(t)}{\partial \nu} = \frac{1}{2}\int_0^t Ji_0(t-x)\, [J_{\nu-1}(x) - J_{\nu+1}(x)]\, dx \tag{4.2.9}$$

$$\operatorname{Re}\nu > 0$$

It comes from the operator s in the Laplace transform domain and is inversed by the Duhamel integral with $J_\nu(0) = 0$ for $\operatorname{Re}\nu > 0$. For the integral Bessel function see paper of Humbert [89].

By arranging the Laplace transforms in different way, van der Pol obtained alternative expressions for the first derivative with respect to the order of the Bessel function of the first kind

$$\frac{\partial J_\nu(t)}{\partial \nu} = \left[\psi\left(\nu - \frac{1}{2}\right) - \ln\left(\frac{t}{2}\right)\right] J_\nu(t) + 2\int_0^t \left(\frac{x}{t}\right)^\nu Ci(t-x) J_{\nu-1}(x)\, dx$$

$$Ci(z) = -\int_z^\infty \frac{\cos t}{t}\, dt \quad ; \quad \operatorname{Re}\nu > 0 \tag{4.2.10}$$

$$\frac{\partial J_\nu(z)}{\partial \nu} = \left[\gamma + \ln\left(\frac{t}{2}\right)\right] J_\nu(t) + \int_0^t \left(\frac{x}{t}\right)^\nu \ln\left(1 - \frac{x^2}{t^2}\right) J_{\nu-1}(x)\, dx \tag{4.2.11}$$

$$\operatorname{Re}\nu > 0$$

and

$$\frac{\partial J_\nu(t)}{\partial \nu} = \frac{1}{2}\,\mathrm{PV}\int_{-\infty}^{\infty} \frac{e^{\nu|x|}}{e^x - 1} J_\nu(te^{-x/2})\,dx \qquad (4.2.12)$$

$$\mathrm{Re}\,\nu > -1$$

After a rather long period of time, only in 1985, the present author and the late Naftali Kravitsky [4] used the Laplace transformation, but this time in a more systematic way to obtain integral representations of derivatives and integrals with respect to the order of the Bessel functions, the integral Bessel function and the Anger function. This was performed by an evaluation of the complex integrals given in (4.2.2) by using different modifications of the Bromwich contour and by applying various operational rules of the Laplace transformation. These operational methods were later extended to the Struve and Kelvin functions during the period 1989–1991 [5, 6]. Results from these investigations and applied mathematical procedures are discussed below.

Let us start with determination of the following integral with respect to the order ν

$$I(\nu,t) = \int_{\nu}^{\infty} J_\lambda(t)\,d\lambda \qquad (4.2.13)$$

The Laplace transform of the Bessel function of the first kind is

$$L\{J_\nu(t)\} = \frac{1}{\sqrt{s^2+1}\,(s+\sqrt{s^2+1})^\nu} = \frac{1}{\sqrt{s^2+1}\,Q^\nu} \qquad (4.2.14)$$

$$Q = (s+\sqrt{s^2+1}) \quad ; \quad \mathrm{Re}\,s > 0 \quad ; \quad \mathrm{Re}\,\nu > -1$$

and by performing integration under the integral sign, we have

$$L\{I(\nu,t)\} = L\left\{\int_{\nu}^{\infty} J_\lambda(t)\,d\lambda\right\}$$

$$= \frac{1}{\sqrt{s^2+1}}\int_{\nu}^{\infty}\frac{d\lambda}{(s+\sqrt{s^2+1})^\lambda} = \frac{1}{\sqrt{s^2+1}\,Q^\nu \ln Q} \qquad (4.2.15)$$

The inverse of (4.2.15) is given by the complex integral (4.2.2)

$$I(\nu,t) = \int_{\nu}^{\infty} J_\lambda(t)\,d\lambda = \frac{1}{2\pi i}\int_{c-i\infty}^{c+i\infty} \frac{e^{st}}{\sqrt{s^2+1}\,Q^\nu \ln Q}\,ds \qquad (4.2.16)$$

$$\mathrm{Re}\,s > c$$

Usually, such complex integrals which have poles, branch points and essential singularities are evaluated with the help of closed, different Bromwich contours, the Cauchy residue theorem and the Jordan lemma. Suitable for complex integration of (4.2.16), the Bromwich contour is presented in Figure 4.1.

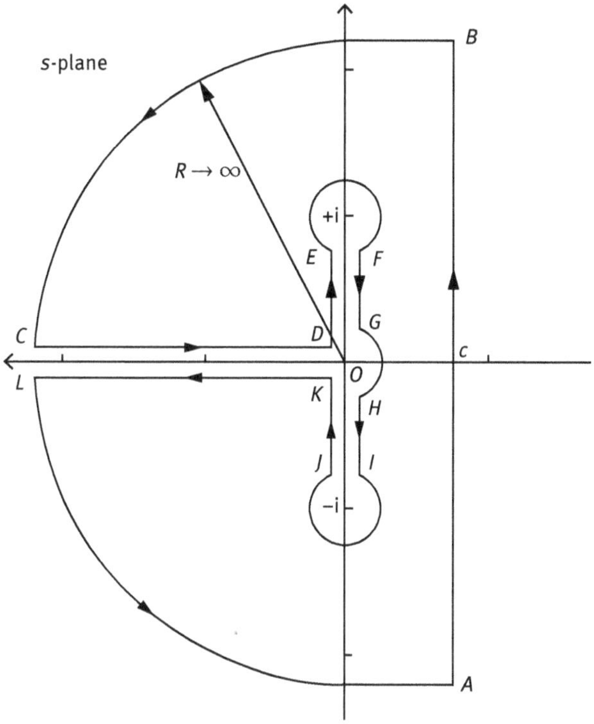

Figure 4.1: The Bromwich contour used in complex integration of equation (4.2.16).

The integrand in (4.2.16) has two conjugate branch points at $s = \pm i$ and a pole at the origin. Contributions to the integral come from CD, DE, EG, HI, JK and KL parts of the contour and a half residue at the pole $s = 0$. The contributions from BC and LA circular arcs with infinite radius R in an imaginary plane vanish according to the Jordan lemma. The sum of these contributions is

$$I(\nu, t) = \int_{\nu}^{\infty} J_{\lambda}(t)\, d\lambda = \frac{1}{2} + \frac{1}{\pi} \int_{0}^{\pi} \sin(t \sin x - \nu x)\, \frac{dx}{x} - \frac{1}{\pi} \int_{0}^{\infty} \frac{e^{-t \sinh x - \nu x} [\pi \cos(\pi \nu) + x \sin(\pi \nu)]}{\pi^2 + x^2}\, dx \tag{4.1.17}$$

For an entire interval, starting from $\nu = 0$, the integral in (4.2.17) reduces to

$$I(0, t) = \int_{0}^{\infty} J_{\lambda}(t)\, d\lambda = \frac{1}{2} + \frac{1}{\pi} \int_{0}^{\pi} \sin(t \sin x)\, \frac{dx}{x} - \int_{0}^{\infty} \frac{e^{-t \sinh x}}{\pi^2 + x^2}\, dx \tag{4.2.18}$$

and it also follows that

$$\int_0^{\nu} J_\lambda(t)\, d\lambda = I(0,t) - I(\nu,t) \tag{4.2.19}$$

The rules of the Laplace transformation permit easily to obtain the limiting behaviour of the integral (4.2.18) for small or large values of t. From (4.2.15) we have

$$L\{I(0,t)\} = L\left\{\int_0^{\infty} J_\lambda(t)\, d\lambda\right\} = \frac{1}{\sqrt{s^2+1}\,\ln\left(s+\sqrt{s^2+1}\right)} \tag{4.2.20}$$

and therefore,

$$L\{I(0,t\to\infty)\} \sim \frac{1}{\ln(1+s)} \sim \frac{1}{s} \quad ; \quad s\to 0 \tag{4.2.21}$$

which means

$$I(0,t\to\infty) \sim L^{-1}\left\{\frac{1}{s}\right\} = 1 \quad ; \quad t\to\infty \tag{4.2.22}$$

Similarly

$$L\{I(0,t\to 0)\} \sim \frac{1}{s\,\ln(2s)} \quad ; \quad s\to\infty \tag{4.2.23}$$

The inverse of (4.2.23) can be expressed in terms of the Volterra function [30]

$$\begin{gathered} I(0,t\to 0) \sim L^{-1}\left\{\frac{1}{s\,\ln(2s)}\right\} = \nu\left(\frac{t}{2}\right) \quad ; \quad t\to 0 \\ \nu\left(\frac{t}{2}\right) = \int_0^{\infty} \frac{\left(\frac{t}{2}\right)^{\lambda}}{\Gamma(\lambda+1)}\, d\lambda \end{gathered} \tag{4.2.24}$$

and its asymptotic expression for small t is known [4]

$$I(0,t\to 0) \sim \frac{1}{\ln(2/t)}\left\{1 - \frac{\psi(1)}{\ln(2/t)} + \frac{[\psi(1)]^2 - \psi'(1)}{[\ln(2/t)]^2} - \dots\right\} \qquad t\to 0 \tag{4.2.25}$$

As expected, a direct differentiation of (4.2.17) leads to the Schlaefli representation of the Bessel function of the first kind

$$J_\nu(t) = -\frac{\partial}{\partial \nu}\left\{\int_\nu^\infty J_\lambda(t)\, d\lambda\right\} = \frac{1}{\pi}\int_0^\pi \cos(t\sin x - \nu x)\, dx \\ -\frac{\sin(\pi\nu)}{\pi}\int_0^\infty e^{-t\sinh x - \nu x}\, dx \tag{4.2.26}$$

Returning to the Laplace transform of derivative given in (4.2.6)

$$L\left\{\frac{\partial J_\nu(t)}{\partial \nu}\right\} = -\frac{\ln\left(s+\sqrt{s^2+1}\right)}{\sqrt{s^2+1}\left(s+\sqrt{s^2+1}\right)^\nu} \\ \operatorname{Re} s > 0 \quad ; \quad \operatorname{Re} \nu > -1 \tag{4.2.27}$$

its inverse is known for $\nu = 0$ [36]

$$\left(\frac{\partial J_\nu(t)}{\partial \nu}\right)_{\nu=0} = L^{-1}\left\{-\frac{\ln\left(s+\sqrt{s^2+1}\right)}{\sqrt{s^2+1}}\right\} = \frac{\pi}{2}Y_0(t) \tag{4.2.28}$$

and the expected result is reached.

The Laplace transforms of higher derivatives with respect to the order can be determined by direct differentiation of (4.2.27), but the operational substitution rule gives a more elegant way to obtain them. If a function *f(t)* has the Laplace transform *F*(*s*) then [31]

$$L\left\{\int_0^\infty f(\lambda)\, J_\lambda(t)\, d\lambda\right\} = \frac{1}{\sqrt{s^2+1}} F\left[\ln\left(s+\sqrt{s^2+1}\right)\right] = \frac{1}{\sqrt{s^2+1}} F(\ln Q) \tag{4.2.29}$$

If in this substitution formula the chosen function is for example *f(t)* = 1, then *F*(*s*) = 1/s, and from (4.2.29) it follows that

$$L\left\{\int_0^\infty J_\lambda(t)\, d\lambda\right\} = \frac{1}{\sqrt{s^2+1}\ln\left(s+\sqrt{s^2+1}\right)} = \frac{1}{\sqrt{s^2+1}\ln Q} \tag{4.2.30}$$

which is the previous result already given in (4.2.20).

The second example is the shifted Dirac function which has the following Laplace transform

$$F(s) = L\{\delta(t-\nu)\} = e^{-\nu s} \tag{4.2.31}$$

Introducing (4.2.31) into (4.2.29) we have

$$L\left\{\int_0^\infty \delta(\lambda-\nu)\,J_\lambda(t)\,d\lambda\right\} = L\{J_\nu(t)\} = \frac{e^{-\nu \ln\left(s+\sqrt{s^2+1}\right)}}{\sqrt{s^2+1}}$$
$$= \frac{1}{\sqrt{s^2+1}\,(s+\sqrt{s^2+1})^\nu} = \frac{1}{\sqrt{s^2+1}\,Q^\nu} \tag{4.2.32}$$

and (4.2.32) is identical with (4.2.14), the Laplace transform of the Bessel function of the first kind.

In order to determine higher derivatives with respect to the order, the chosen functions $f(t)$ should be the derivatives of the shifted Dirac functions

$$F(s) = \left\{\delta^{(n)}(t-\nu)\right\} = (-1)^n s^n e^{-\nu s} \tag{4.2.33}$$

and using (4.2.29) we have

$$L\left\{\int_0^\infty \delta^{(n)}(\lambda-\nu)\,J_\lambda(t)\,d\lambda\right\} = L\left\{\frac{\partial^n J_\nu(t)}{\partial \nu^n}\right\}$$
$$= (-1)^n \frac{e^{-\nu \ln\left(s+\sqrt{s^2+1}\right)}\left[\ln\left(s+\sqrt{s^2+1}\right)\right]^n}{\sqrt{s^2+1}}$$
$$= (-1)^n \frac{\left[\ln\left(s+\sqrt{s^2+1}\right)\right]^n}{\sqrt{s^2+1}\left(s+\sqrt{s^2+1}\right)^\nu} = (-1)^n \frac{(\ln Q)^n}{\sqrt{s^2+1}\,Q^\nu}$$
$$n = 0, 1, 2, 3, \ldots \tag{4.2.34}$$

Thus, formally from (4.2.34) it follows that the higher derivatives with respect to the order of the Bessel function of the first kind are given by determination of the Laplace transform inverses, i.e. by consecutive evaluation of complex integrals

$$\frac{\partial^n J_\nu(t)}{\partial \nu^n} = \frac{(-1)^n}{2\pi i}\int_{c-i\infty}^{c+i\infty} \frac{e^{st}\left[\ln(s+\sqrt{s^2+1})\right]^n}{\sqrt{s^2+1}\,(s+\sqrt{s^2+1})^\nu}\,ds$$
$$= \frac{(-1)^n}{2\pi i}\int_{c-i\infty}^{c+i\infty} \frac{e^{st}\,(\ln Q)^n}{\sqrt{s^2+1}\,Q^\nu}\,ds \tag{4.2.35}$$
$$n = 0, 1, 2, 3, \ldots \quad ; \quad \operatorname{Re} s > c$$

In particular case of $\nu = 0$, the expression (4.2.35) reduces to

$$
\begin{aligned}
\left(\frac{\partial^n J_\nu(t)}{\partial \nu^n}\right)_{\nu=0} &= \frac{(-1)^n}{2\pi i}\int_{c-i\infty}^{c+i\infty}\frac{e^{st}\left[\ln(s+\sqrt{s^2+1})\right]^n}{\sqrt{s^2+1}}\,ds \\
&= \frac{(-1)^n}{2\pi i}\int_{c-i\infty}^{c+i\infty}\frac{e^{st}\,(\ln Q)^n}{\sqrt{s^2+1}}\,ds \\
& n=0,1,2,3,\ \dots\ ;\quad \operatorname{Re} s > c
\end{aligned}
\tag{4.2.36}
$$

Operational calculus permits expressing the first derivative with respect to the order ν differently than that was given by van der Pol in (4.2.8). The Laplace transform in (4.2.34) with $n = 1$ is

$$
\begin{aligned}
L\left\{\frac{\partial J_\nu(t)}{\partial \nu}\right\} &= -\frac{\ln(s+\sqrt{s^2+1})}{\sqrt{s^2+1}\,(s+\sqrt{s^2+1})^\nu} \\
&= -\frac{\ln(s+\sqrt{s^2+1})}{\sqrt{s^2+1}}\cdot\frac{1}{(s+\sqrt{s^2+1})^\nu}
\end{aligned}
\tag{3.2.37}
$$

but [36]

$$
L^{-1}\left\{\frac{\ln(s+\sqrt{s^2+1})}{\sqrt{s^2+1}}\right\} = -\frac{\pi}{2}\,Y_0(t) \tag{4.2.38}
$$

$$
L^{-1}\left\{\frac{1}{(s+\sqrt{s^2+1})^\nu}\right\} = \frac{\nu}{t}\,J_\nu(t)\quad;\quad \operatorname{Re}\nu > 0 \tag{4.2.39}
$$

and from the convolution theorem we have

$$
\begin{aligned}
\frac{\partial J_\nu(t)}{\partial \nu} &= \frac{\pi\nu}{2}\,Y_0(t)*\frac{J_\nu(t)}{t} = \frac{\pi\nu}{2}\int_0^t Y_0(t-x)\,J_\nu(x)\,\frac{dx}{x} \\
& \operatorname{Re}\nu > 0
\end{aligned}
\tag{4.2.40}
$$

The convolution integral can be converted to definite integral by introducing a new variable $x = t\,(\cos\theta)^2$

$$
\begin{aligned}
\frac{\partial J_\nu(t)}{\partial \nu} &= \pi\nu\int_0^{\pi/2}\tan\theta\; Y_0[t\,(\sin\theta)^2]\,J_\nu[t\,(\cos\theta)^2]\,d\theta \\
& \operatorname{Re}\nu > 0
\end{aligned}
\tag{4.2.41}
$$

The integrals in (4.2.41) can be written in a closed form only for $\nu = 1/2$ and $\nu = 1$ because derivatives with respect to the order are known. For other values of order ν, they should be evaluated numerically. If the same procedure is applied for $n = 2$, we have the product of three convolutions

$$L\left\{\frac{\partial^2 J_\nu(t)}{\partial\nu^2}\right\} = \frac{\left[\ln(s+\sqrt{s^2+1})\right]^2}{\sqrt{s^2+1}\,(s+\sqrt{s^2+1})^\nu} = \frac{\ln(s+\sqrt{s^2+1})}{\sqrt{s^2+1}} \cdot \frac{\ln(s+\sqrt{s^2+1})}{s} \cdot \frac{s}{(s+\sqrt{s^2+1})^\nu} \tag{4.2.42}$$

which is equivalent to

$$\frac{\partial^2 J_\nu(t)}{\partial\nu^2} = -\frac{\pi\nu^2}{2}\, Y_0(t) {}^* Ji_0(t) {}^* \frac{d}{dt}\left\{\frac{J_\nu(t)}{t}\right\} \qquad \operatorname{Re}\nu > 0 \tag{4.2.43}$$

However, if (4.2.42) is presented in different form

$$L\left\{\frac{\partial^2 J_\nu(t)}{\partial\nu^2}\right\} = \frac{\left[\ln(s+\sqrt{s^2+1})\right]^2}{\sqrt{s^2+1}\,(s+\sqrt{s^2+1})^\nu} = \frac{\left[\ln(s+\sqrt{s^2+1})\right]^2}{s} \cdot \frac{s}{\sqrt{s^2+1}\,(s+\sqrt{s^2+1})^\nu} \tag{4.2.44}$$

and taking into account (4.2.32) and [36]

$$L^{-1}\left\{\frac{\left[\ln(s+\sqrt{s^2+1})\right]^2}{s}\right\} = -\pi\, Yi_0(t) \tag{4.2.45}$$

we have

$$\frac{\partial^2 J_\nu(t)}{\partial\nu^2} = -\pi\, Yi_0(t) {}^* \frac{d}{dt}\{J_\nu(t)\} \qquad \operatorname{Re}\nu > 0 \tag{4.2.46}$$

or explicitly

$$\frac{\partial^2 J_\nu(t)}{\partial\nu^2} = -\frac{\pi}{2}\int_0^t Yi_0(t-x)\,[J_{\nu-1}(x) - J_{\nu+1}(x)]\,dx \tag{4.2.47}$$

In general case of the nth-derivative with respect to the order, the expression (4.2.34) is

$$L\left\{\frac{\partial^n J_\nu(t)}{\partial \nu^n}\right\} = (-1)^n \frac{\left[\ln(s+\sqrt{s^2+1})\right]^n}{\sqrt{s^2+1}\,(s+\sqrt{s^2+1})^\nu}$$

$$= \left(-\frac{\ln(s+\sqrt{s^2+1})}{s}\right)^n \cdot \frac{s^n}{\sqrt{s^2+1}\,(s+\sqrt{s^2+1})^\nu} \tag{4.2.48}$$

$$n = 0, 1, 2, 3, \ldots \quad ; \quad \operatorname{Re}\nu > -1$$

and the compact form of inverse in (4.2.44) is

$$\frac{\partial^n J_\nu(t)}{\partial \nu^n} = (-1)^n [Ji_0(t)]^{*n} * J_\nu^{(n)}(t) \tag{4.2.49}$$

$$n = 0, 1, 2, 3, \ldots$$

The above expression contains the product of n convolutions of the integral Bessel function of order zero and the *nth* derivative with respect to argument t of the Bessel function of the first kind of the order ν. The formula in (4.2.45) is inconvenient for numerical determination of derivatives with $n > 2$, but operational rules of the Laplace transformation permit to obtain behaviour of derivatives for small and large values of argument. For $t \to \infty$, (4.2.48) becomes

$$L\left\{\frac{\partial^n J_\nu(t)}{\partial \nu^n}\right\} = (-1)^n \frac{\left[\ln(s+\sqrt{s^2+1})\right]^n}{\sqrt{s^2+1}\,(s+\sqrt{s^2+1})^\nu} \sim (-1)^n \frac{[\ln(1+s)]^n}{(1+s)^\nu}$$

$$\sim (-1)^n \frac{s^n}{(1+s)^\nu} \quad ; \quad s \to 0 \tag{4.2.50}$$

the inverse of (4.2.50) is [34]

$$\frac{\partial^n J_\nu(t)}{\partial \nu^n} = (-1)^n \frac{n!}{\Gamma(\nu)}\, t^{\nu-n-1} e^{-t} L_n^{(\nu-n-1)}(t) \quad ; \quad t \to \infty$$

$$\operatorname{Re}\nu > 0 \quad ; \quad n = 0, 1, 2, 3, \ldots \tag{4.2.51}$$

where the generalized Laguerre polynomials are defined by the Rodrigues' formula as

$$L_n^{(\alpha)}(t) = t^{-\alpha} \frac{e^t}{n!} \frac{d}{dt} [t^{n+\alpha} e^{-t}] \tag{4.2.52}$$

For $t \to 0$, the approximation of (4.2.48) is

$$L\left\{\frac{\partial^n J_\nu(t)}{\partial \nu^n}\right\} = (-1)^n \frac{\left[\ln(s+\sqrt{s^2+1})\right]^n}{\sqrt{s^2+1}\,(s+\sqrt{s^2+1})^\nu} \sim (-1)^n \frac{2\,[\ln(2s)]^n}{(2s)^{\nu+1}} \tag{4.2.53}$$

$$s \to \infty$$

In tables of the Laplace transforms, the inverse of (4.2.53) is known only for $n = 1$ and $n = 2$ [36]

$$\frac{\partial J_\nu(t)}{\partial \nu} = L^{-1}\left\{-\frac{2\ln(2s)}{(2s)^{\nu+1}}\right\} = \frac{\left(\frac{t}{2}\right)^\nu}{\Gamma(\nu+1)}\left[\ln\left(\frac{t}{2}\right) - \psi(\nu+1)\right] \tag{4.2.54}$$

$$t \to 0 \quad ; \quad \operatorname{Re}\nu > -1$$

and

$$\frac{\partial^2 J_\nu(t)}{\partial \nu^2} = L^{-1}\left\{\frac{2[\ln(2s)]^2}{(2s)^{\nu+1}}\right\}$$
$$= \frac{\left(\frac{t}{2}\right)^\nu}{\Gamma(\nu+1)}\left\{\left[\ln\left(\frac{t}{2}\right) - \psi(\nu+1)\right]^2 - \psi'(\nu+1)\right\} \tag{4.2.55}$$

$$t \to 0 \quad ; \quad \operatorname{Re}\nu > -1$$

Fényes [52, 76] considered the Laplace transforms of the moments of Bessel functions of the first kind (see Section 4.1)

$$F_n(s) = L\{M_n(t)\} = L\left\{\int_0^\infty \nu^n J_\nu(t)\, d\nu\right\} \tag{4.2.56}$$

$$t, s > 0 \quad ; \quad n = 0, 1, 2, 3, \ldots$$

where

$$F_n(s) = \int_0^\infty\int_0^\infty e^{-st}\nu^n J_\nu(t)\, d\nu\, dt = \int_0^\infty \nu^n\left[\int_0^\infty e^{-st}J_\nu(t)\, dt\right] d\nu$$
$$= \int_0^\infty \frac{\nu^n}{\sqrt{s^2+1}\,(s+\sqrt{s^2+1})^\nu}\, d\nu = \frac{1}{\sqrt{s^2+1}}\int_0^\infty \nu^n e^{-\nu(s+\sqrt{s^2+1})} d\nu \tag{4.2.57}$$
$$= \frac{n!}{\sqrt{s^2+1}\left[\ln(s+\sqrt{s^2+1})\right]^{n+1}} = \frac{n!}{\sqrt{s^2+1}\,(\ln Q)^{n+1}}$$

and

$$F_{n+1}(s) = \frac{(n+1)}{\ln(s+\sqrt{s^2+1})} F_n(s) \tag{4.2.58}$$

By introducing $f(t) = t^n$ and its transform $F(s) = n!/s^{n+1}$, the Laplace transform in (4.2.57) can directly be derived by using the operational rule given in (4.2.29). In the case $n = 0$, the Laplace transform was already discussed in (4.2.30).

The limiting behaviour of moments $M_n(t)$ for small and large values of t can be predicted using operational rules. The case of $t \to \infty$, was already considered by Fényes [52] (see (4.2.29) and (4.2.30)). In operational treatment of the problem we have

$$F_n(s) = \frac{n!}{\sqrt{s^2+1}\left[\ln(s+\sqrt{s^2+1})\right]^{n+1}} \sim \frac{n!}{[\ln(s+1)]^{n+1}} \sim \frac{n!}{s^{n+1}} \tag{4.2.59}$$

$$s \to 0$$

and the inverse transform is

$$L^{-1}\{F_n(s \to 0)\} = L^{-1}\left\{\frac{n!}{s^{n+1}}\right\} = t^n \quad ; \quad t \to \infty \tag{4.2.60}$$

Thus, this approximation gives only the dominant power of t, (from polynomials determined by Fényes), but a very small value of the oscillatory terms in (4.1.29) are omitted.

Fényes did not consider the case $t \to 0$, but it is possible to show that for $s \to \infty$ the Laplace transform can be approximated by

$$F_n(s) = \frac{n!}{\sqrt{s^2+1}[\ln(s+\sqrt{s^2+1})]^{n+1}} \sim \frac{2n!}{(2s)[\ln(2s)]^{n+1}} \quad ; \quad s \to \infty \tag{4.2.61}$$

and its inverse can be expressed in terms of the Volterra function

$$L^{-1}\{F_n(s \to \infty)\} = L^{-1}\left\{\frac{2n!}{(2s)[\ln(2s)]^{n+1}}\right\} = \mu\left(\frac{t}{2}, n\right) \tag{4.2.62}$$

$$t \to 0$$

where [30]

$$\mu(z,\beta) = \int_0^\infty \frac{z^x x^\beta}{\Gamma(\beta+1)\,\Gamma(x+1)}\,dx \quad ; \quad \operatorname{Re}\beta > -1 \tag{4.2.63}$$

and therefore,

$$M_n(t) \sim \frac{1}{n!}\int_0^\infty \frac{\left(\frac{t}{2}\right)^x x^n}{\Gamma(x+1)}\,dx \quad ; \quad t \to 0 \tag{4.2.64}$$

when this integral tends to zero if $t \to 0$.

As shown in Chapter 3, there is the substitution rule of the Laplace transformation which is connected with integrals or derivatives with respect to the order [30]

$$F(s) = L\{f(t)\} = \int_0^\infty e^{-st} f(t)\,dt$$

$$L^{-1}\left\{\frac{1}{(s^2+1)}F\left[\ln(s^2+1)\right]\right\} = \sqrt{\pi}\int_0^\infty \left(\frac{t}{2}\right)^{\nu+1/2} \frac{J_{\nu+1/2}(t)}{\Gamma(\nu+1)} f(\nu)\,d\nu \tag{4.2.65}$$

This expression can be used if inverses of logarithmic functions are known. Unfortunately, they are rare, but the Dirac delta function or its derivatives are suitable for such operations. However, this leads to rather lengthy differentiations of integrands in (4.2.65), but the final results of such calculations are quite attractive.

There is a similar substitution rule for the modified Bessel function [30]

$$F(s) = L\{f(t)\} = \int_0^\infty e^{-st} f(t)\, dt$$

$$L^{-1}\left\{\frac{1}{(s^2-1)} F[\ln(s^2-1)]\right\} = \sqrt{\pi} \int_0^\infty \left(\frac{t}{2}\right)^{\nu+1/2} \frac{I_{\nu+1/2}(t)}{\Gamma(\nu+1)} f(\nu)\, d\nu \tag{4.2.66}$$

and analogous expressions

$$F(s) = L\{f(t)\} = \int_0^\infty e^{-st} f(t)\, dt$$

$$L^{-1}\{F[\ln(s^2+1)]\} = \sqrt{\frac{2\pi}{t}} \int_0^\infty \left(\frac{t}{2}\right)^{\nu} \frac{J_{\nu-1/2}(t)}{\Gamma(\nu)} f(\nu)\, d\nu$$

$$L^{-1}\{F[\ln(s^2-1)]\} = \sqrt{\frac{2\pi}{t}} \int_0^\infty \left(\frac{t}{2}\right)^{\nu} \frac{I_{\nu-1/2}(t)}{\Gamma(\nu)} f(\nu)\, d\nu \tag{4.2.67}$$

4.3 Integration and Differentiation with Respect to the Order of the Modified Bessel Function of the First Kind by Applying the Laplace Transform Approach

Laplace transforms of the Bessel functions of the second kind $Y_\nu(t)$ and the modified Bessel functions of the second kind $K_\nu(t)$ exist only for a limited interval of ν values. This and their behaviour near the origin (the first function tends to minus infinity and the second function to plus infinity) makes the operational calculus less adaptable to use. On the other side, the modified Bessel function of first kind $I_\nu(t)$ is interrelated with $J_\nu(t)$ (the argument t is then replaced with it) and therefore the Laplace transform operations are very similar to those applied in Section 4.2.

The modified Bessel function of the first kind has the following Laplace transform [36]

$$L\{I_\nu(t)\} = \frac{e^{-\nu \ln(s+\sqrt{s^2-1})}}{\sqrt{s^2-1}} = \frac{1}{\sqrt{s^2-1}\,(s+\sqrt{s^2-1})^\nu} = \frac{1}{\sqrt{s^2-1}\, q^\nu} \tag{4.3.1}$$

$$q = (s+\sqrt{s^2-1}) \quad ; \quad \operatorname{Re} s > 1 \quad ; \quad \operatorname{Re} \nu > -1$$

Similarly, as in (4.2.13), let us evaluate the integral

$$I(\nu,t)=\int_{\nu}^{\infty} I_\lambda(t)\,d\lambda \tag{4.3.2}$$

The integration under integral sign gives

$$\begin{aligned} L\{I(\nu,t)\} &= L\left\{\int_{\nu}^{\infty} I_\lambda(t)\,d\lambda\right\} = \frac{1}{\sqrt{s^2-1}}\int_{\nu}^{\infty}\frac{d\lambda}{(s+\sqrt{s^2-1})^{\lambda}} \\ &= \frac{1}{\sqrt{s^2-1}\,q^{\nu}\ln q} \end{aligned} \tag{4.3.3}$$

The inverse of (4.3.3) is

$$I(\nu,t)=\int_{\nu}^{\infty} I_\lambda(t)\,d\lambda=\frac{1}{2\pi i}\int_{c-i\infty}^{c+i\infty}\frac{e^{st}}{\sqrt{s^2-1}\,q^{\nu}\ln q}ds \quad ; \quad \operatorname{Re} s>1 \tag{4.3.4}$$

In this case, the Bromwich contour of complex integration, used in Figure 4.1 should be switched by the angle $\pi/2$ in the clockwise direction (Figure 4.2). The integrand in (4.3.4) has two branch points, but this time on the real axis, $s = \pm 1$. The final result of integration is [4]

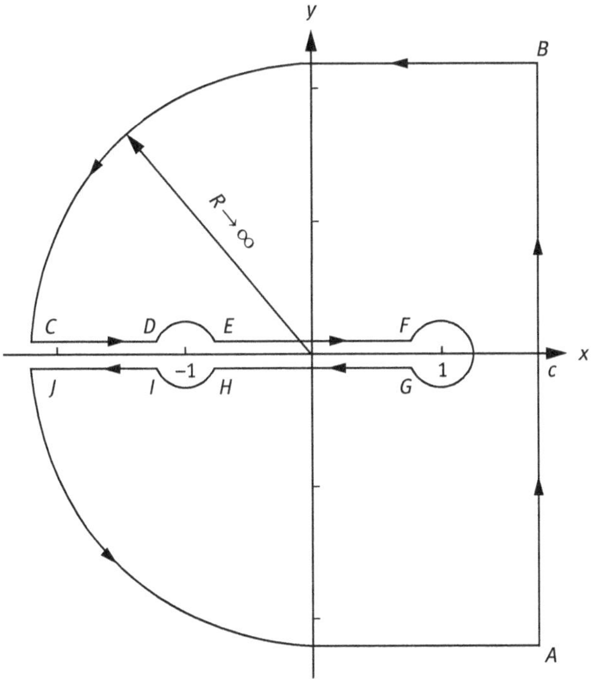

Figure 4.2: The Bromwich contour used in complex integration of equation (4.3.4).

$$\int_{\nu}^{\infty} I_{\lambda}(t)\,d\lambda = \frac{e^t}{2} - \frac{1}{\pi}\int_0^{\pi} e^{t\cos x}\sin(\nu x)\,\frac{dx}{x} - \frac{1}{\pi}\int_0^{\infty} e^{-\nu x - t\cosh x}\,\frac{[x\sin(\pi\nu) + \pi\cos(\pi\nu)]}{\pi^2 + x^2}\,dx \tag{4.3.5}$$

and in particular case of $\nu = 0$ it reduces to

$$\int_0^{\infty} I_{\lambda}(t)\,d\lambda = \frac{e^t}{2} - \int_0^{\infty} \frac{e^{-t\cosh x}}{\pi^2 + x^2}\,dx \tag{4.3.6}$$

The asymptotic behaviour of integral (4.3.5) for $t \to 0$ comes from

$$L\{I(\nu, t \to 0)\} = \frac{1}{\sqrt{s^2-1}\,q^{\nu}\ln q} \sim \frac{2}{(2s)^{\nu+1}\ln(2s)} \qquad s \to \infty \tag{4.3.7}$$

which has the inverse in terms of the Volterra function [30]

$$L^{-1}\left\{\frac{2}{(2s)^{\nu+1}\ln(2s)}\right\} = \nu\left(\frac{t}{2}, \nu\right) = \int_0^{\infty} \frac{\left(\frac{t}{2}\right)^{\lambda+\nu}}{\Gamma(\lambda+\nu+1)}\,d\lambda \qquad t \to 0 \tag{4.3.8}$$

and

$$\begin{aligned} \nu\left(\frac{t}{2}, \nu\right) &\sim \frac{\left(\frac{t}{2}\right)^{\nu}}{\Gamma(\nu+1)\,\ln(2/t)}\, I(t,\nu) \\ I(t,\nu) &= \left\{1 - \frac{\psi(\nu+1)}{\ln(2/t)} + \frac{[\psi(\nu+1)]^2 - \psi'(\nu+1)}{[\ln(2/t)]^2} - \dots\right\} \qquad t \to 0 \end{aligned} \tag{4.3.9}$$

In the case of $t \to \infty$, it follows from (4.3.3) that

$$L\{I(\nu, t \to \infty)\} = \frac{1}{\sqrt{s^2-1}\,q^{\nu}\ln q} \sim \frac{1}{(s^2-1)^{\nu+1}} \quad ; \quad s \to 1 \tag{4.3.10}$$

and its inverse is [36]

$$L^{-1}\left\{\frac{1}{(s^2-1)^{\nu+1}}\right\} = \frac{\sqrt{\pi}\,t^{\nu+1/2}}{2^{\nu+1/2}\Gamma(\nu+1)}\, I_{\nu+1/2}(t) \quad ; \quad t \to \infty \tag{4.3.11}$$

The exponential character of (4.3.11) is evident if the case $\nu = 0$ is considered

$$L^{-1}\left\{\frac{1}{(s^2-1)}\right\} = \sinh t \sim \frac{e^t}{2} \quad ; \quad t \to \infty \tag{4.3.12}$$

This is consistent with (4.3.6) where the contribution from the integral is negligible for large values of argument t. It is worthwhile to note, that the Laplace transform of (4.3.4) with the lower integration limit being zero, can also be derived by applying the operational rule analogous to (4.2.29)

$$L\left\{\int_0^\infty f(\lambda)\, I_\lambda(t)\, d\lambda\right\} = \frac{1}{\sqrt{s^2-1}} F\left[\ln(s+\sqrt{s^2-1})\right] = \frac{1}{\sqrt{s^2-1}} F(\ln q) \tag{4.3.13}$$

As expected, the differentiations of (4.3.5) with respect to the order gives the integral representation of the modified Bessel function of the first kind and its derivatives

$$I_\nu(t) = \frac{1}{\pi}\int_0^\pi e^{t\cos x}\cos(\nu x)\, dx - \frac{\sin(\nu x)}{\pi}\int_0^\infty e^{-\nu x - t\cosh x}\, dx \tag{4.3.14}$$

and

$$\begin{aligned} \frac{\partial I_\nu(t)}{\partial \nu} &= -\frac{1}{\pi}\int_0^\pi x\, e^{t\cos x}\sin(\nu x)\, dx \\ &+ \frac{1}{\pi}\int_0^\infty e^{-\nu x - t\cosh x}\left[x\sin(\pi\nu) - \pi\cos(\pi\nu)\right] dx \end{aligned} \tag{4.3.15}$$

In the case of $\nu = 0$ we have

$$\left(\frac{\partial I_\nu(t)}{\partial \nu}\right)_{\nu=0} = -\int_0^\infty e^{-t\cosh x}\, dx = -K_0(t) \tag{4.3.16}$$

Differentiation of (4.3.1) with respect to the order gives the Laplace transform

$$\begin{aligned} L\left\{\frac{\partial I_\nu(t)}{\partial \nu}\right\} &= -\frac{e^{-\nu\ln(s+\sqrt{s^2-1})}\ln\left(s+\sqrt{s^2-1}\right)}{\sqrt{s^2-1}} \\ &= \frac{\ln\left(s+\sqrt{s^2-1}\right)}{\sqrt{s^2-1}\left(s+\sqrt{s^2-1}\right)^\nu} = \frac{\ln q}{\sqrt{s^2-1}\, q^\nu} \quad ; \quad \operatorname{Re} s > 1 \quad ; \quad \operatorname{Re}\nu > -1 \end{aligned} \tag{4.3.17}$$

The function $K_0(t)$ is represented by the following inverse Laplace transform [36]

$$L^{-1}\left\{\frac{\ln\left(s+\sqrt{s^2-1}\right)}{\sqrt{s^2-1}}\right\}=K_0(t) \tag{4.3.18}$$

and using

$$L^{-1}\left\{\frac{1}{\left(s+\sqrt{s^2-1}\right)^{\nu}}\right\}=\frac{\nu}{t}I_\nu(t) \tag{4.3.19}$$

$$\operatorname{Re}\nu>0$$

the convolution theorem gives

$$\frac{\partial I_\nu(t)}{\partial \nu}=-K_0(t)*\frac{\nu}{t}I_\nu(t)=-\nu\int_0^t K_0(t-x)\,I_\nu(x)\,\frac{dx}{x} \tag{4.3.20}$$

$$\operatorname{Re}\nu>0$$

By introducing $x = t(\cos\theta)^2$, the convolution integral (4.2.20) can be transformed to the following trigonometric integral

$$\frac{\partial I_\nu(t)}{\partial \nu}=-2\nu\int_0^{\pi/2}\tan\theta\,K_0[t(\sin\theta)^2]\,I_\nu([t(\sin\theta)^2])\,d\theta \tag{4.3.21}$$

$$\operatorname{Re}\nu>0$$

Since the derivatives with respect to the order for $\nu = 1/2$ and $\nu = 1$ are known, (4.2.20) can be written also in the following forms

$$\begin{aligned}-K_0(t)*\frac{\sinh t}{t^{3/2}}&=-\int_0^t K_0(t-x)\,\frac{\sinh x}{x^{3/2}}\,dx\\&=\frac{1}{\sqrt{t}}\left[e^tEi(-2t)+e^{-t}Ei(2t)\right]\end{aligned} \tag{4.3.22}$$

and

$$K_0(t)*\frac{I_1(t)}{t}=\int_0^t K_0(t-x)\,\frac{I_1(x)}{x}\,dx=\frac{I_0(t)}{t}-K_1(t) \tag{4.3.23}$$

Similarly as in (4.2.29), the Laplace transforms of higher derivatives with respect to the order can be derived by introducing the shifted Dirac functions from (4.2.13) into (4.2.13)

$$L\left\{\int_0^\infty \delta^{(n)}(\lambda-\nu)\, I_\lambda(t)\, d\lambda\right\} = L\left\{\frac{\partial^n I_\nu(t)}{\partial \nu^n}\right\}$$

$$= (-1)^n \frac{e^{-\nu \ln(s+\sqrt{s^2-1})}\left[\ln(s+\sqrt{s^2-1})\right]^n}{\sqrt{s^2-1}}$$

$$= (-1)^n \frac{\left[\ln(s+\sqrt{s^2-1})\right]^n}{\sqrt{s^2-1}\,(s+\sqrt{s^2-1})^\nu} \tag{4.3.24}$$

$$= (-1)^n \frac{(\ln q)^n}{\sqrt{s^2-1}\, q^\nu} \quad ; \quad n = 0, 1, 2, 3, \ldots \quad \operatorname{Re} s > 1$$

and therefore higher derivatives with respect to the order are given as the Laplace transform inverses from

$$\frac{\partial^n I_\nu(t)}{\partial \nu^n} = \frac{(-1)^n}{2\pi i} \int_{c-i\infty}^{c+i\infty} \frac{e^{st}\left[\ln(s+\sqrt{s^2-1})\right]^n}{\sqrt{s^2+1}\,(s+\sqrt{s^2-1})^\nu}\, ds$$

$$= \frac{(-1)^n}{2\pi i} \int_{c-i\infty}^{c+i\infty} \frac{e^{st}\,(\ln q)^n}{\sqrt{s^2-1}\, q^\nu}\, ds \tag{4.3.25}$$

$$n = 0, 1, 2, 3, \ldots \quad ; \quad \operatorname{Re} s > 1$$

For $\nu = 0$ we have

$$\left(\frac{\partial^n I_\nu(t)}{\partial \nu^n}\right)_{\nu=0} = \frac{(-1)^n}{2\pi i} \int_{c-i\infty}^{c+i\infty} \frac{e^{st}\left[\ln(s+\sqrt{s^2-1})\right]^n}{\sqrt{s^2-1}}\, ds$$

$$= \frac{(-1)^n}{2\pi i} \int_{c-i\infty}^{c+i\infty} \frac{e^{st}\,(\ln q)^n}{\sqrt{s^2-1}}\, ds \quad ; \quad n = 0, 1, 2, 3, \ldots \quad ; \quad \operatorname{Re} s > 1 \tag{4.3.26}$$

The Laplace transform of the second derivative with respect to the order is

$$L\left\{\frac{\partial^2 I_\nu(t)}{\partial \nu^2}\right\} = \frac{\left[\ln(s+\sqrt{s^2-1})\right]^2}{\sqrt{s^2-1}\,(s+\sqrt{s^2-1})^\nu}$$

$$= \frac{\left[\ln(s+\sqrt{s^2-1})\right]^2}{s} \cdot \frac{s}{(s+\sqrt{s^2-1})^\nu} \tag{4.3.27}$$

but

$$L^{-1}\left\{\frac{\left[\ln(s+\sqrt{s^2-1})\right]^2}{s}\right\}=Ki_0(t)-\frac{\pi^2}{4} \tag{4.3.28}$$

and therefore from the convolution theorem we have

$$\begin{aligned}\frac{\partial^2 I_\nu(t)}{\partial\nu^2}&=\left[2Ki_0(t)-\frac{\pi^2}{4}\right]*I_\nu'(t)\\ I_\nu'(t)&=\frac{1}{2}[I_{\nu-1}(t)+I_{\nu+1}(t)]\end{aligned} \tag{4.3.29}$$

It is difficult to find an analogous formula to that in (4.2.49) for derivatives with respect to the order of the modified Bessel functions $I_\nu(t)$. However, operational rules give at least the behaviour of derivatives for small and large values of argument t.

For $t \to 0$, both functions $J_\nu(t)$ and $I_\nu(t)$ behave similarly because (4.3.24) is identical with (4.2.53) for $s \to \infty$

$$L\left\{\frac{\partial^n I_\nu(t)}{\partial\nu^n}\right\}=(-1)^n\frac{\left[\ln(s+\sqrt{s^2-1})\right]^n}{\sqrt{s^2-1}\,(s+\sqrt{s^2-1})^\nu}\sim(-1)^n\frac{2\,[\ln(2s)]^n}{(2s)^{\nu+1}} \qquad s\to\infty \tag{4.3.30}$$

The Laplace transform inverses of (4.3.30) are known only for $n = 1$ and $n = 2$ (see (4.2.54) and (4.2.55)). For the first and second derivatives, with respect to the order, it is possible to obtain corresponding expressions of the Laplace transforms for $s \to 1$

$$L\left\{\frac{\partial I_\nu(t)}{\partial\nu}\right\}=-\frac{\ln(s+\sqrt{s^2-1})}{\sqrt{s^2-1}\,(s+\sqrt{s^2-1})^\nu}\sim-\frac{1}{s\,(s+\sqrt{s^2-1})^\nu}\sim-\frac{1}{s^{\nu+1}} \qquad s\to 1 \tag{4.3.31}$$

which has the simple inverse

$$\frac{\partial I_\nu(t)}{\partial\nu}\sim-\frac{t^\nu}{\Gamma(\nu+1)}\quad;\quad t\to\infty \qquad \operatorname{Re}\nu>0 \tag{4.3.32}$$

In the case of second derivative we have

$$L\left\{\frac{\partial^2 I_\nu(t)}{\partial\nu^2}\right\} = \frac{\left[\ln(s+\sqrt{s^2-1})\right]^2}{\sqrt{s^2-1}\,(s+\sqrt{s^2-1})^\nu} \sim \frac{2\ln(s+\sqrt{s^2-1})}{s\,(s+\sqrt{s^2-1})^\nu} \sim \frac{2\ln s}{s^{\nu+1}} \qquad s\to 1 \tag{4.3.33}$$

and

$$\frac{\partial^2 I_\nu(t)}{\partial\nu^2} \sim \frac{2\,t^\nu}{\Gamma(\nu+1)}\left[\psi(\nu+1)-\ln t\right] \quad ; \quad t\to\infty \qquad \text{Re}\,\nu>0 \tag{4.3.34}$$

The Laplace transforms of moments of modified Bessel functions of the first kind were considered by Fényes in [52, 76]

$$G_n(s) = L\{m_n(t)\} = L\left\{\int_0^\infty \nu^n I_\nu(t)\,d\nu\right\} \quad ; \quad n=0,1,2,3,\ldots \tag{4.3.35}$$

where

$$\begin{aligned} G_n(s) &= \int_0^\infty\int_0^\infty e^{-st}\nu^n I_\nu(t)\,d\nu\,dt = \int_0^\infty \nu^n\left[\int_0^\infty e^{-st} I_\nu(t)\,dt\right]d\nu \\ &= \int_0^\infty \frac{\nu^n}{\sqrt{s^2-1}\,(s+\sqrt{s^2-1})^\nu}\,d\nu = \frac{1}{\sqrt{s^2-1}}\int_0^\infty \nu^n e^{-\nu(s+\sqrt{s^2-1})}\,d\nu \\ &= \frac{n!}{\sqrt{s^2-1}\left[\ln(s+\sqrt{s^2-1})\right]^{n+1}} = \frac{n!}{\sqrt{s^2-1}\,(\ln q)^{n+1}} \quad ; \quad \text{Re}\,s>1 \end{aligned} \tag{4.3.36}$$

with

$$G_{n+1}(s) = \frac{(n+1)}{\ln\left(s+\sqrt{s^2-1}\right)}\,G_n(s) \tag{4.3.37}$$

The expression (4.3.36) follows directly from (4.1.13) considering that for $f(t) = t^n$ we have $F(s) = n!/s^{n+1}$. The Laplace transform in the case $n = 0$ was already discussed in (4.3.3).

The limiting behaviour of moments $m_n(t)$ for large values of t was already considered by Fényes (see (4.1.31)). The Laplace transform of (4.3.36) for $s \to 1$ can be approximated by

$$G_n(s) = \frac{n!}{\sqrt{s^2-1}\left[\ln(s+\sqrt{s^2-1})\right]^{n+1}}$$
$$\sim \frac{n!}{(s^2-1)^{n/2+1}} \sim \frac{n!}{2^{n/2+1}(s-1)^{n/2+1}} \quad ; \quad s \to 1 \tag{4.3.38}$$

and therefore,

$$L^{-1}\{G_n(s\to 1)\} = L^{-1}\left\{\frac{n!}{2^{n/2+1}(s-1)^{n/2+1}}\right\} = \frac{n!\,t^{n/2}e^t}{2^{n/2+1}\Gamma\left(\frac{n}{2}+1\right)} \tag{4.3.39}$$
$$t \to \infty$$

This result is consistent with derived Fényes expressions in (4.1.31). In the case of small values of t, it is possible to show that the Laplace transform can be approximated by

$$G_n(s) = \frac{n!}{\sqrt{s^2-1}\left[\ln(s+\sqrt{s^2-1})\right]^{n+1}} \sim \frac{2n!}{(2s)\,[\ln(2s)]^{n+1}} \tag{4.3.40}$$
$$s \to \infty$$

which is identical with (4.2.61) and therefore

$$m_n(t) \sim \frac{1}{n!}\int_0^\infty \frac{\left(\frac{t}{2}\right)^x x^n}{\Gamma(x+1)}dx \quad ; \quad t \to 0 \tag{4.3.41}$$

4.4 Differentiation with Respect to the Order of the Anger Function and the Weber Function by Applying the Laplace Transform Approach

Since the Anger function is closely related to the Bessel function of the first kind,

$$\mathbf{J}_\nu(t) = \frac{1}{\pi}\int_0^\pi \cos(\nu x - t\sin x)\,dx = J_\nu(t) + \frac{\sin(\pi\nu)}{\pi}\int_0^\infty e^{-t\sinh x - \nu x}dx \tag{4.4.1}$$

an additional integration with respect to the argument t of the infinite integral in (4.4.1) gives the Laplace transform

$$L\{\mathbf{J}_\nu(t)\} = L\{J_\nu(t)\} + \frac{\sin(\pi\nu)}{\pi}\int_0^\infty e^{-st}\left[\int_0^\infty e^{-t\sinh x-\nu x}dx\right]dt$$
$$= \frac{1}{\sqrt{s^2+1}\left[s+\sqrt{s^2+1}\right]^\nu} + \frac{\sin(\pi\nu)}{\pi}\int_0^\infty \frac{e^{-\nu x}}{s+\sinh x}dx \quad ; \quad \operatorname{Re} s>0 \tag{4.4.2}$$

Introducing integral with respect to the order, similar to that in (4.2.13)

$$\mathrm{I}(\nu,t) = \int_\nu^\infty \mathbf{J}_\lambda(t)\,d\lambda = \int_\nu^\infty J_\lambda(t)\,d\lambda + \frac{1}{\pi}\int_\nu^\infty \sin(\pi\nu)\left[\int_0^\infty e^{-t\sinh x-\nu x}dx\right]d\nu \tag{4.4.3}$$

and changing the order of integration in the second integral we have

$$\mathrm{I}(\nu,t) = \int_\nu^\infty \mathbf{J}_\lambda(t)\,d\lambda = I(\nu,t)$$
$$+\frac{1}{\pi}\int_\nu^\infty \frac{e^{-t\sinh x-\nu x}\left[x\sin(\pi\nu)+\pi\cos(\pi\nu)\right]}{\pi^2+x^2}dx \tag{4.4.4}$$
$$\int_{\nu_1}^{\nu_2} \mathbf{J}_\lambda(t)\,d\lambda = \mathrm{I}(\nu_2,t) - \mathrm{I}(\nu_1,t)$$

Combining (4.1.17) with (4.4.4), it follows that $I(\nu;t)$ becomes

$$\mathrm{I}(\nu,t) = \int_\nu^\infty \mathbf{J}_\lambda(t)\,d\lambda = \frac{1}{2} + \frac{1}{\pi}\int_0^\pi \sin(t\sin x-\nu x)\,\frac{dx}{x} \tag{4.4.5}$$

and for $\nu = 0$ it takes the form

$$\mathrm{I}(0,t) = \int_0^\infty \mathbf{J}_\lambda(t)\,d\lambda = \frac{1}{2} + \frac{1}{\pi}\int_0^\pi \sin(t\sin x)\,\frac{dx}{x} \tag{4.4.6}$$

The Laplace transform of this infinite integral is

$$L\{\mathrm{I}(0,t)\} = L\left\{\int_0^\infty \mathbf{J}_\lambda(t)\,d\lambda\right\} = L\{I(0,t)\} + \int_0^\infty e^{-st}\left[\int_0^\infty \frac{e^{-t\sinh x}}{\pi^2+x^2}dx\right]dt$$
$$= \frac{1}{\sqrt{s^2+1}\ln(s+\sqrt{s^2+1})} + \int_0^\infty \frac{dx}{(\pi^2+x^2)(s+\sinh x)} \tag{4.4.7}$$

The asymptotic expressions of this integral for small and large argument t are [4]

$$\mathrm{I}(0, t \to 0) \sim \frac{1}{2} + \frac{Si(\pi)}{\pi} t \quad ; \quad Si(\pi) = 1.85155...$$
$$\mathrm{I}(0, t \to \infty) \sim 1 + \frac{\left(\frac{2}{\pi}\right)^{3/2}}{\sqrt{t}} \sin\left(t - \frac{\pi}{4}\right) \tag{4.4.8}$$

The difference between the integrals (4.4.6) and (4.4.5) gives

$$\int_0^{\nu} \mathrm{J}_{\lambda}(t)\, d\lambda = \mathrm{I}(0, t) - \mathrm{I}(\nu, t) = \frac{2}{\pi} \int_0^{\pi} \cos\left(t \sin x - \frac{\nu x}{2}\right) \sin\left(\frac{\nu x}{2}\right) \frac{dx}{x} \tag{4.4.9}$$

It is worthwhile to compare behaviour of infinite integrals with respect to the order of the Bessel function of the first kind, the modified Bessel function of the first kind and the Anger function as a function of argument t. These three infinite integrals are plotted together in Figure 4.3. As can be observed, the oscillatory character of infinite integrals of the Bessel function of the first kind A and the Anger function C is nearly the same. The contribution coming from the integration of the second term in (4.4.1) is very small. Integrals A and C tend to unity as argument t tends to infinity. In the case of the modified Bessel function of the first kind, the exponential rise of infinite integral B is consistent with its behaviour predicted in (4.3.12).

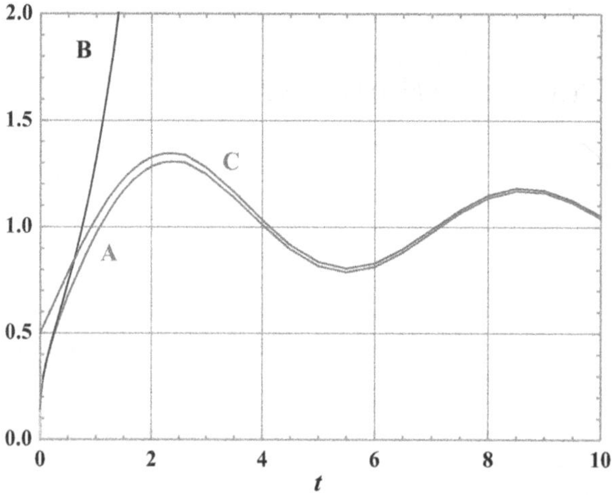

Figure 4.3: Values of infinite integrals with respect to the order as a function of argument t of the Bessel function of the first kind A – equation (4.2.18); the modified Bessel function of the first kind B – equation (4.3.6); and the Anger function C – equation (4.4.6).

Higher derivatives of the Anger function with respect to the order can easily be derived if (4.4.1) is written in complex form

$$\mathbf{J}_\nu(t) = J_\nu(t) + \frac{1}{\pi}\int_0^\infty \frac{e^{-t\sinh x}\left[e^{-(x-\pi i)\nu} - e^{-(x+\pi i)\nu}\right]}{2i}\,dx \tag{4.4.10}$$

and therefore,

$$\frac{\partial^n \mathbf{J}_\nu(t)}{\partial \nu^n} = \frac{\partial^n J_\nu(t)}{\partial \nu^n} + \frac{(-1)^n}{2\pi i}\int_0^\infty e^{-t\sinh x - \nu x}\left[(x-\pi i)^n e^{\pi i\nu} - (x+\pi i)^n e^{-\pi i\nu}\right] dx \tag{4.4.11}$$

For $\nu = 0$, the expression (4.4.11) becomes

$$\left(\frac{\partial^n \mathbf{J}_\nu(t)}{\partial \nu^n}\right)_{\nu=0} = \left(\frac{\partial^n J_\nu(t)}{\partial \nu^n}\right)_{\nu=0} + \frac{(-1)^n}{2\pi i}\int_0^\infty e^{-t\sinh x}\left[(x-\pi i)^n - (x+\pi i)^n\right] dx \tag{4.4.12}$$

From (4.4.12), the first derivative with respect to the order for $\nu = 0$ is

$$\left(\frac{\partial \mathbf{J}_\nu(t)}{\partial \nu}\right)_{\nu=0} = \left(\frac{\partial J_\nu(t)}{\partial \nu}\right)_{\nu=0} + \int_0^\infty e^{-t\sinh x}\,dx \tag{4.4.13}$$

and taking into account that the limiting expression of the Bessel and Anger functions is available, it is possible to present (4.4.13) in the following form

$$\frac{\pi}{2}\,\mathrm{H}_0(t) = \frac{\pi}{2}\,Y_0(t) + \int_0^\infty e^{-t\sinh x}\,dx = \frac{1}{\pi}\int_0^{\pi/2} \sin(t\sin x)\,dx \tag{4.4.14}$$

Similarly, the second derivative from (4.4.12) is

$$\left(\frac{\partial^2 \mathbf{J}_\nu(t)}{\partial \nu^2}\right)_{\nu=0} = \left(\frac{\partial^2 J_\nu(t)}{\partial \nu^2}\right)_{\nu=0} - 2\int_0^\infty x e^{-t\sinh x}\,dx \tag{4.4.15}$$

where

$$\left(\frac{\partial^2 \mathbf{J}_\nu(t)}{\partial \nu^2}\right)_{\nu=0} = \left\{\left[-\frac{\pi^2}{2} + (\ln t + \gamma + \ln 2)\right]^2 - 2(\ln t + \gamma + \ln 2)^2\right\} J_0(t)$$
$$+ \pi(\ln t + \gamma + \ln 2)\, Y_0(t) + \frac{2}{\pi}\int_0^1 \frac{\left[\ln(1-x^2)\right]^2}{\sqrt{1-x^2}} \cos(tx)\, dx \tag{4.4.16}$$

An analogous definition of moments, as in (4.1.22) and (4.1.23) for the Anger function is

$$\mathbf{M}_n(t) = \int_0^\infty \nu^n \mathbf{J}_\nu(t)\, d\nu = \int_0^\infty \nu^n J_\nu(t)\, d\nu + \int_0^\infty \nu^n \left[\frac{\sin(\pi\nu)}{\pi}\int_0^\infty e^{-t\sinh x - \nu x} dx\right] d\nu \tag{4.4.17}$$
$$t > 0 \quad ; \quad n = 0.1.2.3....$$

Changing the order of integration we have

$$\mathbf{M}_n(t) = M_n(t) + \frac{1}{\pi}\int_0^\infty e^{-t\sinh x}\left[\int_0^\infty e^{-\nu x}\nu^n \sin(\pi\nu)\right] d\nu\, dx \tag{4.4.18}$$

where $M_n(t)$ moments are given in (4.1.24).

The inner integral in (4.1.26) can be treated as a known Laplace transform, and therefore,

$$\mathbf{M}_n(t) = M_n(t) + \frac{n!}{\pi}\int_0^\infty e^{-t\sinh x} g_n(x)\, dx$$
$$g_n(x) = (-1)^n \frac{d^n}{dx^n}\left(\frac{\pi}{(\pi^2 + x^2)}\right) \quad ; \quad n = 0, 1, 2, 3, ... \tag{4.4.19}$$

and

$$L\{\mathbf{M}_n(t)\} = \frac{n!}{\sqrt{s^2+1}\left[s + \sqrt{s^2+1}\right]^{n+1}} + \frac{(-1)^n n!}{\pi}\int_0^\infty \frac{g_n(x)}{(s + \sinh x)} dx \tag{4.4.20}$$

The contribution coming from the integral in (4.4.20) is small, and therefore the expected asymptotic behaviour of the Anger function moments resembles those of the Bessel function of the first kind.

The Weber function represented by the following expression:

$$\mathbf{E}_\nu(t) = -\ Y_\nu(t) - \frac{1}{\pi}\int_0^\infty e^{-t\sinh x}[e^{\nu x} + e^{-\nu x}\cos(\pi\nu)]\,dx \quad ; \quad t>0 \tag{4.4.21}$$

is less suitable for the operational treatment because the Bessel function of second kind $Y_\nu(t)$ has the Laplace transform only for limited interval of the orders ν. However the Anger and Weber functions are interrelated in the following way [18]

$$\mathbf{J}_\nu(t) \pm i\,\mathbf{E}_\nu(t) = \frac{1}{\pi}\int_0^\pi e^{\pm i(\nu x - t\sinh x)}\,dx \tag{4.4.22}$$

and therefore,

$$\begin{aligned} \mathbf{E}_\nu(t) &= \frac{1}{\pi}\int_0^\pi \sin(\nu x - t\sinh x)\,dx \\ &= \frac{1}{\pi}\int_0^\pi [\sin(\nu x)\cos(t\sinh x) - \cos(\nu x)\sin(t\sinh x)]\,dx \end{aligned} \tag{4.4.23}$$

The Laplace transform of the Weber function from (4.4.23) is

$$\begin{aligned} L\{\mathbf{E}_\nu(t)\} &= \frac{1}{\pi}\int_0^\pi \sin(\nu x)\left[\int_0^\infty e^{-st}\cos(t\sinh x)\,dt\right]dx \\ &\quad - \frac{1}{\pi}\int_0^\pi \cos(\nu x)\left[\int_0^\infty e^{-st}\sin(t\sinh x)\,dt\right]dx \\ &= \frac{1}{\pi}\int_0^\pi \frac{s\sin(\nu x) - \cos(\nu x)\sinh x}{s^2 + (\sinh x)^2}\,dx \end{aligned} \tag{4.4.24}$$

and the corresponding moments are

$$\begin{aligned} \mathbf{m}_n(t) &= \int_0^\infty \nu^n \mathbf{E}_\nu(t)\,d\nu = \frac{1}{\pi}\int_0^\pi \cos(t\sinh x)\left[\int_0^\infty \nu^n\sin(\pi\nu)\,d\nu\right]dx \\ &\quad - \frac{1}{\pi}\int_0^\pi \sin(t\sinh x)\left[\int_0^\infty \nu^n\cos(\pi\nu)\,d\nu\right]dx \\ &\quad t>0 \quad ; \quad n = 0.1.2.3.... \end{aligned} \tag{4.4.25}$$

The inner integrals in (4.4.25), as suggested by Fényes, can be evaluated using the Laplace transforms of trigonometric functions

$$\begin{aligned} g_n(s) &= \int_0^\infty e^{-sv} v^n \sin(\pi v)\, dv = (-1)^n \frac{d^n}{ds^n}\left(\frac{\pi}{\pi^2+s^2}\right) \\ h_n(s) &= \int_0^\infty e^{-sv} v^n \cos(\pi v)\, dv = (-1)^n \frac{d^n}{ds^n}\left(\frac{s}{\pi^2+s^2}\right) \\ n &= 0.1.2.3.... \end{aligned} \tag{4.4.26}$$

In the limit, by putting $s \to 0$, we have

$$\begin{aligned} g_n(s\to 0) &= (-1)^n \lim_{s\to 0}\left[\frac{d^n}{ds^n}\left(\frac{\pi}{\pi^2+s^2}\right)\right] = \int_0^\infty v^n \sin(\pi v)\, dv \\ g_{2n+1}(s\to 0) &= 0 \\ h_n(s\to 0) &= (-1)^n \lim_{s\to 0}\left[\frac{d^n}{ds^n}\left(\frac{s}{\pi^2+s^2}\right)\right] = \int_0^\infty v^n \cos(\pi v)\, dv \\ h_{2n}(s\to 0) &= 0 \end{aligned} \tag{4.4.27}$$

4.5 Integration and Differentiation with Respect to the Order of the Struve Functions by Applying the Laplace Transform Approach

Derivatives with respect to the order of the Struve function can be derived starting from the Laplace transform-inverse pair [36]

$$\begin{aligned} L\left\{t^{v/2} H_v(z\sqrt{t})\right\} &= \int_0^\infty e^{-st} t^{v/2} H_v(z\sqrt{t})\, dt \\ &= -\frac{i}{s}\left(\frac{z}{2s}\right)^v e^{-z^2/4s} erf\left(\frac{iz}{2\sqrt{s}}\right) \\ &\qquad \operatorname{Re} v > -\frac{3}{2} \quad ; \quad \operatorname{Re} s > 0 \end{aligned} \tag{4.5.1}$$

Differentiation of the left side of (4.5.1) with respect to v gives

$$\frac{\partial}{\partial v}\left\{t^{v/2} H_v(z\sqrt{t})\right\} = \frac{t^{v/2}}{2} \ln t\, H_v(z\sqrt{t}) + t^{v/2}\frac{\partial H_v(z\sqrt{t})}{\partial v} \tag{4.5.2}$$

and of the Laplace transform of (4.5.1) is

$$\frac{\partial}{\partial \nu}\left\{L\left\{t^{\nu/2} H_\nu(z\sqrt{t})\right\}\right\} = -\, i\left(\frac{z}{2s}\right)^{\nu} e^{-z^2/4s} \operatorname{erf}\left(\frac{iz}{2\sqrt{s}}\right) \cdot \frac{\ln\left(\frac{z}{2s}\right)}{s} \tag{4.5.3}$$

Taking into account the following inverses

$$\begin{aligned} L^{-1}\left\{\frac{i}{s}\left(\frac{z}{2s}\right)^{\nu} e^{-z^2/4s} \operatorname{erf}\left(\frac{iz}{2\sqrt{s}}\right)\right\} &= -\frac{z}{2}\, t^{(\nu-1)/2} H_{\nu-1}(z\sqrt{t}) \\ L^{-1}\left\{\frac{\ln s}{s}\right\} &= -\,(\gamma + \ln t) \end{aligned} \tag{4.5.4}$$

from the convolution theorem we have

$$\begin{aligned} \frac{\partial}{\partial \nu}\left\{t^{\nu/2} H_\nu(z\sqrt{t})\right\} = \frac{z}{2}\int_0^t x^{(\nu-1)/2} \ln(t-x)\, H_{\nu-1}(z\sqrt{t})\, dx \\ + \frac{z\gamma}{2}\int_0^t x^{(\nu-1)/2} H_{\nu-1}(z\sqrt{t})\, dx + t^{\nu/2} \ln\left(\frac{z}{2}\right) H_\nu(z\sqrt{t}) \end{aligned} \tag{4.5.5}$$

Considering that

$$\int_0^t x^{\nu} H_\nu(x)\, dx = t^{\nu} H_\nu(t) \tag{4.5.6}$$

and introducing $t = 1$ into (4.5.2) and (4.5.5), the derivative with respect to the order of the Struve function is

$$\begin{aligned} \frac{\partial H_\nu(z)}{\partial \nu} &= \left[\ln\left(\frac{z}{2}\right) + \gamma\right] H_\nu(z) + z\int_0^1 x^{\nu} \ln(1-x^2)\, H_{\nu-1}(zx)\, dx \\ &\operatorname{Re}\nu > -\frac{1}{2} \end{aligned} \tag{4.5.7}$$

For some values of ν, the derivatives with respect to the order are known and therefore the integral in (4.5.7) can be evaluated. The behaviour of the inverse Laplace transforms in (4.5.3) for small and large values of z gives the asymptotic formulas for the considered derivatives (for $\nu = 0$ and $\nu = 1$ see [5]).

The Laplace transformation technique also permits to evaluate the following integral with respect to the order

$$I(z, t, \nu) = \int_0^{\nu} t^{\lambda/2} H_\lambda(z\sqrt{t})\, d\lambda \tag{4.5.8}$$

by using the Laplace transform in (4.5.1)

$$L\{I(z,t,\nu)\} = \frac{ie^{-z^2/4s}}{s\ \ln\left(\frac{2s}{z}\right)}\, erf\left(\frac{iz}{2\sqrt{s}}\right)\left[\left(\frac{z}{2s}\right)^{\nu} - 1\right] \tag{4.5.9}$$

The inverse of (4.5.9) can be determined by applying the convolution theorem in several ways, by using relevant transform-inverse pairs [36]

$$\begin{gathered} L^{-1}\left\{-\frac{i}{s^{\nu}}e^{-z^2/4s} erf\left(\frac{iz}{2\sqrt{s}}\right)\right\} = \left(\frac{z}{2}\right)^{1-\nu} t^{(\nu-1)/2} H_{\nu-1}(z\sqrt{t}) \\ L^{-1}\left\{-\frac{i}{\sqrt{s}}e^{-z^2/4s} erf\left(\frac{iz}{2\sqrt{s}}\right)\right\} = \frac{\sin(z\sqrt{t})}{\sqrt{\pi t}} \\ L^{-1}\left\{\frac{1}{(sa)^{\alpha+1}\ \ln(sa)}\right\} = \frac{1}{a}\,\nu\left[\left(\frac{t}{a}\right), \alpha\right] \\ \nu(t,\alpha) = \int_0^{\infty} \frac{t^{x+\alpha}}{\Gamma(x+\alpha+1)}\,dx \quad ; \quad a>0 \quad ; \quad \operatorname{Re}\alpha > -1 \end{gathered} \tag{4.5.10}$$

Thus, the desired integral can be expressed in terms of the associated Volterra functions $\nu(t,\alpha)$ [30] as

$$\begin{aligned} &\int_0^{\nu} t^{\lambda/2} H_{\lambda}(z\sqrt{t})\,d\lambda \\ &= \sqrt{\frac{z}{2\pi t}}\,\sin(z\sqrt{t}) * \left\{\nu\left(\frac{zt}{2}, -\frac{1}{2}\right) - \nu\left(\frac{zt}{2}, \nu-\frac{1}{2}\right)\right\} \end{aligned} \tag{4.5.11}$$

or

$$\begin{aligned} &\int_0^{\nu} t^{\lambda/2} H_{\lambda}(z\sqrt{t})\,d\lambda \\ &= \frac{z}{2}\,\sin(z\sqrt{t})\,\nu\left(\frac{zt}{2}\right) * \left\{\frac{2-\pi H_1(z\sqrt{t})}{\pi\sqrt{t}} - t^{(\nu-1)/2} H_{\nu-1}(z\sqrt{t})\right\} \end{aligned} \tag{4.5.12}$$

In particular case, if the upper limit of integration is equal to infinity, the explicit form of (4.5.11) is

$$\int_0^{\infty} t^{\lambda/2} H_{\lambda}(z\sqrt{t})\,d\lambda = \sqrt{\frac{z}{2\pi}} \int_0^{t} \frac{\sin(z\sqrt{t-x})}{\sqrt{t-x}}\,\nu\left(\frac{zx}{2}, -\frac{1}{2}\right) dx \tag{4.5.13}$$

For $t = 1$ and $x = 1 - w^2$, (4.5.13) becomes

$$\int_0^\infty H_\lambda(z)\, d\lambda = \sqrt{\frac{2z}{\pi}} \int_0^1 \sin(zw)\, \nu\left[\frac{z}{2}(1-w^2), -\frac{1}{2}\right] dw$$

$$= z \int_0^1 H_{-1}(zw)\, \nu\left[\frac{z}{2}(1-w^2)\right] dw \tag{4.5.14}$$

$$H_{-1}(zw) = \frac{2}{\pi} - H_1(zw)$$

The above integral can also be written as trigonometric integral by introducing $w = \cos\theta$

$$\int_0^\infty \mathrm{H}_\lambda(z)\, d\lambda = \sqrt{\frac{2z}{\pi}} \int_0^{\pi/2} \sin(z\cos\theta)\, \nu\left[\frac{z}{2}(\sin\theta)^2, -\frac{1}{2}\right] \sin\theta\, d\theta \tag{4.5.15}$$

Using the Struve function expansion (2.9.9)

$$H_\mu(z) = \sum_{k=0}^{\infty} (-1)^k \frac{\left(\frac{z}{2}\right)^{2k+\mu+1}}{\Gamma\left(k+\frac{3}{2}\right)\Gamma\left(k+\mu+\frac{3}{2}\right)} \tag{4.5.16}$$

and changing the order of integration and summation, the integrals with respect to the order in (4.5.13) and (4.5.14) can be developed in the series of the Volterra functions

$$\int_0^\infty t^{\lambda/2} H_\lambda(z\sqrt{t})\, d\lambda = \sqrt{\frac{2z}{\pi}} \sum_{k=0}^{\infty} (-1)^k \frac{z^k \nu\left[\left(\frac{zt}{2}\right), k+\frac{1}{2}\right]}{1\cdot 3\cdot 5\cdot \ldots (2k+1)} \tag{4.5.17}$$

and

$$\int_0^\infty H_\lambda(z)\, d\lambda = \sqrt{\frac{2z}{\pi}} \sum_{k=0}^{\infty} (-1)^k \frac{z^k \nu\left[\left(\frac{z}{2}\right), k+\frac{1}{2}\right]}{1\cdot 3\cdot 5\cdot \ldots (2k+1)} \tag{4.5.18}$$

Derivatives and integrals with respect to the order of the modified Struve function $L_\nu(z)$ can be determined in a similar way as demonstrated for $H_\nu(z)$, both functions are interrelated

$$L_\nu(z) = -ie^{-\pi i\nu/2} H_\nu(iz) \tag{4.5.19}$$

and therefore,

$$\frac{\partial L_\nu(z)}{\partial \nu} = -\frac{\pi i}{2} L_\nu(z) - ie^{-\pi i\nu/2} \frac{\partial H_\nu(iz)}{\partial \nu} \tag{4.5.20}$$

Using (4.5.7) we have

$$\frac{\partial L_\nu(z)}{\partial \nu} = \left[\ln\left(\frac{z}{2}\right) + \gamma\right] L_\nu(z) + z \int_0^1 x^\nu \ln(1-x^2)\, L_{\nu-1}(zx)\, dx \tag{4.5.21}$$

$$\operatorname{Re}\nu > -\frac{1}{2}$$

The recurrence relation between the modified Struve functions

$$L_{\nu-1}(z) - L_{\nu+1}(z) = \frac{2\nu}{z} L_\nu(z) + \frac{\left(\frac{z}{2}\right)^\nu}{\sqrt{\pi}\,\Gamma\left(\nu+\frac{3}{2}\right)} \tag{4.5.22}$$

gives

$$\frac{\partial L_{\nu-1}(z)}{\partial \nu} - \frac{\partial L_{\nu+1}(z)}{\partial \nu} = \frac{2\nu}{z}\frac{\partial L_\nu(z)}{\partial \nu} + \frac{2}{z} L_\nu(z) + \frac{\left(\frac{z}{2}\right)^\nu}{\sqrt{\pi}\,\Gamma\left(\nu+\frac{3}{2}\right)} \left[\ln\left(\frac{z}{2}\right) - \psi\left(\nu+\frac{3}{2}\right)\right] \tag{4.5.23}$$

The derivative with respect to the of the modified Struve function can be also presented in the following way [30]

$$\begin{aligned} \frac{\partial L_\nu(z)}{\partial \nu} &= \left[\ln\left(\frac{z}{2}\right) - \psi\left(\nu+\frac{1}{2}\right)\right] L_\nu(z) \\ &+ \frac{2\left(\frac{z}{2}\right)^\nu}{\sqrt{\pi}\,\Gamma\left(\nu+\frac{1}{2}\right)} \left[\ln\left(\frac{z}{2}\right) - \psi\left(\nu+\frac{3}{2}\right)\right] I(\nu,z) \\ I(\nu,z) &= \int_0^1 (1-x^2)^{\nu-1/2} \ln(1-x^2)\, \sinh(zx)\, dx \end{aligned} \tag{4.5.24}$$

Using (4.5.21) and

$$
\begin{aligned}
Ci(iz) &= \frac{1}{2}\,[Ei(z) - E_1(z)] + \frac{\pi i}{2} \\
Si(iz) &= \frac{i}{2}\,[Ei(z) + E_1(z)] \\
L_{-1/2}(z) &= \sqrt{\frac{2}{\pi z}}\,\sinh z \\
L_{1/2}(z) &= \sqrt{\frac{2}{\pi z}}\,(\cosh z - 1)
\end{aligned}
\tag{4.5.25}
$$

it is possible to obtain the derivative with respect to the order for $\nu = -1/2$

$$
\left(\frac{\partial L_\nu(z)}{\partial \nu}\right)_{\nu=-1/2} = \sqrt{\frac{2}{\pi z}}\left\{\frac{e^{-z}}{2}\,[Ei(2z) - 2Ei(z)] + \frac{e^{z}}{2}\,[E_1(2z) - 2E_1(z)]\right\}
\tag{4.5.26}
$$

and for $\nu = 1/2$

$$
\begin{aligned}
&\left(\frac{\partial L_\nu(z)}{\partial \nu}\right)_{\nu=1/2} \\
&= \sqrt{\frac{2}{\pi z}}\left\{\frac{e^{-z}}{2}\,[Ei(2z) - 2Ei(z)] - \frac{e^{z}}{2}\,[E_1(2z) - 2E_1(z)] - \left[\gamma + \ln\left(\frac{z}{2}\right)\right]\right\}
\end{aligned}
\tag{4.5.27}
$$

The last expression (4.5.27) permits to evaluate the integral in (4.5.24) for $\nu = 1/2$

$$
\begin{aligned}
\int_0^1 \ln(1 - x^2)\,\sinh(zx)\,dx = &-\frac{1}{z}\left[\gamma + \ln\left(\frac{z}{2}\right)\right]\,\cosh z \\
&+ \left\{\frac{e^{z}}{2}\,[E_1(2z) - 2E_1(z)] - \frac{e^{-z}}{2}\,[Ei(2z) - 2Ei(z)]\right\}
\end{aligned}
\tag{4.5.28}
$$

Starting in the same way as in (4.5.1), the order derivatives of the modified Struve function can be derived by using the following the Laplace transform [36]

$$
\begin{aligned}
&\mathcal{L}\left\{t^{\nu/2}\,L_\nu(z\sqrt{t})\right\} = -\frac{1}{s}\left(\frac{z}{2s}\right)^{\nu} e^{z^2/4s}\,\mathrm{erf}\left(\frac{z}{2\sqrt{s}}\right) \\
&\qquad \operatorname{Re}\nu > -\frac{3}{2} \quad ; \quad \operatorname{Re}s > 0
\end{aligned}
\tag{4.5.29}
$$

The analogous expression to that in (4.5.7) is

$$\frac{\partial L_\nu(z)}{\partial \nu} = \left[\ln\left(\frac{z}{2}\right) + \gamma\right] L_\nu(z) + z \int_0^1 x^\nu \ln(1-x^2)\, L_{\nu-1}(zx)\, dx$$
$$\operatorname{Re}\nu > -\frac{1}{2} \tag{4.5.30}$$

and to that in (4.5.11) is

$$I(z,t,\nu) = \int_0^\nu t^{\lambda/2} L_\lambda(z\sqrt{t})\, d\lambda$$
$$= \sqrt{\frac{z}{2\pi t}} \sinh(z\sqrt{t}) * \left\{ \nu\left(\frac{zt}{2}, -\frac{1}{2}\right) - \nu\left(\frac{zt}{2}, \nu - \frac{1}{2}\right) \right\} \tag{4.5.31}$$

If the upper limit of integration is equal to infinity and $t = 1$, the explicit form of (4.5.31) is

$$\int_0^\infty L_\lambda(z)\, d\lambda = \sqrt{\frac{2z}{\pi}} \int_0^1 \sinh(zw)\, \nu\left[\frac{z}{2}(1-w^2), -\frac{1}{2}\right] dw \tag{4.5.32}$$

Integrals with respect to the order in (4.5.31) can be developed in the series of the associated Volterra functions from the modified Struve function expansion (2.9.10)

$$L_\mu(z) = \sum_{k=0}^\infty \frac{\left(\frac{z}{2}\right)^{2k+\mu+1}}{\Gamma\left(k+\frac{3}{2}\right)\Gamma\left(k+\mu+\frac{3}{2}\right)} \tag{4.5.33}$$

The inversion of the order of integration and summation in (4.5.33) gives

$$\int_0^\infty t^{\lambda/2} L_\lambda(z\sqrt{t})\, d\lambda = \sqrt{\frac{2z}{\pi}} \sum_{k=0}^\infty \frac{z^k \nu\left[\left(\frac{zt}{2}\right), k+\frac{1}{2}\right]}{1\cdot 3\cdot 5\cdot \ldots (2k+1)} \tag{4.5.34}$$

and

$$\int_0^\infty L_\lambda(z)\, d\lambda = \sqrt{\frac{2z}{\pi}} \sum_{k=0}^\infty \frac{z^k \nu\left[\left(\frac{z}{2}\right), k+\frac{1}{2}\right]}{1\cdot 3\cdot 5\cdot \ldots (2k+1)} \tag{4.5.35}$$

The asymptotic behaviour of the order derivatives or integrals of the modified Struve function as a function of argument z is discussed for a number of particular values of ν in [5]. The Laplace transform technique, which is used to obtain derivatives and integrals with respect to the order of the Struve functions (see (4.5.1) and (4.5.30)) can easily be extended to the Bessel functions because their Laplace transforms are similar, even in simpler form [36]

$$L\left\{t^{\nu/2} J_\nu(z\sqrt{t})\right\} = \frac{1}{s}\left(\frac{z}{2s}\right)^{\nu} e^{-z^2/4s} \quad ; \quad \operatorname{Re} s > 0 \tag{4.5.36}$$

$$L\left\{t^{\nu/2} I_\nu(z\sqrt{t})\right\} = \frac{1}{s}\left(\frac{z}{2s}\right)^{\nu} e^{z^2/4s} \quad ; \quad \operatorname{Re} s > 0 \tag{4.5.37}$$

Differentiation of (4.5.36) and (4.5.37) with respect to the order gives

$$\begin{aligned} \frac{\partial}{\partial \nu} L\left\{t^{\nu/2} J_\nu(z\sqrt{t})\right\} &= \frac{1}{s}\left(\frac{z}{2s}\right)^{\nu} e^{-z^2/4s} \ln\left(\frac{z}{2s}\right) \\ &= \left\{\frac{1}{s}\left(\frac{z}{2s}\right)^{\nu-1} e^{-z^2/4s}\right\}\left\{-\left(\frac{z}{2s}\right) \ln\left(\frac{2s}{z}\right)\right\} \end{aligned} \tag{4.5.38}$$

and

$$\begin{aligned} \frac{\partial}{\partial \nu} L\left\{t^{\nu/2} I_\nu(z\sqrt{t})\right\} &= \frac{1}{s}\left(\frac{z}{2s}\right)^{\nu} e^{z^2/4s} \ln\left(\frac{z}{2s}\right) \\ &= \left\{\frac{1}{s}\left(\frac{z}{2s}\right)^{\nu-1} e^{z^2/4s}\right\}\left\{-\left(\frac{z}{2s}\right) \ln\left(\frac{2s}{z}\right)\right\} \end{aligned} \tag{4.5.39}$$

Using (4.5.7), (4.5.36) and (4.5.37), derivatives with respect to the order of the Bessel functions are given in terms of the following convolution integrals

$$\frac{\partial}{\partial \nu}\left\{t^{\nu/2} J_\nu(z\sqrt{t})\right\} = \frac{z}{2}\, t^{(\nu-1)/2} J_{\nu-1}(z\sqrt{t}) * \left[\gamma + \ln\left(\frac{zt}{2}\right)\right] \tag{4.5.40}$$

$$\frac{\partial}{\partial \nu}\left\{t^{\nu/2} I_\nu(z\sqrt{t})\right\} = \frac{z}{2}\, t^{(\nu-1)/2} I_{\nu-1}(z\sqrt{t}) * \left[\gamma + \ln\left(\frac{zt}{2}\right)\right] \tag{4.5.41}$$

If the same procedure is extended to integrals with respect to the order of the Bessel functions, then we have

$$L\left\{\int_0^{\nu} t^{\lambda/2} J_\lambda(z\sqrt{t})\, d\lambda\right\} = \frac{1}{s \ln\left(\frac{z}{2s}\right)}\, e^{-z^2/4s}\left[\left(\frac{z}{2s}\right)^{\nu} - 1\right] \tag{4.5.42}$$

$$L\left\{\int_0^{\nu} t^{\lambda/2} I_\lambda(z\sqrt{t})\, d\lambda\right\} = \frac{1}{s \ln\left(\frac{z}{2s}\right)}\, e^{z^2/4s}\left[\left(\frac{z}{2s}\right)^{\nu} - 1\right] \tag{4.5.43}$$

and taking into account (4.5.10), the logarithmic term will lead to the convolution integrals including the Volterra and Bessel functions.

$$\int_0^{\nu} t^{\lambda/2} J_\lambda(z\sqrt{t})\, d\lambda = \sqrt{\frac{z}{2t}} J_{-1/2}(z\sqrt{t}) * \nu\left(\frac{zt}{2}, -\frac{1}{2}\right) - \frac{z}{2} t^{(\nu-1)/2} J_{\nu-1}(z\sqrt{t}) * \nu\left(\frac{zt}{2}\right) \tag{4.5.44}$$

$$\int_0^{\nu} t^{\lambda/2} I_\lambda(z\sqrt{t})\, d\lambda = \sqrt{\frac{z}{2t}} I_{-1/2}(z\sqrt{t}) * \nu\left(\frac{zt}{2}, -\frac{1}{2}\right) - \frac{z}{2} t^{(\nu-1)/2} I_{\nu-1}(z\sqrt{t}) * \nu\left(\frac{zt}{2}\right) \tag{4.5.45}$$

Thus, expressions of derivatives and integrals with respect to the order in terms of convolution integrals are based on the product theorem and manipulations with Laplace transforms of the Bessel and related functions. These convolution integrals can be presented in different forms because they depend on how Laplace transforms are arranged in the products for inversion processes.

4.6 Differentiation with Respect to the Order of the Kelvin Functions by Applying the Laplace Transform Approach

Functions introduced by Kelvin, $\mathrm{ber}_\nu(z)$ and $\mathrm{bei}_\nu(z)$, represent the real part of the Bessel function $J_\nu(z)$ of order ν and of argument $i^{3/2}z$ and the imaginary part of the Bessel function $J_\nu(z)$ of order ν and of argument $i^{-3/2}z$ respectively. They can be treated by using the corresponding formulas which are applied to the Bessel function of first kind. However, the Kelvin functions are usually investigated separately, because they occur in a special type of mathematical problems arising in the theory of electrical current, fluid mechanics and elasticity. Similarity of the Kelvin functions with the Bessel functions is evident when operations with transforms and inverses in the Laplace transformation are performed. The Laplace transform of the Kelvin function of the first kind is [36]

$$L\left\{t^{\nu/2}\,\mathrm{ber}_\nu(z\sqrt{t})\right\} = \frac{1}{s}\left(\frac{z}{2s}\right)^{\nu} \cos\left(\frac{z^2}{4s} + \frac{3\pi\nu}{4}\right) \tag{4.6.1}$$

$$\mathrm{Re}\,\nu > -1 \quad ; \quad \mathrm{Re}\,s > 0$$

and its derivative with respect to the order is

$$L\left\{\frac{\partial\left[t^{\nu/2}ber_\nu(z\sqrt{t})\right]}{\partial\nu}\right\}$$

$$=L\left\{t^{\nu/2}\frac{\partial\, ber_\nu(z\sqrt{t})}{\partial\nu}+t^{\nu/2}\ln(\sqrt{t})\,ber_\nu(z\sqrt{t})\right\} \tag{4.6.2}$$

$$=\frac{1}{s}\left(\frac{z}{2s}\right)^\nu\ln\left(\frac{z}{2s}\right)\cos\left(\frac{z^2}{4s}+\frac{3\pi\nu}{4}\right)-\frac{3\pi}{4s}\left(\frac{z}{2s}\right)^\nu\sin\left(\frac{z^2}{4s}+\frac{3\pi\nu}{4}\right)$$

The inverse of the second term in (4.6.2) is known

$$L^{-1}\left\{\frac{3\pi}{4s}\left(\frac{z}{2s}\right)^\nu\sin\left(\frac{z^2}{4s}+\frac{3\pi\nu}{4}\right)\right\}=\frac{3\pi}{4}t^{\nu/2}bei_\nu(z\sqrt{t}) \tag{4.6.3}$$

and taking into account (4.6.1) and (4.5.4), the final result which is based on the convolution theorem is

$$\frac{\partial\left[t^{\nu/2}ber_\nu(z\sqrt{t})\right]}{\partial\nu}=-\frac{3\pi}{4}t^{\nu/2}\,bei_\nu(z\sqrt{t})+t^{\nu/2}\ln\left(\frac{z}{2}\right)\,ber_\nu(z\sqrt{t})$$
$$-\frac{\sqrt{2}z}{4}\int_0^t x^{(\nu-1)/2}[\ln(t-x)+\gamma]\left\{ber_{\nu-1}(z\sqrt{t})+bei_{\nu-1}(z\sqrt{t})\right\}dx$$
$$\operatorname{Re}\nu>0 \tag{4.6.4}$$

Introducing $t=1$ and changing value of parameter z, the expression for the derivative with respect to the order of the Kelvin function of the first kind becomes

$$\frac{\partial\, ber_\nu(z\sqrt{2})}{\partial\nu}=-\frac{3\pi}{4}\,bei_\nu(z\sqrt{2})+\ln\left(\frac{z}{\sqrt{2}}\right)\,ber_\nu(z\sqrt{2})$$
$$-\frac{z}{2}\int_0^1 x^{(\nu-1)/2}[\ln(1-x)+\gamma]\left\{ber_{\nu-1}(z\sqrt{2x})+bei_{\nu-1}(z\sqrt{2x})\right\}dx$$
$$\operatorname{Re}\nu>0 \tag{4.6.5}$$

In the case of $\nu=0$, the Laplace transform should be treated in a different way to obtain inversion of the product

$$\frac{\ln s}{s}\cos\left(\frac{z^2}{4s}\right)=\frac{\ln s}{\sqrt{s}}\cdot\frac{1}{\sqrt{s}}\cos\left(\frac{z^2}{4s}\right) \tag{4.6.6}$$

From

$$L^{-1}\left\{\frac{\ln s}{\sqrt{s}}\right\} = -\frac{1}{\sqrt{\pi t}}\left(\ln t + \gamma + 2\ln 2\right)$$
$$L^{-1}\left\{\frac{1}{\sqrt{s}}\cos\left(\frac{z^2}{4s}\right)\right\} = \frac{1}{\sqrt{\pi t}}\cos\left(z\sqrt{\frac{t}{2}}\right)\cosh\left(z\sqrt{\frac{t}{2}}\right) \tag{4.6.7}$$

the final result after few more steps is

$$\left(\frac{\partial\, ber_\nu(\sqrt{2}z)}{\partial\nu}\right)_{\nu=0} = -\frac{3\pi}{4}\, bei(z\sqrt{2}) + \ln\left(\frac{z}{\sqrt{2}}\right) ber(z\sqrt{2})$$
$$+\frac{1}{\pi}\int_0^1 \frac{[\ln(1-x)+\gamma+2\ln 2]}{\sqrt{x(1-x)}}\cos(z\sqrt{x})\cosh(z\sqrt{x})\,dx \tag{4.6.8}$$

Changing the integration variable in (4.6.8), it is possible to obtain the convolution integral as the finite trigonometric integral

$$\left(\frac{\partial\, ber_\nu(\sqrt{2}z)}{\partial\nu}\right)_{\nu=0} = -\frac{3\pi}{4}\, bei(z\sqrt{2}) + \ln\left(\frac{z}{\sqrt{2}}\right) ber(z\sqrt{2})$$
$$+\frac{4}{\pi}\int_0^{\pi/2} \ln(2C\sin\theta)\cos(z\sqrt{2}\cos\theta)\cosh(z\sqrt{2}\cos\theta)\,d\theta$$
$$\gamma = 2\ln C \quad ; \quad C = 1.3346682\ldots \tag{4.6.9}$$

Following the same procedure, the equivalent expression for the Kelvin function of the second kind is

$$\frac{\partial\, bei_\nu(z\sqrt{2})}{\partial\nu} = \frac{3\pi}{4}\, ber_\nu(z\sqrt{2}) + \ln\left(\frac{z}{\sqrt{2}}\right) bei_\nu(z\sqrt{2})$$
$$-\frac{z}{2}\int_0^1 x^{(\nu-1)/2}[\ln(1-x)+\gamma]\left\{ber_{\nu-1}(z\sqrt{2x}) - bei_{\nu-1}(z\sqrt{2x})\right\}dx$$
$$\operatorname{Re}\nu > 0 \tag{4.6.10}$$

and for $\nu = 0$ we have

$$\left(\frac{\partial\, bei_\nu(\sqrt{2}z)}{\partial\nu}\right)_{\nu=0} = \frac{3\pi}{4}\, ber(z\sqrt{2}) + \ln\left(\frac{z}{\sqrt{2}}\right) bei(z\sqrt{2})$$
$$+\frac{1}{\pi}\int_0^1 \frac{[\ln(1-x)+\gamma+2\ln 2]}{\sqrt{x(1-x)}}\sin(z\sqrt{x})\sinh(z\sqrt{x})\,dx \tag{4.6.11}$$

or in the form of the trigonometric integral

$$\left(\frac{\partial\, bei_\nu(\sqrt{2}z)}{\partial\nu}\right)_{\nu=0} = \frac{3\pi}{4}\, ber(z\sqrt{2}) + \ln\left(\frac{z}{\sqrt{2}}\right) bei(z\sqrt{2})$$
$$+\frac{4}{\pi}\int_0^{\pi/2} \ln(2C\sin\theta)\,\sin(z\sqrt{2}\cos\theta)\,\sinh(z\sqrt{2}\cos\theta)\,d\theta$$
$$\gamma = 2\ln C \quad ; \quad C = 1.3346682... \tag{4.6.12}$$

4.7 Differentiation with Respect to the Order of the Integral Bessel Functions by Applying the Laplace Transform Approach

The integral Bessel function of the first kind is defined in (2.1.68) as

$$Ji_\nu(t) = -\int_t^\infty \frac{J_\nu(x)}{x}\,dx \tag{4.7.1}$$

and the Laplace transform of it resembles that of the Bessel function

$$L\{Ji_\nu(t)\} = \frac{1}{\nu s}\left\{\frac{1}{\left(s+\sqrt{s^2+1}\right)^\nu} - 1\right\}$$
$$\mathrm{Re}\,\nu > 0 \tag{4.7.2}$$

If the complex inversion formula is applied to (4.7.2)

$$Ji_\nu(t) = \frac{1}{2\pi i}\int_{c-i\infty}^{c+i\infty} \frac{e^{st}}{s}\left[\frac{1}{\left(s+\sqrt{s^2+1}\right)^\nu} - 1\right] ds$$
$$\mathrm{Re}\,s > c \tag{4.7.3}$$

the integral representation of the integral Bessel function is derived by using the Bromwich contour plotted in Figure 4.1. The result of complex integration in (4.7.3) is [4]

$$Ji_\nu(t) = -\frac{1}{2\nu} - \frac{1}{\pi\nu}\int_0^\pi \cot x\,\sin(t\sin x - \nu x)\,dx$$
$$+\frac{\sin(\pi\nu)}{\pi\nu}\int_0^\infty e^{-t\sinh x - \nu x}\coth x\,dx \tag{4.7.4}$$

The equivalent form of integral representation of $Ji_\nu(t)$ can be derived starting with the Schlaefli representation of $J_\nu(t)$

$$J_\nu(t) = \frac{1}{\pi}\int_0^{\pi} \cos(t\sin x - \nu x)\,dx - \frac{\sin(\pi\nu)}{\pi}\int_0^{\infty} e^{-t\sinh x - \nu x}\,dx \qquad \mathrm{Re}\,\nu > -1 \tag{4.7.5}$$

The integral Bessel function is reached by an integration of (4.7.5)

$$\begin{aligned} Ji_\nu(t) = &-\int_t^{\infty} \frac{J_\nu(u)}{u}\,du = -\frac{1}{\pi}\int_t^{\infty}\left[\int_0^{\pi} \cos(u\sin x - \nu x)\,dx\right]\frac{du}{u} \\ &+ \frac{\sin(\pi\nu)}{\pi}\int_t^{\infty}\left[\int_0^{\infty} e^{-u\sinh x - \nu x}\,dx\right]\frac{du}{u} \end{aligned} \tag{4.7.6}$$

Changing the order of integration in (4.7.6) and taking into account that

$$\begin{aligned} si(t) &= -\int_t^{\infty} \frac{\sin x}{x}\,dx \\ Ci(t) &= -\int_t^{\infty} \frac{\cos x}{x}\,dx \\ E_1(t) &= \int_t^{\infty} \frac{e^{-x}}{x}\,dx \end{aligned} \tag{4.7.7}$$

we have

$$\begin{aligned} Ji_\nu(t) = &\frac{1-\cos(\pi\nu)}{2\nu} + \frac{\sin(\pi\nu)}{\pi}\int_0^{\infty} e^{-\nu x}E_1(t\sinh x)\,dx \\ &+ \frac{1}{\pi}\int_0^{\pi}[\cos(\nu x)\,Ci(t\sin x) + \sin(\nu x)\,si(t\sin x)]\,dx \end{aligned} \tag{4.7.8}$$

For $\nu = 0$, the expected result is obtained by applying l'Hôpital's rule

$$Ji_0(t) = \frac{1}{\pi}\int_0^{\pi} Ci(t\sin x)\,dx \tag{4.7.9}$$

If the order ν is odd or even positive integer, the expression (4.7.8) can be represented by

$$Ji_{2n+1}(t) = \frac{1}{2n+1} + \frac{1}{\pi}\int_0^{\pi} \sin[(2n+1)x]\,si(t\sin x)\,dx \tag{4.7.10}$$

because

$$\int_0^{\pi} \cos[(2n+1)x]\, Ci(t\sin x)\, dx = 0 \qquad (4.7.11)$$
$$n = 0, 1, 2, 3, \ldots$$

and

$$Ji_{2n}(t) = \frac{1}{\pi}\int_0^{\pi} \sin(2nx)\, Ci(t\sin x)\, dx \qquad (4.7.12)$$

because

$$\int_0^{\pi} \sin(2nx)\, si(t\sin x)\, dx = 0 \qquad (4.7.13)$$
$$n = 1, 2, 3, \ldots$$

These results can be written differently, if odd and even values of ν are introduced into (4.7.4)

$$Ji_{2n+1}(t) = -\frac{1}{2(2n+1)} - \frac{1}{(2n+1)\pi}\int_0^{\pi} \cot x\, \cos[(2n+1)x]\, \sin(t\sin x)\, dx \qquad (4.7.14)$$

because

$$\int_0^{\pi} \cot x\, \sin[(2n+1)x]\, \cos(t\sin x)\, dx = 0 \qquad (4.7.15)$$
$$n = 0, 1, 2, 3, \ldots$$

and

$$Ji_{2n}(t) = -\frac{1}{4n} + \frac{1}{2n\pi}\int_0^{\pi} \cot x\, \sin(2nx)\, \cos(t\sin x)\, dx \qquad (4.7.16)$$

because

$$\int_0^{\pi} \cot x\, \cos(2nx)\, \sin(t\sin x)\, dx = 0 \qquad (4.7.17)$$
$$n = 1, 2, 3, \ldots$$

Equalities of involved integrals can be established if (4.7.10) is compared with (4.7.14) and (4.7.12) with (4.7.16).

Direct differentiation with respect to the order of the integral representation in (4.7.8) gives

$$\frac{\partial Ji_\nu(t)}{\partial\nu} = \frac{1-\cos(\pi\nu)}{2\nu^2} - \frac{\pi\sin(\pi\nu)}{2\nu} + \frac{\sin(\pi\nu)}{\pi}\int_0^\infty e^{-\nu x}E_1(t\sinh x)\,[x\sin(\pi\nu)-\pi\cos(\pi\nu)]\,dx + \frac{1}{\pi}\int_0^\pi x\,[\sin(\nu x)\,Ci(t\sin x) - \cos(\nu x)\,si(t\sin x)]\,dx \tag{4.7.18}$$

For $\nu = 0$, by applying l'Hôpital's rule we have

$$-\left(\frac{\partial Ji_\nu(t)}{\partial\nu}\right)_{\nu=0} = \frac{\pi^2}{4} + \int_0^\infty E_1(t\sinh x)\,dx + \frac{1}{\pi}\int_0^\pi x\,si(t\sin x)]\,dx \tag{4.7.19}$$

The first derivative with respect to the order the integral Bessel function can be derived also by using the Laplace transform method. From (4.7.2) it follows that

$$L\left\{\frac{\partial Ji_\nu(t)}{\partial\nu}\right\} = \left\{\frac{1-[s+\sqrt{s^2+1}]^{-\nu}}{\nu^2 s} - \frac{\ln[s+\sqrt{s^2+1}]}{\nu s\,[s+\sqrt{s^2+1}]^\nu}\right\} \qquad \operatorname{Re}\nu>0 \tag{4.7.20}$$

and using (4.7.3) it is possible to present the inverse of (4.7.20) as

$$\frac{\partial Ji_\nu(t)}{\partial\nu} = \frac{Ji_\nu(t)}{\nu} - \frac{1}{2\pi i\nu}\int_{c-i\infty}^{c+i\infty}\frac{e^{st}}{s}\,\frac{\ln(s+\sqrt{s^2+1})}{(s+\sqrt{s^2+1})^\nu}\,ds \qquad \operatorname{Re}s>c\ ;\ \operatorname{Re}\nu>0 \tag{4.7.21}$$

Solving the complex integral with the help of Bromwich contour (Figure 4.1), the final result of the inverse is [4]

$$\frac{\partial Ji_\nu(t)}{\partial\nu} = \frac{Ji_\nu(t)}{\nu} + \frac{1}{\pi\nu}\int_0^\pi x\cot x\,\cos(t\sin x-\nu x)\,dx - \frac{1}{\pi\nu}\int_0^\infty e^{-t\sinh x-\nu x}\coth x\,[x\sin(\pi\nu)-\pi\cos(\pi\nu)]\,dx \qquad \operatorname{Re}\nu>0 \tag{4.7.22}$$

This integral representation can be present in a more compact form taking into account the inverse Laplace transforms [36]

$$L^{-1}\left\{\frac{\ln(s+\sqrt{s^2+1})}{s}\right\}=Ji_0(t)$$
$$L^{-1}\left\{\frac{1}{(s+\sqrt{s^2+1})}\right\}=\frac{\nu J_\nu(t)}{t} \tag{4.7.23}$$

which permit to express the integral in (4.7.21) as the convolution integral

$$\frac{\partial Ji_\nu(t)}{\partial \nu}=\frac{Ji_\nu(t)}{\nu}-\int_0^t Ji_0(t-x)\,J_\nu(x)\,\frac{dx}{x} \qquad \operatorname{Re}\nu>0 \tag{4.7.24}$$

or by changing the variable of integration we have the finite trigonometric integral

$$\frac{\partial Ji_\nu(t)}{\partial \nu}=\frac{Ji_\nu(t)}{\nu}-2\int_0^{\pi/2}\tan\theta\, Ji_0[t\,(\sin\theta)^2]\,J_\nu[t\,(\sin\theta)^2]\,d\theta \qquad \operatorname{Re}\nu>0 \tag{4.7.25}$$

4.8 Zeros of the Bessel and Related Functions As a Function of Their Order

In two investigations performed in the period 1918–1919, Watson [77] examined transmission of electrical waves around the Earth. For the first time, as a solution of this physical problem he found the necessity to determine zeros (roots) of functional expressions (algebraic sums of products) when the order of the Bessel and Hankel functions is a variable. In 1936, as the next step, Coulomb [47] postulated that in the case of the Bessel function of the first kind, if argument z is real and positive, its zeros are real, simple and asymptotically close to the negative integers for large nth zeros. About two, three decades later, in various physical and technological studies (quantum-mechanical potential scattering and in numerous diffraction phenomena), similar transcendental functions were encountered. The unknown quantities to be determined were series of orders $\nu_1(z)$, $\nu_2(z)$, $\nu_3(z)$, $\nu_4(z)$ ... which were associated with prescribed values of arguments z. Values of these orders were physically identified as poles of scattering amplitudes resulting from interaction of various kinds of waves with spheres and cylinders. Thus, various physical or engineering problems, in different geometric situations, were linked with determination of zeros of the Bessel and related functions as a function of their order. And therefore, this topic found some kind of importance in the theory of electromagnetic waves and also in the solution of non-self-adjoint boundary value problems.

The most significant results from this subject are reported briefly here by using a uniform notation. They are presented in the final form, without giving derivations and proofs. In the mathematical literature [61, 64–68, 90–95], three equations, with an increasing complexity, were solved. The first from these equations is

$$\begin{aligned} &J_\nu(z) = 0 \\ &J_{\nu_n}(z) = 0 \quad , \quad n = 1, 2, 3, \ldots \end{aligned} \tag{4.8.1}$$

where the argument z is fixed.

Using the asymptotic expansions of the Bessel function for particular values of argument z and order ν, Cohen [64] solved (4.8.1) and he found that for $0 < | z | \ll 1$, and/or $| \nu | \gg | z | + 1$

$$\nu_n(z) = -n + \frac{\left(\frac{z}{2}\right)^{2n}}{n!(n-1)!} - \frac{2n\left(\frac{z}{2}\right)^{2n+2}}{(n-1)(n-1)!(n+1)!} + O\left(\frac{z^{2n+4}e^{2n}}{n^{2n+1}}\right) \tag{4.8.2}$$

and this result is consistent with the Coulomb prediction [47]. If order and argument have similar value $z \sim \nu$, but $| z - \nu |$ is large, then

$$\begin{aligned} &\nu_n(z) = z - \frac{q_n z^{1/3}}{6^{1/3}} - \frac{3 q_n^2}{10\, 6^{2/3} z^{1/3}} + O(z^{-1}) \\ &Ai(q_n) = 0 \end{aligned} \tag{4.8.3}$$

where q_n are zeros of the Airy function and they are tabulated in [9].

If order is considerably smaller than argument $| z | \gg | \nu |$ and $| \arg(z) | < \pi$, the nth zero of (4.8.1) is

$$\begin{aligned} &\nu_n(z) = \nu_0(z) + \frac{[4\nu_0(z)^2 - 1]}{4\pi z} + \frac{[4\nu_0(z)^2 - 1][4\nu_0(z)^2 - 25]}{192\pi z^3} + O(z^{-5}) \\ &\nu_0(z) = -2n - \frac{1}{2} + \frac{2z}{\pi} \quad ; \quad n \approx \frac{z}{\pi} \end{aligned} \tag{4.8.4}$$

In the next equation to solve, the derivative of the Bessel function with regard to the argument z is considered

$$\frac{dJ_\nu(z)}{dz} = 0 \tag{4.8.5}$$

Cohen found that solutions of (4.8.5) are similar to those derived for the solution in (4.8.1). For $0 < | z | \ll 1$, and /or $| \nu | \gg | z | + 1$, his result was

$$v_n(z) = -n - \frac{\left(\frac{z}{2}\right)^{2n}}{n!(n-1)!} + \frac{2(n^2-2)\left(\frac{z}{2}\right)^{2n+2}}{(n^2-1)(n!)^2} + O\left(\frac{z^{2n+4}e^{2n}}{n^{2n+1}}\right) \tag{4.8.6}$$

In the case of $z \sim v$, $|z - v|$ is large, he found that

$$v_n(z) = z - \frac{q'_n z^{1/3}}{6^{1/3}} - \frac{\left[3(q'_n)^2 - \frac{2}{q'_n}\right]}{10\, 6^{2/3} z^{1/3}} + O(z^{-1}) \tag{4.8.7}$$
$$Ai'(q'_n) = 0$$

where the zeros of derivatives of the Airy function are considered to be known. For $|z| \gg |v|$ and $|\arg(z)| < \pi$, the expression is

$$v_n(z) = v_0(z) + \frac{4v_0(z)^2 + 3}{4\pi z} + \frac{16 v_0(z)^4 + 181 v_0(z)^2 - 63}{192\pi z^3} + O(z^{-5})$$
$$v_0(z) = -2n - \frac{1}{2} + \frac{2z}{\pi} \quad ; \quad n \approx \frac{z}{\pi} \tag{4.8.8}$$

In the third equation to solve, the Bessel function and its derivative with regard to argument z appear in the form

$$\frac{dJ_v(z)}{dz} + \alpha J_v(z) = 0 \tag{4.8.9}$$

where α is given real or complex constant. This type of equation frequently emerges in a solution of boundary value problems.

Cohen obtained, for this case that if $0 < |z| \ll 1$ and/or $|v| \gg |z| + 1$, the following expression for the zeros of equation (4.8.9)

$$v_n(z) = -n - \frac{\left(\frac{z}{2}\right)^{2n}}{n!(n-1)!} - \frac{4\alpha\left(\frac{z}{2}\right)^{2n+2}}{(n!)^2} + O\left(\frac{z^{2n+4}e^{2n}}{n^{2n+1}}\right)$$
$$|v| > |\alpha| \tag{4.8.10}$$

If $z \sim v$ and $|z - v|$ is large, the solution is represented by

$$v_n(z) = z - \frac{q'_n z^{1/3}}{6^{1/3}} - \frac{\left[3(q'_n)^2 - \frac{2}{q'_n}\right]}{10\, 6^{2/3} z^{1/3}} + O(z^{-1})$$
$$Ai'(q'_n) + \alpha\left(\frac{z}{6}\right)^{1/3} Ai(q'_n) = 0 \tag{4.8.11}$$

In the case $|z| \gg |v|$ and $|\arg(z)| < \pi$, Cohen found two expressions valid for small and large values of parameter α

$$
\begin{aligned}
\nu_n(z) = \nu_0(z) - &\frac{(\alpha^2-1)\,[4\,\nu_0(z)^2+3]+\alpha^2(\alpha^2-1)\,[4\,\nu_0(z)^2-1]}{4\pi z} \\
&-\frac{3\alpha(1-\alpha^2)[4\,\nu_0(z)^2-1][4\,\nu_0(z)^2+15]}{96\pi z^2} - \\
&-\frac{2\alpha\{[4\,\nu_0(z)^2+3]+\alpha^2[4\,\nu_0(z)^2-1]\}^2}{96\pi z^2} + O(z^{-3}) \\
\nu_0(z) = -2n - \frac{1}{2} &+ \frac{2z}{\pi} - \frac{2\alpha\left(1-\frac{\alpha^2}{3}\right)}{\pi} \quad ; \quad n \approx \frac{z}{\pi} \quad ; \quad |\alpha| \ll 1
\end{aligned}
\tag{4.8.12}
$$

and

$$
\begin{aligned}
\nu_n(z) = \nu_0(z) - &\frac{(1-3\alpha^2)\,\{[4\,\nu_0(z)^2+3]+\alpha^2\,[4\,\nu_0(z)^2-1]\}}{12\pi z\alpha^4} \\
&-\frac{\alpha^2(1+3\alpha^2)[4\,\nu_0(z)^2-1][4\,\nu_0(z)^2+15]}{96\pi z^2\alpha^5} - \\
&-\frac{[4\,\nu_0(z)^2+3]+\alpha^2[4\,\nu_0(z)^2-1]}{96\pi z^2\alpha^5} + O(z^{-3}) \\
\nu_0(z) = -2n + \frac{1}{2} &+ \frac{2z}{\pi} - \frac{2\left(\frac{1}{\alpha}-\frac{1}{3\alpha^3}\right)}{\pi} \quad ; \quad n \approx \frac{z}{\pi} \quad ; \quad |\alpha| \gg 1
\end{aligned}
\tag{4.8.13}
$$

Starting from the mid-fifties of the previous century, a number of investigations were directed to finding zeros of the Hankel functions, the derivatives with respect to the argument z and the linear combination of them all as a function of the order [91–95]. The simplest equations to solve are

$$
\begin{aligned}
&H_\nu^{(1)}(z) = 0 \quad ; \quad H_\nu^{(2)}(z) = 0 \\
&H_{\nu_n}^{(1)}(z) = 0 \quad ; \quad n = 1, 2, 3, \ldots
\end{aligned}
\tag{4.8.14}
$$

However, in this case, Magnus and Kotin [68] observed that only one Hankel function should be taken into account, because for a real argument, zeros of these functions are conjugates and the Hankel functions are interrelated in the following way

$$
\begin{aligned}
&\overline{H_{\nu^*}^{(1)}(z)} = H_\nu^{(2)}(z^*) \\
&\overline{H_\nu^{(1)}(x)} = H_{\nu^*}^{(2)}(x) \quad ; \quad x > 0 \\
&\overline{H_{-\nu^*}^{(1)}(x)} = -H_\nu^{(1)}(e^{\pi i}x) \\
&H_{-\nu}^{(1)}(z) = e^{\pi\nu i} H_\nu^{(1)}(z) \\
&z = x + iy \quad ; \quad z^* = x - iy \quad ; \quad \nu = \alpha + i\beta \quad ; \quad \nu^* = \alpha - i\beta
\end{aligned}
\tag{4.8.15}
$$

where a bar denotes the complex conjugate of the indicated Hankel function.

The Magnus and Kotin paper is important because it includes detailed analysis about the number of zeros and their location as a function of complex and real arguments. Magnus and Kotin evaluated only large zeros which are practically important. They demonstrated that if $0 < \arg\nu < \pi$, and ν is sufficiently large, the real and imaginary parts of the large zeros of the Hankel function of the first kind are

$$\begin{aligned}
\alpha_n &= \pi\left(\frac{\pi}{2}-\theta\right)\left(n-\frac{3}{4}\right)\left|\ln\left[\frac{\pi(4n-1)}{2er}\right]\right|^{-2}(1+\varepsilon_n)\\
\beta_n &= \pi\left(n-\frac{1}{4}\right)\left|\ln\left[\frac{\pi(4n-1)}{2er}\right]\right|^{-1}(1+\delta_n)\\
\varepsilon_n &= O\left|\frac{\ln(\ln n)}{\ln n}\right| \quad ; \quad \delta_n \sim \varepsilon_n \quad ; \quad n \gg 1
\end{aligned} \tag{4.8.16}$$

In this expression, some misprints which were mentioned in [66] are corrected. Magnus and Kotin [68] found that for real, positive large argument x, a much simpler relation exists

$$\nu_n(x) \sim x + \frac{x^{1/3}}{2}\left[3\pi\left(n+\frac{3}{4}\right)\right]^{2/3} e^{\pi i/3} \quad ; \quad x \gg 1 \tag{4.8.17}$$

Following the Cohen method which was used in determination of zeros of the Bessel function of the first kind, Keller, Rubinow and Goldstein [66] derived corresponding expressions for specific ranges of the order and argument. For $0 < |z| \ll 1$ or $|\nu| \gg 1 + |z|^2$, their result for zeros of the equation (4.8.14) is

$$\begin{aligned}
\nu_n(z) &= -\frac{\pi n i}{\ln\left(\frac{r}{2}\right)} F(r,\theta)\\
F(r,\theta) &= 1 + \frac{\left[i\left(\frac{\pi}{2}-\theta\right)-\gamma\right]}{\ln\left(\frac{r}{2}\right)} + \frac{\left[i\left(\frac{\pi}{2}-\theta\right)-\gamma\right]^2}{\left[\ln\left(\frac{r}{2}\right)\right]^2}\\
&\quad + \frac{\left[i\left(\frac{\pi}{2}-\theta\right)-\gamma\right]^3 - \frac{\zeta(3)\pi^2 n^2}{3}}{\left[\ln\left(\frac{r}{2}\right)\right]^3} + O\left(\frac{z^2}{\ln(z/2)}\right)\\
z &= re^{i\theta} \quad ; \quad |z| \ll 1
\end{aligned} \tag{4.8.18}$$

In the case of z fixed and ν large, the expression (4.8.16) slightly differs by a numerical coefficient from that given below

$$\alpha_n = \pi\left(\frac{\pi}{2} - \theta\right)\left(n - \frac{1}{4}\right)\left|\ln\left[\frac{\pi(4n-1)}{2er}\right]\right|^{-2}\left[1 + O\left(\frac{\ln(\ln n)}{\ln n}\right)\right]$$
$$\beta_n = \pi\left(n - \frac{1}{4}\right)\left|\ln\left[\frac{\pi(4n-1)}{2er}\right]\right|^{-1}\left[1 + O\left(\frac{\ln(\ln n)}{\ln n}\right)\right] \quad ; \quad n \gg 1 \tag{4.8.19}$$

If the absolute values of the argument and order are large, Keller, Rubinow and Goldstein [66] quoted the Franz [79] result

$$\nu_n(z) = z + \frac{q_n z^{1/3} e^{\pi i/3}}{6^{1/3}} + \frac{6^{1/3}(q_n)^2 e^{2\pi i/3}}{180 z^{1/3}} + O\left(z^{-1}\right) \quad ; \quad |z| \gg n > 0$$
$$Ai(q_n) = 0 \tag{4.8.20}$$

Zeros of the derivatives with respect to the argument of the Hankel function of the first kind are symmetric about the origin, and therefore only solutions of

$$\frac{dH_\nu^{(1)}(z)}{dz} = 0 \tag{4.8.21}$$

are needed for real values of ν and different ranges of the argument z. For small values of z, the final result resembles (4.8.18)

$$\nu_n(z) = -\frac{\pi i\left(n - \frac{1}{2}\right)}{\ln\left(\frac{r}{2}\right)} F(r, \theta)$$
$$F(r, \theta) = 1 + \frac{\left[i\left(\frac{\pi}{2} - \theta\right) - \gamma\right]}{\ln\left(\frac{r}{2}\right)} + \frac{\left[i\left(\frac{\pi}{2} - \theta\right) - \gamma\right]^2}{\left[\ln\left(\frac{r}{2}\right)\right]^2}$$
$$+ \frac{\left[i\left(\frac{\pi}{2} - \theta\right) - \gamma\right]^3 - \frac{\zeta(3)\pi^2 n^2}{3}}{\left[\ln\left(\frac{r}{2}\right)\right]^3} + O\left(\frac{z^2}{\ln(z/2)}\right) \tag{4.8.22}$$
$$z = re^{i\theta} \quad ; \quad |z| \ll 1$$

In case of $|\nu| \gg 1 + |z|^2$, the real and imaginary components of ν, which are associated with (4.8.21) are

$$\alpha_n = \pi\left(\frac{\pi}{2} - \theta\right)\left(n - \frac{3}{4}\right)\left|\ln\left[\frac{\pi(4n-3)}{2er}\right]\right|^{-2}\left[1 + O\left(\frac{\ln(\ln n)}{\ln n}\right)\right]$$
$$\beta_n = \pi\left(n - \frac{3}{4}\right)\left|\ln\left[\frac{\pi(4n-3)}{2er}\right]\right|^{-1}\left[1 + O\left(\frac{\ln(\ln n)}{\ln n}\right)\right] \quad ; \quad n \gg 1 \tag{4.8.23}$$

For $|z|$ large and n fixed, the derivative of the Airy function is involved

$$\nu_n(z) = z + \frac{q'_n z^{1/3} e^{\pi i/3}}{6^{1/3}} + \frac{6^{1/3} e^{2\pi i/3}}{z^{1/3}} \left[\frac{(q'_n)^2}{180} + \frac{1}{10 q'_n} \right] + O(z^{-1})$$
$$Ai'(q'_n) = 0 \quad ; \quad |z| \gg n > 0 \tag{4.8.24}$$

In addition, Keller, Rubinow and Goldstein [66] considered the linear combination of (4.8.14) and (4.8.21)

$$\frac{dH_\nu^{(1)}(z)}{dz} + i\alpha H_\nu^{(1)}(z) = 0 \tag{4.8.25}$$

However, their results are less convenient for numerical determinations, e.g. for $|\nu| \gg 1 + |z|^2$ or $|z| \ll 1$, they were able to give zeros of (4.8.25) only as a solution of the rather complicated transcendental equation

$$\nu_n(z) = \frac{\pi i\left(\frac{1}{2} - n\right)}{\left[\ln\left(\frac{z}{2}\right) - \frac{\pi i}{2}\right]} + \frac{\ln\left[\frac{\Gamma(1+\nu_n(z))}{\Gamma(1-\nu_n(z))}\right]}{2\left[\ln\left(\frac{z}{2}\right) - \frac{\pi i}{2}\right]} + O\left(\frac{\alpha z}{\nu_n(z)}\right) + O\left(\frac{z^2}{\nu_n(z)}\right) \tag{4.8.26}$$

The solution of (4.8.25) for $|z|$ large and n fixed was also derived by Levy and Keller [67]

$$\nu_n(z) = z + \frac{q_n(\alpha z^{1/3})\, z^{1/3}}{6^{1/3}} + O\left(z^{-1/3}\right)$$
$$lptAi'(q_n) - \frac{e^{5\pi i/6} \alpha z^{1/3}}{6^{1/3}} Ai(q_n) = 0 \quad ; \quad |z| \gg n > 0 \tag{4.8.27}$$

The importance of determination of zeros with respect to the order of the Hankel function, its derivative and the linear combination of them led Cochran, [65] in 1965, to reconsider solutions of (4.8.14), (4.8.21) and (4.8.25). His attention was directed to large values of the order ν and unrestricted values of argument z. In order to obtain solution for such conditions, Cochran derived an asymptotic expansion of the Hankel function of the first kind directly in terms of the Airy functions and not as usually in terms of the Bessel functions. He postulated that for large $|z|$, the ν – zeros of equation

$$H_\nu^{(1)}(z) = 0 \tag{4.8.28}$$

are all negative real numbers, and they can be approximated by

$$\nu_n(z) = z + \frac{q_n z^{1/3} e^{-2\pi i/3}}{2^{1/3}} + \frac{1}{z^{1/3}} O(1) \quad ; \quad |z| \gg 1$$

$$Ai(q_n) = 0 \tag{4.8.29}$$

$$q_n \approx -\left(\frac{3\pi n}{2}\right)^{2/3} \left[1 + \frac{1}{n} O(1)\right]$$

and for fixed z, the large order zeros of the Hankel function can be evaluated from

$$\nu_n(z) = \frac{\pi n i}{i\left(\frac{\pi}{2} - \arg z\right) + \ln\left(\frac{3\pi n}{e|z|}\right)} [1 + \frac{\ln(\ln n)}{\ln n} O(1)] \quad ; \quad n \gg 1 \tag{4.8.30}$$

In the case of the ν – zeros of derivatives of the Hankel function of the first kind

$$\frac{dH_\nu^{(1)}(z)}{dz} = 0 \tag{4.8.31}$$

Cochran found for large $|z|$ that

$$\nu_n(z) = z + \left(\frac{z}{2}\right)^{1/3} q'_n e^{-2\pi i/3} + \frac{e^{-4\pi i/3}}{(2z)^{1/3}} \left[\frac{(q'_n)^2}{30} - \frac{1}{5q'_n}\right] + \frac{1}{z} O(1) \tag{4.8.32}$$

$$|z| \gg 1$$

and for large ν

$$\nu_n(z) = \frac{z}{2} e^{2w/3} q'_n e^{-2\pi i/3} \left[1 + \frac{[\ln(q'_n)]^2}{(q'_n)^3} O(1)\right] \quad ; \quad |n| > > 1$$

$$w\, e^{2w/3} = \frac{2(q'_n)^{3/2} e^{-\pi i}}{ez} \tag{4.8.33}$$

$$q'_n = q_n \left[1 + \frac{1}{n} O(1)\right]$$

The result given in (4.8.33) is also nearly valid for solution of

$$\frac{dH_\nu^{(1)}(z)}{dz} + i\alpha H_\nu^{(1)}(z) = 0 \tag{4.8.34}$$

when large ν – zeros of the Hankel function are considered. For large values of the argument z, Cochran derived the following expression:

$$v_n(z) = z + \left(\frac{z}{2}\right)^{1/3} q'_n e^{-2\pi i/3} + \frac{(q'_n)^2 e^{-4\pi i/3}}{2^{2/3}\, 30\, z^{1/3}} + \frac{1}{z} O(1) \quad ; \quad |z| >> 1$$

$$Ai'(q'_n) - \alpha \left(\frac{z}{2}\right)^{1/3} e^{-\pi i/6} Ai(q'_n) = 0 \tag{4.8.35}$$

$$q'_n = q_n + \frac{2^{1/3} e^{\pi i/6}}{\alpha z^{1/3}} + \frac{1}{z^{2/3}} O(1)$$

He also gave the estimated accuracy of large order zeros when evaluated by his approximated expressions.

From studies devoted to determination of the order zeros, it is worthwhile to mention also two papers of Streifer and Kodis [70, 71]. They investigated diffraction effects in cylindrical geometry by using the Hankel functions of the first kind. However, the solution of physical problem involved significantly more complicated equations, in the form of the following quotients:

$$\frac{\left(z \dfrac{dH_\nu^{(1)}(z)}{dz}\right)}{H_\nu^{(1)}(z)} - \frac{\left(w \dfrac{dH_\nu^{(2)}(w)}{dw}\right)}{H_\nu^{(2)}(w)} = 0 \quad ; \quad z, w >> 1 \tag{4.8.36}$$

and

$$\frac{\left(z \dfrac{dH_\nu^{(1)}(z)}{dz}\right)}{H_\nu^{(1)}(z)} - \frac{\left(w \dfrac{dJ_\nu(w)}{dw}\right)}{J_\nu(w)} = 0 \tag{4.8.37}$$

where z and w denote prescribed positive values of arguments. Solutions for particular values of z and w are very long and complex to be presented here, they are available in the Streifer and Kodis papers [24, 71, 72].

From mathematical point of view, the most elaborate are two investigations performed by Nagase [96, 97] in 1954. He studied the diffraction of elastic waves by a sphere and found the necessity to determine zeros with respect to the order of the linear homogenous forms with the Hankel functions and their derivatives

$$f_1(\lambda, z)\, H_\nu^{(2)}(z) + f_2(\lambda, z) \frac{dH_\nu^{(2)}(z)}{dz} = 0 \tag{4.8.38}$$

$$\lambda = \nu^2$$

where the argument z is real positive number and the order ν is a complex variable. In the prescribed functions f_1 and f_2, it was convenient to use $\lambda = \nu^2$ and not ν as a variable. Nagase gave general expressions for estimation of the zeros of (4.8.38) by using asymptotic expansions of the Hankel functions in forms which are valid for a concerned region of argument. He devoted much attention to the case when the argument z and the order ν are nearly the same. Nagase also involved the Nicolson integral for an estimation of his solutions which were expressed in the form of infinite integrals. He discussed in a detail two particular cases of (4.3.38) which are important in mathematical physics

$$\frac{\alpha}{z} H_\nu^{(2)}(z) + \frac{dH_\nu^{(2)}(z)}{dz} = 0 \tag{4.8.39}$$

where α is a real constant, and the following bilinear form

$$\begin{aligned} &f_1(\lambda, z, \alpha)\, H_\nu^{(2)}(z)\, H_\nu^{(2)}(\alpha z) + f_2(\lambda, z, \alpha)\, \frac{dH_\nu^{(2)}(z)}{dz}\, H_\nu^{(2)}(\alpha z) \\ &+ f_3(\lambda, z, \alpha)\, H_\nu^{(2)}(z)\, \frac{dH_\nu^{(2)}(\alpha z)}{d(\alpha z)} \\ &+ f_4(\lambda, z, \alpha)\, \frac{dH_\nu^{(2)}(z)}{d(\alpha z)}\, \frac{dH_\nu^{(2)}(\alpha z)}{d(\alpha z)} = 0 \\ &\alpha < 1 \quad ; \quad \lambda = \nu^2 \end{aligned} \tag{4.8.40}$$

where f_1, f_2, f_3 and f_4 are any given functions.

Nagase illustrated, with an example, the proposed method to determine zeros in (4.8.40) with the following set of prescribed functions:

$$\begin{aligned} f_1(\lambda, z, \alpha) &= 1 - \frac{\lambda}{z^2} + \frac{4\lambda}{z^4}\left(\lambda - \frac{9}{4}\right) \quad ; \quad \lambda = \nu^2 \\ f_2(\lambda, z, \alpha) &= \frac{2}{z}\left[1 + \frac{1}{z^2}\left(\lambda - \frac{9}{4}\right)\right] \\ f_3(\lambda, z, \alpha) &= -\frac{4\alpha}{z}\left[1 - \frac{1}{2z^2}\left(\lambda - \frac{9}{4}\right)\right] \quad ; \quad \alpha < \frac{1}{\sqrt{2}} \\ f_4(\lambda, z, \alpha) &= -\frac{4\alpha}{z^2}\left(\lambda - \frac{9}{4}\right) \end{aligned} \tag{4.8.41}$$

He extensively used an estimation of the Hankel functions based on their representation coming from the Nicolson integral [7]

$$\begin{aligned} H_\nu^{(1)}(z)\, H_\nu^{(2)}(z) &= [J_\nu(z)]^2 + [Y_\nu(z)]^2 \\ &= \frac{8}{\pi^2} \int_0^\infty K_0(2z \sinh \xi) \cosh(2\nu \xi)\, d\xi \end{aligned} \tag{4.8.42}$$

and from the value of the Wronskian

$$\left\{H_\nu^{(1)}(z), H_\nu^{(2)}(z)\right\} = -\frac{4i}{\pi z} \tag{4.8.43}$$

From differentiating (4.8.42) with respect to the argument z he derived

$$H_\nu^{(1)}(z)\frac{dH_\nu^{(2)}(z)}{dz} + H_\nu^{(2)}(z)\frac{dH_\nu^{(1)}(z)}{dz} = -\frac{16}{\pi^2}\int_0^\infty K_1(2z\sinh\xi)\sinh\xi\cosh(2\nu\xi)\,d\xi \tag{4.8.44}$$

and after few manipulations by using (4.8.42) – (4.8.44) also

$$\frac{\left(\frac{dH_\nu^{(2)}(z)}{dz}\right)}{H_\nu^{(2)}(z)} + \frac{\left(\frac{dH_\nu^{(1)}(z)}{dz}\right)}{H_\nu^{(1)}(z)} = -\frac{2\int_0^\infty K_1(2z\sinh\xi)\sinh\xi\cosh(2\nu\xi)\,d\xi}{\int_0^\infty K_0(2z\sinh\xi)\cosh(2\nu\xi)\,d\xi} \tag{4.8.45}$$

and

$$\frac{\left(\frac{dH_\nu^{(2)}(z)}{dz}\right)}{H_\nu^{(2)}(z)} - \frac{\left(\frac{dH_\nu^{(1)}(z)}{dz}\right)}{H_\nu^{(1)}(z)} = -\frac{\left(\frac{\pi i}{2z}\right)}{\int_0^\infty K_0(2z\sinh\xi)\cosh(2\nu\xi)\,d\xi} \tag{4.8.46}$$

These two equations permit to separate quotients of the derivative and Hankel function and to introduce the corresponding integrals into (4.8.39) and (4.8.40). The estimation of integral expressions for different values of argument z allowed Nagase to determine approximated values of the order zeros.

4.9 Zeros of the Bessel Functions with Respect to the Argument As a Function of Their Order

In 15.6 section of his treatise, Watson [7] discussed the mode of variation of the zeros of the Bessel functions of the first and second kind when their orders are also variable. At present, a huge number of investigations exist in the mathematical literature [60, 90–92, 98–153] which are devoted to the problem of finding and evaluation of zeros at fixed value of the order. If we start only from seventies of the previous century, it is possible to mention a number of leading mathematicians who are associated with this topic, their names arranged in alphabetic order are Á. Barich, T.H. Boyer, F. Calogero, Á. Elbert, L. Gatteschi, E.K. Ifantis, A. Laforgia, M.K. Kerimov, L. Lorch (also civil rights activist) M.E. Muldoon, F.W.J. Olver, T. Pálmal, T.K. Pogány, P.D. Siafarikas and J. Steining. In References, it is possible to find few others, who also contributed to this subject.

The theory and applied mathematical methods, the conditions for monotonicity, concavity and convexity linked with zeros of the Bessel functions were discussed in 1981 by Muldoon [69]. These new results and available numerical methods were reviewed in 2001 by Elbert [109]. A more recent survey, from the period 2014–2018, was presented by Kerimov [125–128]. In a series of papers, he discussed available inequalities and estimations for zeros of the Bessel functions and the Euler-Rayleigh sums.

Zeros of the Bessel and related functions were investigated not only by mathematicians, but they were also of significant interest in mathematical physics and engineering sciences. Quite different cases which are connected with finding zeros of the Bessel functions illustrate such physical problems. The first is determination of bound states associated with solutions of the Schrödinger equation and the second one is the application of linear viscoelastic models to establish retardation and relaxation times in solid bodies [154].

In this section only a short overview on the subject is given, it includes materials collected from the literature. Presented formulas, inequalities and approximations are of an unequal value and importance, and their verification can be found in original papers.

Following the Watson notation, for real, nonnegative values of the order, $\nu \geq 0$, the kth positive zero (root) of the Bessel function of the first and second kind are

$$\begin{aligned} &J_\nu(j_{\nu,k}) = 0 \quad ; \quad k = 1, 2, 3, \ldots \\ &Y_\nu(y_{\nu,k}) = 0 \end{aligned} \tag{4.9.1}$$

The orders ν can be also extended to negative values (for details, see [7]). The zeros of derivatives of the Bessel functions are denoted in a similar way $j'_{\nu,k}, j''_{\nu,k}, j'''_{\nu,k}, \ldots$ $y'_{\nu,k}, y''_{\nu,k}, y'''_{\nu,k} \ldots$

If a more general form of cylindrical functions is considered, the zeros are denoted as $c_{\nu k}$

$$J_\nu(c_{\nu,k}) \cos\alpha - Y_\nu(c_{\nu,k}) \sin\alpha = 0 \quad ; \quad k = 1, 2, 3, \ldots \tag{4.9.2}$$

where α lies in the interval $0 \leq \alpha < \pi$ and $c_{\nu,k}$ can take also negative values. In the case $\alpha = \pi/2$ in (4.9.2), the positive zeros of the Bessel function of the second kind are denoted as $y_{\nu k}$. Three equations from (4.9.1) and (4.9.2) have infinite number of zeros and their values depend also on the order ν.The positive zeros of $J_\nu(z)$ are interlaced with those of $J_{\nu+1}(z)$ and this is true in the general case of cylindrical functions.

$$j_{\nu,1} < j_{\nu+1,1} < j_{\nu,2} < j_{\nu+1,2} < j_{\nu,3} < j_{\nu+1,3} < \dots \quad ; \quad \nu > -1 \tag{4.9.3}$$

$$y_{\nu,1} < y_{\nu+1,1} < y_{\nu,2} < y_{\nu+1,2} < y_{\nu,3} < y_{\nu+1,3} < \dots \quad ; \quad \nu > -1 \tag{4.9.4}$$

In the case of zeros of functions and derivatives it is possible to write the following inequality sequences:

$$\nu \geq j'_{\nu,1} < y_{\nu,1} < y'_{\nu,1} < j_{\nu,1} < j'_{\nu,2} < y_{\nu,2} < y'_{\nu,2} < i_{\nu,2} < j_{\nu,2} < j'_{\nu,3} < \dots \tag{4.9.5}$$

$$j'_{\nu,1} < j'_{\nu+1,1} < j'_{\nu,2} < j'_{\nu+1,2} < j'_{\nu,3} < j'_{\nu+1,3} < \dots \tag{4.9.6}$$

$$y'_{\nu,1} < y'_{\nu+1,1} < y'_{\nu,2} < y'_{\nu+1,2} < y'_{\nu,3} < y'_{\nu+1,3} < \dots \tag{4.9.7}$$

$$\begin{aligned} & j_{\nu+1,k} < j'_{\nu,k+1} \quad ; \quad k = 1, 2, 3, \dots \\ & j_{\nu+1,k} < j'_{\nu,k} \end{aligned} \tag{4.9.8}$$

$$j''_{\nu,k} > j'_{\nu,k} \quad ; \quad 0 < \nu \leq 1 \quad ; \quad k = 1, 2, 3, \dots \tag{4.9.9}$$

$$j''_{\nu,k} > j_{\nu,k} \quad ; \quad 0 < \nu \leq 1 \quad ; \quad k = 2, 3, 4 \dots \tag{4.9.10}$$

$$j'_{\nu,k} < j''_{\nu,k} < j_{\nu,k} \quad ; \quad \nu > 1 \quad ; \quad k = 1, 2, 3 \dots \tag{4.9.11}$$

$$j'''_{\nu,1} < j''_{\nu,1} < j'''_{\nu,2} < j''_{\nu,2} < j'''_{\nu,3} < j''_{\nu,3} < \dots \quad ; \quad \nu \geq 2 \tag{4.9.12}$$

$$j''_{\nu,1} < j'_{\nu,1} < j''_{\nu,2} < j'_{\nu,2} < j''_{\nu,3} < j'_{\nu,3} < \dots \quad ; \quad \nu \geq 1 \tag{4.9.13}$$

The Schläfli, Schafheitlin, Watson, Ifantis and Siafarikas integral and summation formulas [7, 122, 123] and the expressions derived from them are useful when behaviour of the zeros of the Bessel functions is considered

$$\frac{dj_{\nu,k}}{d\nu} = \frac{2\nu}{j_{\nu,k}\,[J_{\nu+1}(j_{\nu,k})]^2} \int_0^{j_{\nu,k}} [J_\nu(t)]^2 \frac{dt}{t} \quad ; \quad \nu > 0 \tag{4.9.14}$$

$$\frac{d(j_{\nu,k})^2}{d\nu} = 4\,(j_{\nu,k})^2 \int_0^\infty e^{-2\nu t} K_0(2j_{\nu,k}\sinh t)\,dt \quad ; \quad \nu > 0 \tag{4.9.15}$$

$$\frac{dj_{\nu,k}}{d\nu} = \frac{j_{\nu,k} \sum_{k=1}^{\infty} [J_{\nu+k}(j_{\nu,k})]^2}{\sum_{k=1}^{\infty} (\nu+k)\,[J_{\nu+k}(j_{\nu,k})]^2} \tag{4.9.16}$$

$$\frac{dj'_{\nu,k}}{d\nu} = \frac{2j_{\nu,k}}{[(j'_{\nu,k})^2 - \nu^2]} \int_0^\infty e^{-2\nu t}[(j'_{\nu,k})^2 \cosh(2t) - \nu^2]\, K_0(2j'_{\nu,k} \sinh t)\, dt \tag{4.9.17}$$

$$j'_{\nu,k} \neq |\nu|$$

$$\frac{dj''_{\nu,k}}{d\nu} = \frac{\nu\left\{\int_0^1 [J_\nu(j''_{\nu,k}x)]^2 \frac{dx}{x} - [J_\nu(j''_{\nu,k})]^2\right\}}{\left\{j''_{\nu,k}\int_0^1 x\,[J_\nu(j''_{\nu,k}x)]^2 dx - [J_\nu(j''_{\nu,k})]^2\right\}} \quad ; \quad \nu > 0 \tag{4.9.18}$$

$$\frac{dj''_{\nu,k}}{d\nu} = \frac{2\nu\left\{\int_0^{j''_{\nu,k}} [J_\nu(j''_{\nu,k})]^2 \frac{dx}{x} - [J_\nu(j''_{\nu,k})]^2\right\}}{(j''_{\nu,k})^2 J_\nu(j''_{\nu,k})\, J'''_\nu(j''_{\nu,k})} \quad ; \quad \nu > 0 \tag{4.9.19}$$

$$\frac{dc_{\nu,k}}{d\nu} = 2c_{\nu,k}\int_0^\infty e^{-2\nu t} K_0(2c_{\nu,k} \sinh t)\, dt \tag{4.9.20}$$

$$\frac{dc_{\nu,k}}{d\nu} = \frac{2c'_{\nu,k}}{(c'_{\nu,k} - \nu^2)}\int_0^\infty e^{-2\nu t}[(c'_{\nu,k})^2 \cosh(2t) - \nu^2] K_0(2c'_{\nu,k} \sinh t)\, dt \tag{4.9.21}$$

These formulas help to establish properties of the zeros of the Bessel functions and their derivatives as a function of the order ν. This permits to establish whether on investigated interval of ν, the function increases, decreases, is monotonic, convex or concave.

Starting from Watson [7], there is a very large number of inequalities for estimation of the zeros of the Bessel functions and their derivatives. In the case of first smallest zeros, the upper bounds are:

$$j_{\nu,1} < \pi\left(\frac{\nu}{2} + \frac{3}{4}\right) \quad ; \quad \nu > \frac{1}{2} \tag{4.9.22}$$

$$j_{\nu,1} < \sqrt{\nu+1}\,[\sqrt{\nu+2} + 1] \tag{4.9.23}$$

$$j_{\nu 1} < \sqrt{(j_{0,1})^2 + 2\nu[\pi^2 - (j_{0,1})^2]} \quad ; \quad 0 < \nu < \frac{1}{2} \tag{4.9.24}$$

$$j_{\nu,1} < j_{0,1} + \frac{\pi\nu}{2} \quad ; \quad \nu > 0 \tag{4.9.25}$$

and the lower bounds are:

$$j_{\nu,1} > \pi(\nu + 1) \quad ; \quad -1 < \nu < -\frac{1}{2} \tag{4.9.26}$$

$$j_{\nu,1} > \frac{\pi}{2}\left(\nu+\frac{3}{2}\right) \quad ; \quad -\frac{1}{2}<\nu<0 \tag{4.9.27}$$

$$j_{\nu,1} > \sqrt{\nu(\nu+2)} \quad ; \quad \nu\geq 0 \tag{4.9.28}$$

$$j_{\nu,1} > \sqrt{\nu^2-19+6\sqrt{2\nu^2+10\nu+17}} \quad ; \quad \nu>0 \tag{4.9.29}$$

$$j_{\nu,1} > \frac{(\nu+1)(\nu^2+12\nu+23)}{\sqrt{\nu+4}} \quad ; \quad \nu>0 \tag{4.9.30}$$

$$j_{\nu,1} > \sqrt{\nu+1}\left\{1+\frac{2}{3}(\nu+1)\right\}^{1/4} \quad ; \quad \nu\geq -1 \tag{4.9.31}$$

$$j_{\nu,1} > \sqrt{(\nu+1)(\nu+5)} \quad ; \quad -1<\nu<\infty \tag{4.9.32}$$

$$j_{\nu,1} > j_{0,1}+2\nu\left(j_{0,1}-\frac{\pi}{2}\right) \quad ; \quad -\frac{1}{2}<\nu<0 \tag{4.9.33}$$

$$j_{\nu 1} > \sqrt{(j_{0,1})^2(1+\nu)} \quad ; \quad -1<\nu<0 \tag{4.9.34}$$

$$j_{\nu 1} > j_{0,1}+2\nu(\pi-j_{0,1}) \quad ; \quad 0<\nu<\frac{1}{2} \tag{4.9.35}$$

$$j_{\nu 1} > \sqrt{(j_{0,1})^2+\nu^2} \quad ; \quad \nu>0 \tag{4.9.36}$$

and in particular

$$\sqrt{\nu(\nu+2)} < j_{\nu,1} < 2\sqrt{(\nu+1)(\nu+3)} \quad ; \quad 0<\nu<1 \tag{4.9.37}$$

$$\sqrt{\nu(\nu+2)} < j_{\nu,1} < 2\sqrt{\frac{4(\nu+1)(\nu+5)}{3}} \quad ; \quad 1<\nu<4 \tag{4.9.38}$$

$$\sqrt{\nu(\nu+3)} < j_{\nu,1} < 2\sqrt{\frac{4(\nu+1)(\nu+5)}{3}} \quad ; \quad \nu>4 \tag{4.9.39}$$

$$2[\nu(\nu+2)]^{1/4} < j_{\nu,1} < \sqrt{2(\nu+1)(\nu+3)} \quad ; \quad \nu>-1 \tag{4.9.40}$$

$$\sqrt{2(\nu+1)} < j_{\nu,1} < \sqrt{\nu+1}(1+\sqrt{\nu+2}) \quad ; \quad \nu>1 \tag{4.9.41}$$

$$\frac{(\nu+1)(\nu^2+12\nu+13)}{\nu+4} < j_{\nu,1} < \sqrt{\frac{2(\nu+1)(\nu+5)(5\nu+1)}{7\nu+19}} \quad ; \quad \nu\geq 0 \tag{4.9.42}$$

The first three zeros of $J_\nu(z)$ and $Y_\nu(z)$ as a function of ν, for large values of the order can be expressed by:

$$j_{\nu,1} \sim \nu + 1.85575708\,\nu^{1/3} + \frac{1.0331502}{\nu^{1/3}} - \frac{0.00397406}{\nu} - \frac{0.0907627}{\nu^{5/3}} + \frac{0.0433385}{\nu^{7/3}} + \ldots \quad ; \quad \nu \gg 1 \tag{4.9.43}$$

$$j_{\nu,2} \sim \nu + 3.2446076\,\nu^{1/3} + \frac{3.958244}{\nu^{1/3}} - \frac{0.08331}{\nu} - \frac{0.8437}{\nu^{5/3}} + \ldots \quad \nu \gg 1 \tag{4.9.44}$$

$$j_{\nu,3} \sim \nu + 4.3816712\,\nu^{1/3} + \frac{5.759713}{\nu^{1/3}} - \frac{0.22607}{\nu} - \frac{2.8039}{\nu^{5/3}} + \ldots \quad \nu \gg 1 \tag{4.9.45}$$

and it is known also that

$$j_{\nu,1} = 2\beta\left[1 + \frac{\beta}{4} - \frac{7\beta^2}{96} + \frac{49\beta^3}{1152} - \frac{78363\beta^4}{276480} + \ldots\right] \quad ; \quad -1 < \nu < 0 \quad \beta = \nu + 1 \tag{4.9.46}$$

In general case the zeros can be determined from [144]

$$j_{\nu,k} \sim \beta - \frac{\mu - 1}{8\beta} - \frac{(\mu-1)(7\mu-31)}{3!(8\beta)^2} - \frac{4(\mu-1)(83\mu^2 - 982\mu + 3779))}{5!8^3\beta^5} + \ldots \quad \mu = 4\nu^2 \quad ; \quad \beta = \pi\left(k + \frac{\nu}{2} - \frac{1}{4}\right) \quad ; \quad k \gg \nu \tag{4.9.47}$$

$$j'_{\nu,k} \sim \beta - \frac{\mu + 3}{8\beta} - \frac{4(7\mu^2 + 82\mu - 9)}{3!(8\beta)^2} - \frac{32(83\mu^3 + 2075\mu^2 - 3039\mu + 3537)}{5!8^3\beta^5} + \ldots \quad \mu = 4\nu^2 \quad ; \quad \beta = \pi\left(k + \frac{\nu}{2} + \frac{1}{4}\right) \quad ; \quad k \gg \nu \tag{4.9.48}$$

$$y_{\nu,k} \sim \beta - \frac{\mu - 1}{8\beta} - \frac{(\mu-1)(7\mu-31)}{3!(8\beta)^2} - \frac{4(\mu-1)(83\mu^2 - 982\mu + 3779)}{5!8^3\beta^5} + \ldots \quad \mu = 4\nu^2 \quad ; \quad \beta = \pi\left(k + \frac{\nu}{2} - \frac{3}{4}\right) \quad ; \quad k \gg \nu \tag{4.9.49}$$

The zeros of the Bessel functions are frequently expressed in the terms of the zeros of the Airy function a_k, these zeros all lie on the negative real axis [150]

$$j_{\nu,k} \sim \nu + \frac{|a_k|}{(2\nu)^{1/3}} + O\left(\frac{1}{\nu^{1/3}}\right) \quad ; \quad \nu \gg 1$$
$$Ai(-a_k) = 0 \quad ; \quad k = 1, 2, 3, \ldots \tag{4.9.50}$$

$$j_{\nu,k} > \nu + \left(\frac{\nu}{2}\right)^{1/3} |a_{k-1}| \quad ; \quad \nu \gg k \quad ; \quad \nu \geq \frac{1}{2}$$
$$Ai(-a_k) = 0 \quad ; \quad k = 1, 2, 3, \ldots \tag{4.9.51}$$

$$j_{\nu,k} \sim \nu + a_k \left(\frac{\nu}{2}\right)^{1/3} + \frac{3}{20} (a_k)^2 \left(\frac{2}{\nu}\right)^{1/3} + \ldots \quad ; \quad \nu > 0$$
$$Ai(-a_k) = 0 \quad ; \quad k = 1, 2, 3, \ldots \tag{4.9.52}$$

$$\nu - a_k \left(\frac{\nu}{2}\right)^{1/3} < j_{\nu,k} < \nu + a_k \left(\frac{\nu}{2}\right)^{1/3} + \frac{3}{20} (a_k)^2 \left(\frac{2}{\nu}\right)^{1/3} \quad ; \quad \nu > 0$$
$$Ai(-a_k) = 0 \quad ; \quad k = 1, 2, 3, \ldots \tag{4.9.53}$$

$$j_{\nu,k} < \frac{\nu}{\left(1 - \frac{|a_k|}{2^{1/3}\nu^{2/3}}\right)} \quad ; \quad \nu > \frac{3\pi}{2^{3/2}} k$$
$$Ai(-a_k) = 0 \quad ; \quad k = 1, 2, 3, \ldots \tag{4.9.54}$$

$$j_{\nu,k} < \frac{\pi\nu}{2} + \frac{2}{3} |a_k|^{3/2} - \frac{\nu^2}{2\left(\frac{\pi\nu}{2} + \frac{2}{3} |a_k|^{3/2}\right)} \quad ; \quad k \gg \nu \quad ; \quad \nu \geq \frac{1}{2}$$
$$Ai(-a_k) = 0 \quad ; \quad k = 1, 2, 3, \ldots \tag{4.9.55}$$

The zeros of the Airy function can be estimated from [93–95]

$$j_{\nu,k} > \nu + \frac{2}{3} |a_{k-1}|^{3/2} \quad ; \quad k \gg \nu \quad ; \quad \nu \geq \frac{1}{2}$$
$$Ai(-a_k) = 0 \quad ; \quad k = 1, 2, 3 \ldots \tag{4.9.56}$$
$$\left[\frac{3\pi}{8}(4k - 1.4)\right]^{2/3} < |a_k| < \left[\frac{3\pi}{8}(4k - 0.965)\right]^{2/3}$$

$$-\left[\frac{3\pi}{8}(4k-1)+\frac{3}{2}\tan^{-1}\left(\frac{5}{18\pi(4k-1)}\right)\right]^{2/3} < a_k < -\left[\frac{3\pi}{8}(4k-1)\right]^{2/3}$$

$$\delta_k \le j_{\nu,k} \le \delta_k - \frac{4\nu^2-1}{8\delta_k} \quad ; \quad \delta_k = \pi\,(k+\frac{\nu}{2}-\frac{1}{4}) \quad ; \quad 0<\nu\le\frac{1}{2} \tag{4.9.57}$$

$$j_{\nu,k} \le \delta_k \quad ; \quad \nu \ge \frac{1}{2} \quad ; \quad k=1,2,3,\ldots$$

The upper bounds for squares of the first zero of the Bessel function $J_\nu(z)$ are:

$$(j_{\nu,1})^2 < 4(\nu+1)\,(\nu+2) \quad ; \quad \nu > -1 \tag{4.9.58}$$

$$(j_{\nu,1})^2 < \frac{2(\nu+1)\,(\nu+5)\,(5\nu+1)}{7\nu+19} \quad ; \quad \nu > -1 \tag{4.9.59}$$

$$(j_{\nu 1})^2 < (j_{0,1})^2(1+\nu) \quad ; \quad -1<\nu<0 \tag{4.9.60}$$

with more inequalities available for the lower bounds

$$(j_{\nu,1})^2 > 2(\nu+1)\,(\nu+3) \quad ; \quad -2<\nu<-1 \tag{4.9.61}$$

$$(j_{\nu,1})^2 > 2^{5/3}(\nu+1)\,[(\nu+2)\,(\nu+5)]^{1/3} \quad ; \quad -2<\nu<-1 \tag{4.9.62}$$

$$(j_{\nu,1})^2 > 4(\nu+1) \tag{4.9.63}$$

$$(j_{\nu,1})^2 > 4(\nu+1)\sqrt{\nu+2} \tag{4.9.64}$$

$$(j_{\nu,1})^2 > (\nu+1)\,(\nu+5) \quad ; \quad \nu > -1 \tag{4.9.65}$$

$$(j_{\nu,1})^2 < 2(\nu+1)\,(\nu+3) \quad ; \quad \nu > -1 \tag{4.9.66}$$

$$(j_{\nu,1})^2 > (\nu+1)\,(\nu+\frac{13}{2}) \quad ; \quad \nu > -1 \tag{4.9.67}$$

$$(j_{\nu,1})^2 > \frac{24(\nu+1)^2}{1-2\nu+\sqrt{(2\nu+1)\,(2\nu+3)}} - 2(\nu^2-1) \quad ; \quad \nu > -1 \tag{4.9.68}$$

$$(j_{\nu,1})^2 > (\nu+1)\,(\nu+\frac{2\pi^2}{3}-\frac{1}{2}) \quad ; \quad \nu > \frac{1}{2} \tag{4.9.69}$$

$$(j_{\nu 1})^2 > (j_{\nu 0})^2 + 4\nu \quad ; \quad \nu > 0 \tag{4.9.70}$$

$$(j_{\nu 1})^2 > (j_{\nu 0})^2 + \nu^2 \quad ; \quad \nu \ge 0 \tag{4.9.71}$$

The inequalities for both sides are:

$$2(\nu+1)\,(\nu+3) < (j_{\nu,1})^2 < 4(\nu+1)\,(\nu+2) \quad ; \quad -2<\nu<-1 \tag{4.9.72}$$

$$4(\nu+1) < (j_{\nu,1})^2 < 4(\nu+1)\sqrt{(\nu+2)} \quad ; \quad -2<\nu<-1 \tag{4.9.73}$$

$$2^{5/2}(\nu+1)\sqrt{\nu+2} < (j_{\nu,1})^2 < 4(\nu+1)\sqrt{\nu+2} \quad ; \quad -2<\nu<-1 \tag{4.9.74}$$

$$4(\nu+1) < (j_{\nu,1})^2 < 4(\nu+1)(\nu+2) \quad ; \quad \nu>-1 \tag{4.9.75}$$

$$4(\nu+1)\sqrt{\nu+2} < (j_{\nu,1})^2 < 2(\nu+1)(\nu+3) \quad ; \quad \nu>-1 \tag{4.9.76}$$

$$2^{5/3}(\nu+1)[(\nu+2)(\nu+2)]^{1/3} < (j_{\nu,1})^2 < \frac{8(\nu+1)(\nu+2)(\nu+4)}{5\nu+11} \tag{4.9.77}$$

$\nu>-1$

The lower and upper bounds for the squares of smallest zeros of first derivatives of the Bessel functions are given by

$$(j'_{\nu 1})^2 > \frac{4\nu(\nu+1)}{\nu+2} \quad ; \quad \nu>0 \tag{4.9.78}$$

$$(j'_{\nu 1})^2 > \frac{4\nu(\nu+1)\sqrt{\nu+2}}{\sqrt{\nu^2+8\nu+8}} \quad ; \quad \nu>0 \tag{4.9.79}$$

$$(j'_{\nu 1})^2 > \frac{2^{5/3}\nu(\nu+1)(\nu+2)^{1/2}(\nu+3)^{1/4}}{\sqrt{\nu^3+16\nu^2+38\nu+21}} \quad ; \quad \nu>0 \tag{4.9.80}$$

$$(j'_{\nu 1})^2 < \frac{4\nu(\nu+1)(\nu+2)}{\nu^2+8\nu+8} \quad ; \quad \nu>0 \tag{4.9.81}$$

$$(j'_{\nu 1})^2 < \frac{2\nu(\nu^2+8\nu+8)(\nu+1)(\nu+3)}{\nu^3+16\nu^2+38\nu+21} \quad ; \quad \nu>0 \tag{4.9.82}$$

and the corresponding inequalities for second derivatives are:

$$j''_{\nu 1} > \sqrt{\nu(\nu+2)} \quad ; \quad 0<\nu\le 1 \tag{4.9.83}$$

$$j''_{\nu 1} > \sqrt{\nu(\nu-1)} \quad ; \quad \nu>1 \tag{4.9.84}$$

$$j''_{\nu 1} > \sqrt{\frac{2\nu(\nu^2-1)}{2\nu-1}} \quad ; \quad \nu>1 \tag{4.9.85}$$

$$j''_{\nu 1} < \sqrt{\nu^2-1} \quad ; \quad \nu>1 \tag{4.9.86}$$

$$j''_{\nu 1} < \sqrt{\frac{\nu(\nu-1)(\nu+2)}{(\nu+1)}} \quad ; \quad \nu>1 \tag{4.9.87}$$

$$j''_{\nu 1} > \frac{j_{\nu 0}}{\sqrt{2}} \quad ; \quad \nu>1 \tag{4.9.88}$$

$$j''_{\nu 1} > j_{\nu 0}\sqrt{\frac{8}{8+(j_{\nu 0})^2}} \quad ; \quad \nu > 1 \tag{4.9.89}$$

$$j''_{\nu 1} > j_{\nu 1}\sqrt{\frac{(2\nu+1)}{2(\nu+1)}} \quad ; \quad 0 < \nu \le 1 \tag{4.9.90}$$

and their squares are:

$$(j''_{\nu 1})^2 > 2\nu(\nu-1) \quad ; \quad -1 < \nu < 0 \tag{4.9.91}$$

$$(j''_{\nu 1})^2 > \frac{4\nu(\nu-1)}{\nu+2} \quad ; \quad \nu > 1 \tag{4.9.92}$$

$$(j''_{\nu 1})^2 > \frac{4\nu(\nu-1)\sqrt{(\nu+1)(\nu+2)}}{\sqrt{\nu^3+13\nu^2+32\nu+8}} \quad ; \quad \nu > 1 \tag{4.9.93}$$

$$(j''_{\nu 1})^2 > \frac{2^{5/3}\nu(\nu-1)[(\nu+1)(\nu+2)(\nu+3)]^{1/3}}{(\nu^4+27\nu^3+138\nu^2+134\nu+24)^{1/3}} \quad ; \quad \nu > 1 \tag{4.9.94}$$

$$(j''_{\nu 1})^2 < \frac{4\nu(\nu-1)(\nu+1)(\nu+2)^2}{\nu^3+13\nu^2+32\nu+8} \quad ; \quad \nu > 1 \tag{4.9.95}$$

$$(j''_{\nu 1})^2 < \frac{2\nu(\nu-1)(\nu+3)(\nu^3+13\nu^2+32\nu+8)}{\nu^4+27\nu^3+138\nu^2+134\nu+24} \quad ; \quad \nu > 1 \tag{4.9.96}$$

There is a large number of inequalities in general case, for any zeros of the Bessel function, for the lower bounds, we have

$$j_{\nu,k} > \nu + k \quad ; \quad -k < \nu < \infty \tag{4.9.97}$$

$$j_{\nu,k} > \nu + k\pi - \frac{\pi}{2} + \frac{1}{2} \quad ; \quad \nu > -\frac{1}{2} \quad ; \quad k = 1, 2, 3, \ldots \tag{4.9.98}$$

$$j_{\nu,k} \ge \pi\left(k + \frac{\nu}{2} - \frac{1}{4}\right) \quad ; \quad |\nu| \le \frac{1}{2} \quad ; \quad k = 1, 2, 3, \ldots \tag{4.9.99}$$

$$j_{\nu,k} > \nu + k\pi - \frac{1}{2} \quad ; \quad \nu > \frac{1}{2} \quad ; \quad k = 1, 2, 3, \ldots \tag{4.9.100}$$

$$j_{\nu,k} > \frac{2(\nu+1)(2\nu+3)}{\pi} \quad ; \quad \nu > \frac{1}{2} \tag{4.9.101}$$

$$j_{\nu,k} > j_{0,k}\left(1 + \frac{\nu}{k}\right) \quad ; \quad -k < \nu < \infty \tag{4.9.102}$$

$$j_{\nu,k} > j_{0,k}(1-2\nu) + 2\nu\pi k \quad ; \quad 0 < \nu < \frac{1}{2} \tag{4.9.103}$$

$$j_{\nu,k} > j_{0,k} + (j_{1,k} - j_{0,k})\nu \quad ; \quad 0 < \nu < 1 \tag{4.9.104}$$

$$j_{\nu,k} > j_{0,k} + \nu \quad ; \quad \nu > 0 \tag{4.9.105}$$

$$j_{\nu,k} > \sqrt{(j_{0,k})^2 + \nu^2} \quad ; \quad \nu > 0 \tag{4.9.106}$$

and for upper bounds, we have

$$\begin{gathered} j_{\nu,k} \le \beta_{\nu,k} - \frac{4\nu^2 - 1}{8\beta_{\nu,k}} \quad ; \quad 0 \le \nu \le \frac{1}{2} \quad ; \quad k = 1, 2, 3, \ldots \\ \beta_{\nu,k} = \pi\left(k + \frac{\nu}{2} - \frac{1}{4}\right) \end{gathered} \tag{4.9.107}$$

$$j_{\nu,k} < j_{0,k} + \frac{\pi\nu}{2} \quad ; \quad k = 1, 2, 3, \ldots \tag{4.9.108}$$

$$(j_{\nu k})^2 < (j_{0,1})^2 + 2\left[\pi^2 k^2 - (j_{0,1})^2\right]\nu \quad ; \quad 0 < \nu < \frac{1}{2} \tag{4.9.109}$$

$$j_{\nu,k} \le j_{0,k} + \frac{\nu}{j_{0,k}[J_1(j_{0,k})]^2} \quad ; \quad -k < \nu < \infty \quad ; \quad \nu \ne 0 \tag{4.9.110}$$

$$j_{\nu,k} < \frac{(\nu - \frac{1}{2})}{j_{0,k}\,[J_1(j_{0,k})]^2} + \pi k \quad ; \quad \nu > \frac{1}{2} \tag{4.9.111}$$

with

$$\begin{gathered} \beta_{\nu,k} - \frac{4\nu^2 - 1}{8\beta_{\nu,k}} - \frac{4(4\nu^2 - 1)(28\nu^2 - 3)}{3(8\beta_{\nu,k})^3} \le j_{\nu,k} \le \beta_{\nu,k} - \frac{4\nu^2 - 1}{8\beta_{\nu,k}} \\ \beta_{\nu,k} = \pi\left(k + \frac{\nu}{2} - \frac{1}{2}\right) \quad ; \quad |\nu| \le \frac{1}{2} \quad ; \quad k = 1, 2, 3, \ldots \end{gathered} \tag{4.9.112}$$

The inequalities for the first and second derivatives are:

$$j'_{\nu,k} \le \pi\left(k + \frac{\nu}{2} - \frac{3}{4}\right) \quad ; \quad \nu \ge 0 \quad ; \quad k = 1, 2, 3, \ldots \tag{4.9.113}$$

$$j'_{\nu,k} < \frac{\left(j_{\nu,k} + \frac{1}{j_{\nu,k}}\right)}{(\nu + 2)} \tag{4.9.114}$$

$$j'_{\nu,k} > \frac{2}{j_{\nu,k}} + 8\frac{(\nu + 1)^2}{(j_{\nu,k})^3} \tag{4.9.115}$$

$$j''_{\nu,k} \le \pi\left(k + \frac{\nu}{2} - \frac{1}{4}\right) \quad ; \quad \nu \ge 0 \quad ; \quad k = 1, 2, 3, \ldots \tag{4.9.116}$$

$$j''_{\nu,k} > \frac{\nu (j'_{\nu,k})^2 - j_{\nu,k} j'_{\nu,k}}{j_{\nu,k}(\nu + j_{\nu,k})} \tag{4.9.117}$$

There is a large number of differential inequalities, their general behaviour can be expressed as

$$\begin{aligned} &\frac{dj_{\nu,k}}{d\nu} < 0 \quad ; \quad \nu \geq 0 \\ &\frac{d^2 j_{\nu,k}}{d\nu^2} < 0 \\ &\frac{d^3 j_{\nu,k}}{d\nu^3} > 0 \end{aligned} \tag{4.9.118}$$

For lower bounds, they are

$$\frac{dj_{\nu,1}}{d\nu} > 1 \quad ; \quad \nu \geq 0 \tag{4.9.119}$$

$$\frac{1}{j_{\nu,1}} \frac{dj_{\nu,1}}{d\nu} > \frac{3\nu + 5}{4(\nu+1)(\nu+4)} \quad ; \quad \nu > 1 \tag{4.9.120}$$

$$\frac{1}{j_{\nu,1}} \frac{dj_{\nu,1}}{d\nu} > \frac{1}{(j_{\nu,1})^2} \left[1 + \sqrt{1 + (j_{\nu,1})^2} \right] \quad ; \quad \nu > -1 \tag{4.9.121}$$

$$\frac{d(j_{\nu k})^2}{d\nu} > 0 \quad ; \quad \nu \geq 3 \tag{4.9.122}$$

$$\frac{dj_{\nu,k}}{d\nu} > 1 \quad ; \quad \nu \geq 0 \tag{4.9.123}$$

$$\frac{dj_{\nu,k}}{d\nu} > \frac{1}{j_{\nu,k}} + \sqrt{1 + \frac{1}{j^2_{\nu,k}}} \quad ; \quad \nu > -1 \quad ; \quad k = 1, 2, 3, \ldots \tag{4.9.124}$$

$$\frac{dj_{\nu,k}}{d\nu} > \frac{2}{j_{\nu,k}} + \frac{8(\nu+1)^2}{(j_{\nu,k})^3} \quad ; \quad \nu > -1 \quad ; \quad k = 1, 2, 3, \ldots \tag{4.9.125}$$

$$\begin{aligned} &\frac{dj_{\nu,k}}{d\nu} > \frac{4}{j_{\nu,k}} - \frac{8(\nu+1)(\nu+3)}{(j_{\nu,k})^3} + \frac{32(\nu+1)^2(\nu+2)^2}{(j_{\nu,k})^5} \\ &\nu > -1 \quad ; \quad k = 1, 2, 3, \ldots \end{aligned} \tag{4.9.126}$$

$$j_{\nu,k} \frac{dj_{\nu,k}}{d\nu} > 2 \left[1 + \left(\frac{2(\nu+1)}{j_{\nu,k}} \right)^2 \right] \quad ; \quad \nu > -1 \quad ; \quad k = 1, 2, 3, \ldots \tag{4.9.127}$$

$$j_{\nu,k}\frac{dj_{\nu,k}}{d\nu}>2\left[1+2-\frac{4(\nu+1)(\nu+3)}{j_{\nu,k}^2}+\left(\frac{4(\nu+1)(\nu+2)}{j_{\nu,k}^2}\right)^2\right] \tag{4.9.128}$$

$\nu>-1 \quad ; \quad k=1,2,3,\ldots$

and for upper bounds, they are

$$\frac{dj_{\nu,k}}{d\nu}<\frac{j_{\nu,k}}{\nu+k} \quad ; \quad k=1,2,3,\ldots \quad ; \quad \nu>-1 \tag{4.9.129}$$

$$\frac{dj_{\nu,k}}{d\nu}<\frac{j_{\nu,k}}{\nu+\frac{1}{2}} \quad ; \quad k=1,2,3,\ldots \quad ; \quad \nu\geq 0 \tag{4.9.130}$$

$$\frac{dj_{\nu,k}}{d\nu}<\frac{\pi j_{\nu,k}}{2j_{\nu,k}+(\pi-2)\nu} \quad ; \quad \nu\geq 0 \quad ; \quad k\geq 0 \tag{4.9.131}$$

$$\frac{dj_{\nu,k}}{d\nu}<\frac{j_{\nu,k}}{(\nu+3)}\left[j_{\nu,k}+\frac{4}{j_{\nu,k}}+\frac{8(\nu+1)^2}{(j_{\nu,k})^2}\right] \tag{4.9.132}$$

$\nu>-1 \quad ; \quad k=1,2,3,\ldots$

$$\frac{dj_{\nu,k}}{d\nu}<\frac{j_{\nu,k}}{(\nu+4)}\left[j_{\nu,k}+\frac{8}{j_{\nu,k}}-\frac{16(\nu+1)^2}{(j_{\nu,k})^3}+\frac{32(\nu+1)^2(\nu+2)^2}{(j_{\nu,k})^5}\right] \tag{4.9.133}$$

$\nu>-1 \quad ; \quad k=1,2,3,\ldots$

The expressions for $y_{\nu,k}$ zeros of the Bessel function of the second kind are given by

$$y_{\nu,1}\sim\nu+0.9315768\nu^{1/3}+\frac{0.230351}{\nu^{1/3}}+\frac{0.01198}{\nu}-\frac{0.0060}{\nu^{5/3}}+\ldots \tag{4.9.134}$$

$\nu\gg 1$

$$y_{\nu,2}\sim\nu+2.5962685\nu^{1/3}+\frac{2.022183}{\nu^{1/3}}-\frac{0.03572}{\nu}-\frac{0.3463}{\nu^{5/3}}+\ldots \tag{4.9.135}$$

$\nu\gg 1$

$$y_{\nu,3}\sim\nu+3.8341592\nu^{1/3}+\frac{4.410233}{\nu^{1/3}}-\frac{0.14676}{\nu}-\frac{1.6444}{\nu^{5/3}}+\ldots \tag{4.9.136}$$

$\nu\gg 1$

$$y_{\nu,k}<y_{0,k}+1.4470201\ldots\nu \quad ; \quad \nu>0 \tag{4.9.137}$$

$$y_{\nu,k}<y_{0,k}+\frac{1}{y_{0,k}[Y_1(y_{0,k})]^2} \quad ; \quad \nu>0 \tag{4.9.138}$$

$$\frac{dy_{\nu,k}}{d\nu} < \frac{\pi}{2} \quad ; \quad \nu > 0 \quad ; \quad k = 1, 2, 3, \ldots \tag{4.9.139}$$

$$\frac{d^2 y_{\nu,k}}{d\nu^2} < 0 \quad ; \quad \nu > 0 \quad ; \quad k = 1, 2, 3, \ldots \tag{4.9.140}$$

and similarly for $c_{\nu,k}$ zeros, the following inequalities hold

$$c_{\nu,k} > c_{0,k} + \nu \quad ; \quad \nu > 0 \quad ; \quad k = 2, 3, 4 \ldots \tag{4.9.141}$$

$$c_{\nu,k} \geq \pi \left(k + \frac{\nu}{2} - \frac{1}{4} \right) - \alpha \quad ; \quad 0 \leq \nu \leq \frac{1}{2} \quad ; \quad k = 1, 2, 3, \ldots \tag{4.9.142}$$

$$c_{\nu,k} < c_{0,k} + \frac{\pi \nu}{2} \quad ; \quad \nu > 0 \quad ; \quad k = 1, 2 \tag{4.9.143}$$

$$c_{\nu,k} \leq \pi \left(k + \frac{\nu}{2} - \frac{1}{4} \right) - \alpha \quad ; \quad \nu \geq \frac{1}{2} \quad ; \quad k = 1, 2, 3, \ldots \tag{4.9.144}$$

$$\frac{dc_{\nu,k}}{d\nu} < \frac{c_{\nu,k}}{\nu + \frac{1}{2}} \quad ; \quad \nu > 0 \quad ; \quad k = 1, 2, 3, \ldots \tag{4.9.145}$$

$$\frac{d^2 c_{\nu,k}}{d\nu^2} < 0 \quad ; \quad \nu > 0 \quad ; \quad k = 1, 2, 3, \ldots \tag{4.9.146}$$

Euler-Rayleigh sums for positive zeros of $J_\nu^{(n)}(z)$, $n = 0,1,2,3, \ldots$, or the Rayleigh function (Lord Rayleigh started investigation of this topic in 1874 [7]) is defined by

$$\sigma_n(\nu) = \sum_{k=1}^{\infty} \frac{1}{(j_{\nu,k})^{2n}} \quad ; \quad n = 1, 2, 3, \ldots \tag{4.9.147}$$

and it gives the following estimation for the square of the Bessel function zeros

$$\frac{1}{[\sigma_n(\nu)]^{1/n}} < (j_{\nu,k})^2 < \frac{\sigma_n(\nu)}{\sigma_{n+1}(\nu)} \quad ; \quad n = 1, 2, 3, \ldots \tag{4.9.148}$$

The sums can be successively evaluated using the Kishore recurrence formula [113, 128, 155–157]

$$\begin{aligned} \sigma_n(\nu) &= \frac{1}{(n+\nu)} \sum_{k=1}^{n-1} \sigma_k(\nu)\, \sigma_{n-k}(\nu) \\ \sigma_{2n+1}(\nu) &= 0 \end{aligned} \tag{4.9.149}$$

The explicit form of first few Rayleigh functions is

$$\sum_{k=1}^{\infty} \frac{1}{(j_{\nu,k})^2} = \frac{1}{4(\nu+1)} \tag{4.9.150}$$

$$\sum_{k=1}^{\infty}\frac{1}{(j_{\nu,k})^4}=\frac{1}{16(\nu+1)^2(\nu+2)} \tag{4.9.151}$$

$$\sum_{k=1}^{\infty}\frac{1}{(j_{\nu,k})^6}=\frac{1}{32(\nu+1)^3(\nu+2)(\nu+3)} \tag{4.9.152}$$

$$\sum_{k=1}^{\infty}\frac{1}{(j_{\nu,k})^8}=\frac{5\nu+11}{256(\nu+1)^4(\nu+2)^2(\nu+3)(\nu+4)} \tag{4.9.153}$$

$$\sum_{k=1}^{\infty}\frac{1}{(j_{\nu,k})^{10}}=\frac{7\nu+19}{512(\nu+1)^5(\nu+2)^2(\nu+3)(\nu+4)(\nu+5)} \tag{4.9.154}$$

The results associated with the zeros of the Bessel function can also be extended to derivatives. If $\nu > n-1$, then nth derivative of the Bessel function of the first kind of the order ν has infinitely many zeros which are all real simple except at the origin. If $\nu > n$, then the positive zeros of nth and $(n+1)$th derivative are interlacing. For the first and second derivatives of the Bessel function, we have the following Euler-Rayleigh sums

$$\sum_{k=1}^{\infty}\frac{1}{(j'_{\nu,k})^2}=\frac{\nu+2}{4\nu(\nu+1)} \tag{4.9.155}$$

$$\sum_{k=1}^{\infty}\frac{1}{(j'_{\nu,k})^4}=\frac{\nu^2+8\nu+8}{[4\nu(\nu+1)]^2(\nu+2)} \tag{4.9.156}$$

$$\sum_{k=1}^{\infty}\frac{1}{(j'_{\nu,k})^6}=\frac{\nu^3+16\nu^2+38\nu+24}{2^5[\nu(\nu+1)]^3(\nu+2)(\nu+3)} \tag{4.9.157}$$

$$\sum_{k=1}^{\infty}\frac{1}{(j''_{\nu,k})^2}=\frac{\nu+2}{4\nu(\nu-1)} \tag{4.9.158}$$

$$\sum_{k=1}^{\infty}\frac{1}{(j''_{\nu,k})^4}=\frac{13\nu^3+19\nu^2+26\nu+8}{[4\nu(\nu-1)]^2(\nu+1)(\nu+2)} \tag{4.9.159}$$

It is worthwhile to mention also the general result for zeros of the nth derivative

$$\sum_{k=1}^{\infty}\frac{1}{(j^{(n)}_{\nu,k})^2}=\frac{\nu+2}{4(\nu-n+1)(\nu-n+2)} \tag{4.9.160}$$

$$\sum_{k=1}^{\infty}\frac{1}{(j^{(n)}_{\nu,k})^4}=\frac{\nu+2}{16(\nu-n+1)(\nu-n+2)}R(\nu,n)$$

$$R(\nu,n)=\left[\frac{(\nu+2)^2}{(\nu-n+1)(\nu-n+2)}-\frac{(\nu+3)(\nu+4)}{(\nu-n+3)(\nu-n+4)}\right] \tag{4.9.161}$$

In special cases of the Bessel functions, $J_{1/2}(z)$ and $J_{-1/2}(z)$, their positive zeros are known $j_{1/2,k} = \pi k$ and $j_{-1/2,k} = \pi(k_{-1/2})$, and therefore the Rayleigh functions are reduced to

$$\sigma_n\left(\frac{1}{2}\right) = \sum_{k=1}^{\infty} \frac{1}{\left(j_{1/2,k}\right)^{2n}} = \frac{\varsigma(2n)}{\pi^{2n}} \quad ; \quad n = 1, 2, 3, \ldots \tag{4.9.162}$$

$$\sigma_n\left(-\frac{1}{2}\right) = \sum_{k=1}^{\infty} \frac{1}{\left(j_{-1/2,k}\right)^{2n}} = \frac{(2^{2n}-1)\,\varsigma(2n)}{\pi^{2n}} \quad ; \quad n = 1, 2, 3, \ldots \tag{4.9.163}$$

where $\zeta(z)$ is the zeta function.

In case of the Airy functions, the following Euler-Rayleigh sums are known

$$\sum_{k=1}^{\infty} \frac{1}{|a'_k|^2} = -\frac{Ai(0)}{Ai'(0)} = \frac{\Gamma\left(\frac{1}{3}\right)}{3^{1/3}\Gamma\left(\frac{2}{3}\right)} \quad ; \quad Ai'(a'_k) = 0$$

$$a'_k = -\varsigma^{2/3}\left\{1 - \frac{7}{48\varsigma^2} + \frac{35}{288\varsigma^4} - \frac{181228}{207360\varsigma^6} + \ldots\right\} \tag{4.9.164}$$

$$\varsigma = \frac{3\pi(4k-3)}{8}$$

$$\sum_{k=1}^{\infty} \frac{1}{|a'_k|^3} = 1 \tag{4.9.165}$$

$$\sum_{k=1}^{\infty} \frac{1}{|a'_k|^4} = \frac{\left[\Gamma\left(\frac{1}{3}\right)\right]^2}{2 \cdot 3^{2/3}\left[\Gamma\left(\frac{2}{3}\right)\right]^2} \tag{4.9.166}$$

$$\sum_{k=1}^{\infty} \frac{1}{|a'_k|^5} = \frac{2\Gamma\left(\frac{1}{3}\right)}{3^{4/3}\Gamma\left(\frac{2}{3}\right)} \tag{4.9.167}$$

$$\sum_{k=1}^{\infty} \frac{1}{|a'_k|^6} = \frac{1}{4}\left\{1 + \frac{\left[\Gamma\left(\frac{1}{3}\right)\right]^3}{3\left[\Gamma\left(\frac{2}{3}\right)\right]^3}\right\} \tag{4.9.168}$$

$$\sum_{k=1}^{\infty} \frac{1}{|a'_k|^7} = \frac{7\left[\Gamma\left(\frac{1}{3}\right)\right]^2}{3^{2/3} \cdot 15\left[\Gamma\left(\frac{2}{3}\right)\right]^2} \tag{4.9.169}$$

$$\sum_{k=1}^{\infty} \frac{1}{|a'_k|^8} = \frac{11\Gamma\left(\frac{1}{3}\right)}{3^{1/3} \cdot 36\Gamma\left(\frac{2}{3}\right)} + \frac{\left[\Gamma\left(\frac{1}{3}\right)\right]^4}{3^{4/3} \cdot 8\left[\Gamma\left(\frac{2}{3}\right)\right]^4} \tag{4.9.170}$$

Contrary to significant interest devoted to zeros of the Bessel functions of the first and the second kind $J_\nu(z)$ and $Y_\nu(z)$, their linear combinations or cross products, only limited attention has been directed to zeros of the Struve functions $h_{\nu,k}$, and even less to zeros of the Lommel functions, $\sigma_{\nu,k}.(\mu)$

In the case of the Struve functions

$$H_\nu(h_{\nu,k})=0 \tag{4.9.171}$$

the positive zeros are all simple and they satisfy the following chains of inequalities

$$\begin{aligned}&j_{\nu,k}<\pi k<h_{\nu,k}<j_{\nu,k+1}<\pi(k+1)<h_{\nu,k+1}<j_{\nu,k+2}<\dots\\&|\nu|<\frac{1}{2}\quad;\quad k=1,2,3,\dots\end{aligned} \tag{4.9.172}$$

$$\begin{aligned}&y_{\nu,k}<y_{\nu,k+1}<h_{\nu,k}<\pi k<h_{\nu,k+1}<y_{\nu,k+2}<\dots\\&|\nu|<\frac{1}{2}\quad;\quad k=1,2,3,\dots\end{aligned} \tag{4.9.173}$$

$$\begin{aligned}&y_{\nu,k}<j_{\nu,k}<y_{\nu,k+1}<h_{\nu,k}<j_{\nu,k+1}<\pi(k+1)<h_{\nu,k+1}<y_{\nu,k+2}<\\&|\nu|<\frac{1}{2}\quad;\quad k=1,2,3,\dots\end{aligned} \tag{4.9.174}$$

and

$$\begin{aligned}&h_{\nu,k}<y_{\nu,k}<j_{\nu,k}<y_{\nu,k+1}<h_{\nu,k+1}<j_{\nu,k+1}<h_{\nu,k+2}<\dots\\&-n-\frac{1}{2}<\nu\le -n\quad;\quad k,n=1,2,3,\dots\end{aligned} \tag{4.9.175}$$

$$\begin{aligned}&j_{\nu,k}<h_{\nu,k}<y_{\nu,k}<j_{\nu,k+1}<y_{\nu,k+1}<h_{\nu,k+1}<j_{\nu,k+2}<\dots\\&-n<\nu\le -n+\frac{1}{2}\quad;\quad k,n=1,2,3,\dots\end{aligned} \tag{4.9.176}$$

For the smallest positive real zeros of the Struve function and its derivatives, the lower and upper bounds are:

$$h_{\nu,1}>\sqrt{3(2\nu+3)}\quad;\quad|\nu|<\frac{1}{2} \tag{4.9.177}$$

$$(h_{\nu,1})^2>3(2\nu+3)\sqrt{\frac{5(2\nu+5)}{7-2\nu}}\quad;\quad|\nu|<\frac{1}{2} \tag{4.9.178}$$

$$h'_{\nu,1}>\sqrt{\frac{3(\nu+1)(2\nu+3)}{\nu+3}}\quad;\quad|\nu|<\frac{1}{2} \tag{4.9.179}$$

$$(h'_{\nu,1})^2>\frac{3(\nu+1)(2\nu+3)}{\nu+3}\quad;\quad|\nu|<\frac{1}{2} \tag{4.9.180}$$

$$(h'_{\nu,1})^2 > 3(\nu+1)(2\nu+3)\sqrt{\frac{5(2\nu+5)}{-2\nu^3-5\nu^2+72\nu+135}} \qquad |\nu| < \frac{1}{2} \tag{4.9.181}$$

$$(h'_{\nu,1})^2 < \frac{15(\nu+1)(\nu+3)(2\nu+3)(2\nu+5)}{-2\nu^3-5\nu^2+72\nu+135} \qquad |\nu| < \frac{1}{2} \tag{4.9.182}$$

$$(h'_{\nu,1})^2 < \frac{21(\nu+1)(2\nu+3)(2\nu+7)(-2\nu^3-5\nu^2+72\nu+135)}{-4\nu^5-132\nu^4-1115\nu^3+621\nu^2+12339\nu+14931} \qquad |\nu| < \frac{1}{2} \tag{4.9.183}$$

$$(h''_{\nu,1})^2 > 3(\nu+1)(2\nu+3)\sqrt{\frac{3\nu(\nu+1)(2\nu+3)}{(\nu+2)(\nu+3)}} \qquad \nu > 0 \tag{4.9.184}$$

$$(h''_{\nu,1})^2 > 3\nu(\nu+1)(2\nu+3)I(\nu)$$
$$I(\nu) = \sqrt{\frac{5(2\nu+5)}{-2\nu^5-13\nu^4+92\nu^3+763\nu^2+1500\nu+900}} \qquad \nu > 0 \tag{4.9.185}$$

The Euler–Rayleigh sums of zeros of the Struve function and its derivatives are:

$$\sum_{k=1}^{\infty} \frac{1}{(h_{\nu,k})^2} = \frac{1}{3(2\nu+3)} \quad ; \quad |\nu| < \frac{1}{2} \tag{4.9.186}$$

$$\sum_{k=1}^{\infty} \frac{1}{(h_{\nu,k})^4} = \frac{7-2\nu}{45(2\nu+3)(2\nu+5)} \quad ; \quad |\nu| < \frac{1}{2} \tag{4.9.187}$$

$$\sum_{k=1}^{\infty} \frac{1}{(h'_{\nu,k})^2} = \frac{\nu+3}{3(\nu+1)(2\nu+3)} \quad ; \quad |\nu| < \frac{1}{2} \tag{4.9.188}$$

$$\sum_{k=1}^{\infty} \frac{1}{(h'_{\nu,k})^4} = \frac{-2\nu^3-5\nu^2+72\nu+135}{45[(\nu+1)(2\nu+3)]^2(2\nu+5)} \quad ; \quad |\nu| < \frac{1}{2} \tag{4.9.189}$$

$$\sum_{k=1}^{\infty} \frac{1}{(h''_{\nu,k})^2} = \frac{(\nu+2)(\nu+3)}{3\nu(\nu+1)(2\nu+3)} \quad ; \quad \nu > 0 \tag{4.9.190}$$

$$\sum_{k=1}^{\infty} \frac{1}{(h''_{\nu,k})^4} = \frac{-2\nu^5 - 13\nu^4 + 92\nu^3 + 763\nu^2 + 1500\nu + 900}{45[\nu(\nu+1)(2\nu+3)]^2(2\nu+5)} \tag{4.9.191}$$
$$\nu > 0$$

In the case of positive zeros of the Lommel function $s_{\mu,\nu}(z)$, only two inequality chains are available

$$\begin{aligned} &0 < j_{\nu,1} < \sigma_{\nu,1}(\mu) < j_{\nu,2} < \sigma_{\nu,2}(\mu) < j_{\nu,3} < \sigma_{\nu,3}(\mu) < \dots \\ &\nu < \mu + 1 \\ &0 < \sigma_{\nu,1}(\mu) < j_{\nu,1} < \sigma_{\nu,2}(\mu) < j_{\nu,2} < \sigma_{\nu,3}(\mu) < j_{\nu,3} < \dots \\ &\nu > \mu + 1 \end{aligned} \tag{4.9.192}$$

More information about zeros of the Struve and the Lommel functions is given in the Steinig [153], Koumandos and Lamprecht [130], and Cho and Chung [104] papers.

4.10 Bessel Functions of Equal or Nearly Equal Order and Argument

Watson, in his book, mentioned that the case of the Bessel functions with equal order and argument (large positive orders) has been discussed by Kelvin, Airy, Stokes and others, but the first result for the Bessel function of the first kind $J_\nu(\nu)$ belongs to Cauchy, who gave, in 1854, the following expression

$$J_\nu(\nu) \sim \frac{\Gamma\left(\frac{1}{3}\right)}{2^{2/3} 3^{1/6} \pi \nu^{1/3}} \tag{4.10.1}$$

Watson extended this result to

$$J_\nu(\nu) \sim \frac{\Gamma\left(\frac{1}{3}\right)}{2^{2/3} 3^{1/6} \pi \nu^{1/3}} - \frac{3^{5/6}\Gamma\left(\frac{5}{3}\right)}{2^{1/3} 140 \pi \nu^{5/3}} + O\left(\frac{1}{\nu^{5/3}}\right) \tag{4.10.2}$$

and showed that at limit we have the Cauchy result

$$\nu^{1/3} J_\nu(\nu) < \lim_{\nu\to\infty} \left[\nu^{1/3} J_\nu(\nu)\right] = \frac{\Gamma\left(\frac{1}{3}\right)}{2^{2/3} 3^{1/6} \pi} = 0.44731\dots \tag{4.10.3}$$

and similarly for the derivative he demonstrated that the dominant term is

$$\nu^{2/3} J'_\nu(\nu) < \lim_{\nu\to\infty} \left[\nu^{2/3} J'_\nu(\nu)\right] = \frac{3^{1/6}\Gamma\left(\frac{2}{3}\right)}{2^{1/3}\pi} = 0.41085\dots.. \tag{4.10.4}$$

Thus, the Cauchy formula is always in error by excess, and Watson gave also the estimation for the Bessel function of the second kind

$$Y_\nu(\nu) \sim -\frac{3^{1/3}\Gamma(\frac{1}{3})}{2^{2/3}\pi\nu^{1/3}} \tag{4.10.5}$$

Considering the monotonicity properties of Bessel functions Lorch [78] studied products of more general type that present in (4.10.3) and (4.10.4), they were $\nu^{1/\alpha}J_\nu(\nu)$, $\nu^{2/\alpha}J'_\nu(\nu)$ and $\nu^{1/\alpha}J'_\nu(\nu)$ with $\alpha > 3$.

Over a long period, the Bessel functions of the same argument and order continue to attract mathematicians and in 1955, Gatteschi [115] proposed the next approximation for the Bessel function of the first kind

$$J_\nu(\nu) \sim \frac{\Gamma\left(\frac{1}{3}\right)}{2^{2/3}3^{1/6}\pi\nu^{1/3}} - \eta\theta$$

$$\eta = \frac{e^{-2\pi/\sqrt{3}} + 0.521\, e^{-(2\pi/\sqrt{3})^3\nu/6}}{\pi\nu} + \frac{1.4}{\pi}\left(\frac{\nu}{6}\right)^{5/3} \quad ; \quad 0 < \theta \le 1 \tag{4.10.6}$$

$$J_\nu(\nu) \sim \frac{\Gamma\left(\frac{1}{3}\right)}{2^{2/3}3^{1/6}\pi\nu^{1/3}} - \frac{9\theta}{\nu^{5/3}} \quad ; \quad 0 < \theta \le 1 \quad ; \quad \nu \ge 6$$

and for the Bessel function of the second kind

$$Y_\nu(\nu) \sim -\frac{3^{1/3}\Gamma\left(\frac{1}{3}\right)}{2^{2/3}\pi\nu^{1/3}} + \rho$$

$$|\rho| \le \frac{0.252}{\pi\nu} \quad ; \quad \nu \ge 1 \tag{4.10.7}$$

$$Y_\nu(\nu) \sim -\frac{3^{1/3}\Gamma\left(\frac{1}{3}\right)}{2^{2/3}\pi\nu^{1/3}} + \frac{\theta}{\nu} \quad ; \quad 0 < \theta < 1 \quad ; \quad \nu \ge 6$$

The corresponding formulas for derivatives of the Bessel functions are:

$$J'_\nu(\nu) \sim \frac{6^{2/3}}{3\,\Gamma\left(\frac{1}{3}\right)\nu^{2/3}} + \frac{6^{4/3}\sqrt{3}\Gamma\left(\frac{1}{3}\right)}{180\pi\nu^{4/3}} + \frac{2\theta}{\nu^2} \quad ; \quad |\theta| < 1$$

$$J'_\nu(\nu) \sim \frac{1}{2\sqrt{3}\pi}\left[\Gamma\left(\frac{1}{3}\right)\left(\frac{6}{\nu}\right)^{2/3} - \frac{\Gamma\left(\frac{1}{3}\right)}{30}\left(\frac{6}{\nu}\right)^{4/3}\right] + \frac{2\theta}{\nu^2} \tag{4.10.8}$$

$$0 < \theta \le 1 \quad ; \quad \nu \ge 6$$

Descriptive properties of the Bessel function of the first kind $J_\nu(\nu x)$ in the $0 < x \le 1$ interval were discussed by Watson [7] who showed that the Carlini formula is the first term in the asymptotic expansion which is valid for large values of the order ν

$$J_\nu(\nu x) \sim \frac{e^{\nu\sqrt{1-x^2}}}{\sqrt{2\pi\nu}\,(1-x^2)^{1/4}}\left[\frac{x}{1+\sqrt{1-x^2}}\right]^\nu \tag{4.10.9}$$

and derived the following inequalities

$$\begin{aligned} J_\nu(\nu x) &\le \frac{e^{\nu\sqrt{1-x^2}}}{\sqrt{2\pi\nu}}\left[\frac{x}{1+\sqrt{1-x^2}}\right]^\nu \quad ; \quad 0 \le x < 1 \quad ; \quad \nu > 0 \\ J'_\nu(\nu x) &\le \frac{e^{\nu\sqrt{1-x^2}}(1+x)^{1/4}}{\sqrt{2\pi\nu x}}\left[\frac{x}{1+\sqrt{1-x^2}}\right]^\nu \end{aligned} \tag{4.10.10}$$

By using the recursion formulas of the Bessel functions, Siegel and Sleator [158] modified these expressions correspondingly for $x > 1$. They also presented the equivalent formulas for the Bessel function of the first kind

$$\begin{aligned} J_\nu(\nu x) &\le \frac{e^{\nu\sqrt{1-x^2}}}{\sqrt{2\pi\nu}}\left[\frac{x}{1+\sqrt{1-x^2}}\right]^\nu \left[1+\left(\frac{1+x^2}{4\pi^2\nu^2}\right)^{1/4}\right] \\ & 0 \le x < 1 \quad ; \quad \nu \gg 1 \end{aligned} \tag{4.10.11}$$

and the formulas for the Bessel function of the second kind $Y_\nu(\nu x)$ [159]

$$\begin{aligned} |Y_\nu(\nu x)| &\le \frac{\Gamma(\frac{1}{3})}{\sqrt{x}}\left[\frac{1}{\pi\nu^{1/3}}\left(\frac{1+2\cdot 3^{-1/2}+|\cot(\pi\nu)|}{2^{2/3}\cdot 3^{1/6}}\right)+\frac{1}{\Gamma(\frac{1}{3})|\sin(\pi\nu)|}\right] \\ & x>1 \quad ; \quad \nu>0 \quad ; \quad \nu \ne 1,2,3,\ldots n \end{aligned} \tag{4.10.12}$$

and

$$\begin{aligned} |Y_n(nx)| &\le \frac{1}{\sqrt{x}}\left[\frac{0.964}{n^{1/3}}+\frac{1}{2\pi n}+1\right] \quad ; \quad n=1,2,3,\ldots \\ |Y_\nu(\nu x)| &\le \frac{1}{\sqrt{x}}\left[\frac{0.964}{\nu^{1/3}}+\frac{|\cos(\pi\nu)|}{2\pi n}+e^{\pi\varepsilon}\right] \\ & \varepsilon=\nu-n \quad ; \quad \varepsilon \ll 1 \end{aligned} \tag{4.10.13}$$

The Siegel and Sleator investigations inspired Gatteschi to derive new expressions for the Bessel function of the first and second kind

$$|J_\nu(\nu x)| < \frac{4}{\pi\sqrt{x}\,\nu^{1/3}} \quad ; \quad x>1 \quad ; \quad \nu \geq \frac{1}{2}$$
$$|J_\nu(\nu x)| < \frac{\Gamma(\frac{1}{3})(3^{1/3}+3^{-1/6})}{2^{3/2}\pi\,\nu^{1/3}\sqrt{x}} + \frac{1}{\nu\sqrt{x}} \quad ; \quad x \geq 1 \quad ; \quad \nu > 6 \tag{4.10.14}$$

and

$$|Y_\nu(\nu x)| < \frac{1}{\sqrt{x}}\left[\frac{3.841}{\pi\,\nu^{1/3}} + \frac{0.252}{\pi\nu}\right] \quad ; \quad x>1 \quad ; \quad \nu>0 \tag{4.10.15}$$

A quite different approach in treating the Bessel functions of nearly equal argument and order was presented by Paris [146]. He derived a number of new inequalities, most of them for the $0 < x \leq 1$ interval

$$J_{\nu+1}(\nu x) < \frac{\nu x}{\nu+2} J_\nu(\nu x) \quad 0<x\leq 1 \quad ; \quad \nu>0 \tag{4.10.16}$$

$$0 < \frac{\nu x}{2\nu+2} < \frac{J_{\nu+1}(\nu x)}{J_\nu(\nu x)} < \frac{\nu x}{\nu+2} < 1 \quad ; \quad 0<x\leq 1 \quad ; \quad \nu>0 \tag{4.10.17}$$

$$0 < \frac{1}{x} - \frac{J'_\nu(\nu x)}{J_\nu(\nu x)} < 1 \quad ; \quad 0<x\leq 1 \quad ; \quad \nu>0 \tag{4.10.18}$$

and by using the exponential functions he demonstrated that

$$1 \leq \frac{J_\nu(\nu x)}{x^\nu J_\nu(\nu)} \leq e^{\nu(1-x)} \quad ; \quad 0<x\leq 1 \quad ; \quad \nu>0 \tag{4.10.19}$$

$$e^{\nu^2(1-x^2)/(4\nu+4)} \leq \frac{J_\nu(\nu x)}{x^\nu J_\nu(\nu)} \leq e^{\nu^2(1-x^2)/(2\nu+2)}$$
$$0<x<1 \quad ; \quad \nu>0 \tag{4.10.20}$$

It is worthwhile to mention some formulas of general character when both the argument x and the order ν are involved in mathematical operations. Usually, they are related to descriptive properties of the Bessel functions in the $0 \leq x \leq 1$ interval and with $\nu > 0$

$$[J_\nu(\nu x)]^2 + [Y_\nu(\nu x)]^2 < \frac{1}{x}\left\{[J_\nu(\nu)]^2 + [Y_\nu(\nu)]^2\right\}$$
$$|J_\nu(\nu x)| < \frac{1}{\sqrt{x}}\{|J_\nu(\nu)| + |Y_\nu(\nu)|\} \tag{4.10.21}$$
$$\nu \geq \frac{1}{2} \quad ; \quad x>1$$

$$\frac{\partial}{\partial \nu}\left\{\nu^{1/3} J_\nu(\nu)\right\} > 0 \tag{4.10.22}$$

$$\frac{\partial}{\partial \nu}\left\{\nu^{1/3} J'_\nu(\nu)\right\} > 0 \tag{4.10.23}$$

and

$$\frac{\partial}{\partial \nu}\left\{\frac{J_\nu(\nu x)}{J_\nu(\nu)}\right\} \le 0 \tag{4.10.22}$$

$$\frac{\partial}{\partial \nu}\left\{\nu^{1/3}\frac{J'_\nu(\nu)}{J_\nu(\nu)}\right\} \ge 0 \tag{4.10.25}$$

$$\frac{\partial}{\partial x}\left\{\left(\frac{\partial J_\nu(\nu x)}{\partial \nu}\right)/J_\nu(\nu x)\right\} \ge 0 \tag{4.10.26}$$

$$\frac{1}{J_\nu(\nu x)}\left(\frac{\partial J_\nu(\nu x)}{\partial x}\right) \le \frac{1}{J_\nu(\nu)}\left(\frac{\partial J_\nu(\nu)}{\partial \nu}\right) \tag{4.10.27}$$

$$J_\nu(\nu x)\frac{\partial^2 J_\nu(\nu x)}{\partial \nu \partial x} - \left(\frac{\partial J_\nu(\nu x)}{\partial \nu}\right)\left(\frac{\partial J_\nu(\nu x)}{\partial x}\right) \ge 0 \tag{4.10.28}$$

The second group of formulas is related to the Nicolson integral (4.8.42)

$$[J_\nu(x)]^2 + [Y_\nu(x)]^2 = \frac{8}{\pi^2}\int_0^\infty K_0(2x\sinh\xi)\cosh(2\nu\xi)\,d\xi \tag{4.10.29}$$

$$\frac{2}{\pi\sqrt{x^2-\nu^2}} > \left\{[J_\nu(x)]^2 + [Y_\nu(x)]^2\right\} < \frac{2}{\pi x} \tag{4.10.30}$$

$$J_\nu(x)\frac{\partial Y_\nu(x)}{\partial \nu} - Y_\nu(x)\frac{\partial J_\nu(x)}{\partial \nu} = -\frac{4}{\pi}\int_0^\infty e^{-2\nu\xi} K_0(2x\sinh\xi)\,d\xi \tag{4.10.31}$$

$$J_\nu(x)\frac{\partial Y_\nu(x)}{\partial \nu} - Y_\nu(x)\frac{\partial J_\nu(x)}{\partial \nu} > 0 \tag{4.10.32}$$

$$\begin{aligned} &J'_\nu(x)\frac{\partial Y'_\nu(x)}{\partial \nu} - Y'_\nu(x)\frac{\partial J'_\nu(x)}{\partial \nu} \\ &= -\frac{4}{\pi x^2}\int_0^\infty e^{-2\nu\xi}[x^2\cosh(2\xi) - \nu^2]\,K_0(2x\sinh\xi)\,d\xi \end{aligned} \tag{4.10.33}$$

4.11 The Asymptotic Limit of the Bessel Function $\nu J_\nu(\nu x)$ Expressed as the Shifted Dirac Function $\delta(x-1)$. Evaluation of Integral Representations of Elementary Functions, Special Functions, Mathematical Constants, Integral Transforms, Asymptotic Relations, Integrals and Limits of Functional Series

A sudden excitation of physical systems is expressed mathematically by the impulse functions. These functions were introduced to mathematics by Fourier, Cauchy, Poisson and Hermite. However, their use was significantly extended when Helmholtz, Kirchhoff, Kelvin and Heaviside started to apply them in solutions of various practical problems in electricity, fluid mechanics and heat transfer [82, 160–162]. The recognition of impulse functions as a powerful tool in many branches of applied mathematics, physics and engineering was firmly established when Dirac introduced them to quantum mechanics in 1926. From that time, the impulse function is named after him, $\delta(x)$ – Dirac delta function. The impulse functions are not ordinary functions in mathematical sense, but functions defined by giving a rule for integrating their product with a continuous function $f(x)$. These functions are approximated by the sequence of functions converging to them (i.e. delta sequences).

$$\delta(x-1) = \lim_{n\to\infty} \delta_n(x-1) \tag{4.11.1}$$

The above sequence is written in the form for the shifted delta function $\delta(x-1)$, which will be used in this section. A number of delta sequences are known in mathematics, three "classical" sequences are presented below

$$\delta(x-1) = \lim_{\nu\to\infty}\left(\frac{\sin[\nu(x-1)]}{\pi(x-1)}\right) \tag{4.11.2}$$

$$\delta(x-1) = \frac{1}{\pi}\lim_{\nu\to\infty}\left(\frac{\nu}{1+\nu^2(x-1)^2}\right) = \frac{1}{\pi}\lim_{\varepsilon\to 0}\left(\frac{\varepsilon}{\varepsilon^2+(x-1)^2}\right) \tag{4.11.3}$$

$$\delta(x-1) = \frac{1}{2}\lim_{\nu\to\infty}\left(\frac{\nu e^{-|\nu(x-1)|^n}}{\Gamma\left(1+\frac{1}{n}\right)}\right) \quad ; \quad n = 1, 2, 3, \ldots \tag{4.11.4}$$

The first sequence was proposed by Fourier, the second sequence by Cauchy and the last by Kelvin, when for $n = 2$, the sequence is called the "heat source." Since our main interest is directed to the Bessel and related functions, the shifted Dirac function $\delta(x-1)$ can be represented by the orthogonality identities for the Bessel functions of the first and second kind and the Struve function of the first kind

$$\delta(x-1) = \int_0^\infty (xt) J_\nu(t) J_\nu(xt)\, dt \qquad x > 0 \quad ; \quad \operatorname{Re}\nu > -1 \tag{4.11.5}$$

and

$$\delta(x-1) = \int_0^\infty (xt)\, Y_\nu(t) H_\nu(xt)\, dt \qquad x > 0 \quad ; \quad \operatorname{Re}\nu > -1 \tag{4.11.6}$$

However, the delta sequence, which is directly related to the order of Bessel function and therefore is of particular interest to us, was derived in 1969 by Lamborn [83]

$$\delta(x-1) = \lim_{\nu\to\infty} [\nu J_\nu(\nu x)] \tag{4.11.7}$$

Using this shifted Dirac delta function, the present author over the 1999–2008 period, in a series of papers, was able to derive many new mathematical relations. They included representations of the mathematical constants, the elementary and special functions, the asymptotic and functional relations, infinite integrals, series and limits [84,85,29]. Following these investigations, Laforgia and Natalini [163] obtained additional limiting relations, mainly for the Bessel functions. The equivalent form of the Lamborn expression in terms of the Bessel-Clifford function is

$$\delta(4x-1) = \frac{1}{4} \lim_{\nu\to\infty} \left[\nu^{\nu+1} x^{\nu-1/2} C_\nu(\nu^2 x)\right] \qquad C_\nu(x) = x^{-\nu/2} J_\nu(2\sqrt{x}) \tag{4.11.8}$$

It is also possible to express the delta function by using the infinite integral with integrand having product of the Airy functions

$$\delta(x-a) = \int_{-\infty}^\infty Ai(t-x)\, Ai(t-a)\, dt \tag{4.11.9}$$

The shifted Dirac delta function $\delta(x-a)$ is defined by

$$\delta(x-a) = \begin{cases} 0 & \text{for all } x \neq a \\ \infty & \text{for } x = a \end{cases} \tag{4.11.10}$$

and its most important property is the so-called sifting property or the sampling property

$$\int_{-\infty}^{\infty} \delta(x-a)f(x)dx = f(a) \tag{4.11.11}$$

The following formula represents derivatives of the delta function

$$\int_{-\infty}^{\infty} \delta^{(n)}(x-a)f(x)dx = (-1)^n f^{(n)}(a) \qquad n = 0, 1, 2, 3\ldots \tag{4.11.12}$$

where it is assumed that $f(x)$ is continuously differentiable in the neighbourhood of point a, $-\infty < a < \infty$. The sifting property in (4.11.11) is independent of the actual values of the limits of integration, it depends only on the behaviour of the integrand near the point a. Thus, the delta function and its derivatives are defined by a rule, by integration of the products appearing in (4.11.11) and (4.11.12).

If the Lamborn delta sequence in (4.11.7) is multiplied by a function $f(tx)$ and integrated from zero to infinity with respect to variable x we have

$$f(t) = \int_0^{\infty} f(tx)\delta(x-1)\,dx = \lim_{\nu\to\infty}\left[\nu\int_0^{\infty} f(tx)J_\nu(\nu x)\,dx\right] \tag{4.11.13}$$

It is assumed here and in other places that inversion of mathematical operations is permissible. Thus, it follows from (4.11.13) that function $f(t)$ is represented by the asymptotic limit of the infinite integral of product of $f(tx)$ and the Bessel function $J_\nu(\nu x)$. If the right-hand integral in (4.11.13) is evaluated in the closed form then the limit can be regarded as the generalization of the l'Hôpital's rule.

$$f(t) = \lim_{\nu\to\infty}[\nu\,\Phi(t,\nu)] \qquad \Phi(t,\nu) = \int_0^{\infty} f(tx)J_\nu(\nu x)\,dx \tag{4.11.14}$$

Since many infinite integrals having the integrand Bessel functions and various $f(tx)$ functions are tabulated, it is possible to present many functions $f(t)$ as the limit given in (4.11.14). A number of examples, mainly with the Bessel functions, will illustrate the applied procedure. Let us start with the Bessel function itself, $f(tx) = J_\mu(tx)$

$$\Phi(t,\nu,\mu) = \int_0^{\infty} J_\mu(tx)J_\nu(\nu x)\,dx \tag{4.11.15}$$

but this is the Weber-Schafheitlin type integral [7]

$$\int_0^\infty J_\nu(\nu x)J_\mu(tx)\,dx = \frac{\Gamma\left(\frac{\mu+\nu+1}{2}\right)t^\mu}{\Gamma(\mu+1)\,\Gamma\left(\frac{\nu-\mu+1}{2}\right)\nu^{\mu+1}}\,{}_2F_1\left(\frac{\mu+\nu+1}{2},\frac{\mu-\nu+1}{2};\mu+1;\frac{t^2}{\nu^2}\right)$$

$$\operatorname{Re}\mu > -1 \quad ; \quad 0 < t < \nu \qquad (4.11.16)$$

Thus, using (4.11.14) and (4.11.16), the Bessel function is represented as the limit of the hypergeometric function

$$J_\mu(t) = \frac{t^\mu}{\Gamma(\mu+1)}\lim_{\nu\to\infty}\left\{\frac{\Gamma\left(\frac{\mu+\nu+1}{2}\right)}{\Gamma\left(\frac{\nu-\mu+1}{2}\right)\nu^\mu}I(t,\mu,\nu)\right\}$$

$$I(t,\mu,\nu) = {}_2F_1\left(\frac{\mu+\nu+1}{2},\frac{\mu-\nu+1}{2};\mu+1;\frac{t^2}{\nu^2}\right) \qquad (4.11.17)$$

$$\operatorname{Re}\mu > -1 \quad ; \quad 0 < t < \nu$$

In particular cases, for $\mu = 0$ and $\mu = 1$, (4.11.17) becomes

$$J_0(t) = \lim_{\nu\to\infty}\left\{{}_2F_1\left(\frac{1+\nu}{2},\frac{1-\nu}{2};1;\frac{t^2}{\nu^2}\right)\right\}$$

$$J_1(t) = \frac{t}{2}\lim_{\nu\to\infty}\left\{{}_2F_1\left(\frac{\nu}{2}+1,1-\frac{\nu}{2};2;\frac{t^2}{\nu^2}\right)\right\} \qquad (4.11.18)$$

In the second example the square of the Bessel function, $f(tx) = [J_\mu(tx)]^2$, is considered

$$\Phi(t,\nu,\mu) = \int_0^\infty [J_\mu(tx)]^2 J_\nu(\nu x)\,dx \qquad (4.11.19)$$

but also in this case the integral in (4.11.19) is known

$$\int_0^\infty [J_\mu(tx)]^2 J_\nu(\nu x)\,dx = \frac{\Gamma\left(\frac{2\mu+\nu+1}{2}\right) t^{2\mu}}{[\Gamma(\mu+1)]^2 \Gamma\left(\frac{2\mu-\nu+1}{2}\right) \nu^{2\mu+1}} I(t,\mu,\nu)$$

$$I(t,\mu,\nu) = \left[{}_2F_1\left(\frac{2\mu-\nu+1}{2}, \frac{2\mu+\nu+1}{2}; \mu+1; \frac{1-\sqrt{1-\left(\frac{4t}{\nu}\right)^2}}{2}\right)\right]^2 \qquad (4.11.20)$$

$$\mathrm{Re}\,(2\mu+\nu)-1$$

and therefore by using (4.11.14) and (4.11.20) we have the limiting form for the square of the Bessel functions

$$[J_\mu(t)]^2 = \frac{1}{[\Gamma(\mu+1)]^2} \lim_{\nu\to\infty} \frac{\Gamma\left(\frac{2\mu+\nu+1}{2}\right)}{\Gamma\left(\frac{2\mu-\nu+1}{2}\right)} \left(\frac{t}{\nu}\right)^{2\mu} I(t,\mu,\nu)$$

$$I(t,\mu,\nu) = \left[{}_2F_1\left(\frac{2\mu-\nu+1}{2}, \frac{2\mu+\nu+1}{2}; \mu+1; \frac{1-\sqrt{1-\left(\frac{4t}{\nu}\right)^2}}{2}\right)\right]^2 \qquad (4.11.21)$$

$$\mathrm{Re}\,(2\mu+\nu)-1$$

For $\mu = 0$ and $\mu = 1$, the limit representations of the square of the Bessel functions are

$$[J_0(t)]^2 = \lim_{\nu\to\infty} \left[{}_2F_1\left(\frac{1-\nu}{2}, \frac{1+\nu}{2}; 1; \frac{1-\sqrt{1-\left(\frac{4t}{\nu}\right)^2}}{2}\right)\right]^2$$

$$\left[\frac{J_1(t)}{t}\right]^2 = \lim_{\nu\to\infty} \left[{}_2F_1\left(1-\frac{\nu}{2}, \frac{1+\nu}{2}; 2; \frac{1-\sqrt{1-\left(\frac{4t}{\nu}\right)^2}}{2}\right)\right]^2 \qquad (4.11.22)$$

If $\mu = 1/2$, the Bessel function is the elementary function

$$J_{1/2}(t) = \sqrt{\frac{2}{\pi t}}\sin t \qquad (4.11.23)$$

and after few steps it is possible to obtain that

$$\left(\frac{\sin t}{t}\right)^2 = \lim_{\nu\to\infty} \left[{}_2F_1\left(-\frac{\nu}{2}, 1+\frac{\nu}{2}; \frac{3}{2}; \frac{1-\sqrt{1-\left(\frac{4t}{\nu}\right)^2}}{2}\right)\right]^2 \qquad (4.11.24)$$

Since the following integral is known [13]

$$\int_0^\infty J_\nu(\nu x) K_\mu(tx)\, dx = \frac{\Gamma\left(\frac{\mu+\nu+1}{2}\right)\Gamma\left(\frac{\nu-\mu+1}{2}\right)\nu^\nu}{2\Gamma(\mu+1)\,\Gamma(\nu+1)\,t^{\nu+1}} I(t,\mu,\nu)$$
$$I(t,\mu,\nu) = {}_2F_1\left(\frac{\mu+\nu+1}{2}, \frac{\nu-\mu+1}{2}; \nu+1; -\frac{\nu^2}{t^2}\right) \tag{4.11.25}$$
$$\mathrm{Re}\,(t+\nu) > 0$$

the limit representations of the modified Bessel function of the second kind is

$$K_\mu(t) = \lim_{\nu\to\infty}\left\{\frac{\Gamma\left(\frac{\mu+\nu+1}{2}\right)\Gamma\left(\frac{\nu-\mu+1}{2}\right)}{\Gamma(\nu+1)}\left(\frac{\nu}{t}\right)^{\nu+1} I(t,\mu,\nu)\right\}$$
$$I(t,\mu,\nu) = \frac{1}{2\Gamma(\mu+1)}\, {}_2F_1\left(\frac{\mu+\nu+1}{2}, \frac{\nu-\mu+1}{2}; \nu+1; -\frac{\nu^2}{t^2}\right) \tag{4.11.26}$$

For $\mu = 0$ and $\mu = 1$, (4.11.26) reduces to

$$K_0(t) = \lim_{\nu\to\infty}\left\{\frac{\sqrt{\pi}\,\Gamma\left(\frac{\nu+1}{2}\right)}{\Gamma\left(\frac{\nu}{2}+1\right)}\left(\frac{\nu}{2t}\right)^{\nu+1} {}_2F_1\left(\frac{\nu+1}{2}, \frac{\nu+1}{2}; \nu+1; -\frac{\nu^2}{t^2}\right)\right\}$$
$$K_1(t) = \lim_{\nu\to\infty}\left\{\frac{\sqrt{\pi}\,\Gamma\left(\frac{\nu}{2}\right)}{\Gamma\left(\frac{\nu+1}{2}\right)}\left(\frac{\nu}{2t}\right)^{\nu+1} {}_2F_1\left(\frac{\nu}{2}+1, \frac{\nu}{2}; \nu+1; -\frac{\nu^2}{t^2}\right)\right\} \tag{4.11.27}$$

From the next integral [13]

$$\int_0^\infty \frac{\left[tx+\sqrt{a^2+t^2x^2}\right]^\mu}{\sqrt{a^2+t^2x^2}} J_\nu(\nu x)\, dx = \frac{a^\mu}{t} I_{(\nu-\mu)/2}\left(\frac{at}{2}\right) K_{(\nu+\mu)/2}\left(\frac{at}{2}\right) \tag{4.11.28}$$
$$\mathrm{Re}\, a > 0 \quad ; \quad \mathrm{Re}\,\nu > -1 \quad ; \quad \mathrm{Re}\,\mu < \frac{3}{2}$$

it is possible to obtain the limit representation of the product of the modified Bessel functions

$$\lim_{\nu\to\infty}\left\{\left(\frac{\nu a^\mu}{t}\right) I_{(\nu-\mu)/2}\left(\frac{a\nu}{2t}\right) K_{(\nu+\mu)/2}\left(\frac{a\nu}{2t}\right)\right\} = \frac{\left(t+\sqrt{a^2+t^2}\right)^\mu}{\sqrt{a^2+t^2}} \tag{4.11.29}$$

Denoting $\lambda = \nu/2$, $\tau = a/t$, and $\rho = \mu/2$, the limit in (4.11.29) takes the form

$$\lim_{\lambda\to\infty}\{\lambda I_{\lambda-\rho}(\lambda\tau)K_{\lambda+\rho}(\lambda\tau)\}=\frac{\left(1+\sqrt{1+\tau^2}\right)^{2p}}{2\tau^{2p}\sqrt{1+\tau^2}} \tag{4.11.30}$$

$$\mathrm{Re}\,\tau>0\quad;\quad \mathrm{Re}\,\rho<\frac{3}{4}$$

and for $\rho = 0$ reduces to

$$\lim_{\lambda\to\infty}\{\lambda I_\lambda(\lambda\tau)K_\lambda(\lambda\tau)\}=\frac{1}{2\sqrt{1+\tau^2}}\quad;\quad \mathrm{Re}\,\tau>0 \tag{4.11.31}$$

The left side expression in (4.11.30) is the first term in the uniform asymptotic expansion of the product of modified Bessel functions for large orders [9]

$$\begin{aligned}
I_\lambda(\lambda\tau)&\sim\frac{e^{\lambda\eta}}{\sqrt{2\pi\lambda}(1+\tau^2)^{1/4}}\{1+\ldots\}\\
K_\lambda(\lambda\tau)&\sim\sqrt{\frac{\pi}{2\lambda}}\frac{e^{-\lambda\eta}}{(1+\tau^2)^{1/4}}\{1+\ldots\}\\
\eta&=\sqrt{1+\tau^2}+\ln\left(\frac{1}{1+\sqrt{1+\tau^2}}\right)
\end{aligned} \tag{4.11.32}$$

By introducing $\tau = 1$ into (4.11.32), we have for large values of λ the expected result

$$\lambda I_\lambda(\lambda\tau)\,K_\lambda(\lambda\tau)\sim\frac{1}{2\sqrt{1+\tau^2}}\left\{1-\frac{1}{2}\frac{(4\lambda^2-1)}{(2\lambda\tau)^2}+\frac{3}{8}\frac{(4\lambda^2-1)(4\lambda^2-9)}{(2\lambda\tau)^4}-\ldots\right\} \tag{4.11.33}$$

$$\lambda\to\infty$$

Using the following integral [13]

$$\int_0^\infty e^{-t^2x^2}J_\nu(\nu x)\,dx=\frac{\sqrt{\pi}}{2t}e^{-\nu^2/8t^2}I_{\nu/2}\left(\frac{\nu^2}{8t}\right) \tag{4.11.34}$$

it is possible, after few steps, to obtain the limit representation of the modified Bessel function of the first kind

$$\lim_{\lambda\to\infty}\left\{\lambda e^{-\lambda^2/2\tau}I_\lambda\left(\frac{\lambda^2}{2\tau}\right)\right\}=\sqrt{\frac{\tau}{\pi}}e^{-\tau} \tag{4.11.35}$$

and using

$$\int_0^\infty e^{-1/tx} J_\nu(\nu x) \frac{dx}{x} = 2J_\nu\left(\sqrt{\frac{2\nu}{t}}\right) K_\nu\left(\sqrt{\frac{2\nu}{t}}\right) \quad ; \quad \nu, t > 0 \tag{4.11.36}$$

the limit of following product of the Bessel functions is

$$\lim_{\nu \to \infty} \left\{\nu J_\nu(\sqrt{\nu\tau}) K_\nu(\sqrt{\nu\tau})\right\} = \frac{e^{-\tau/2}}{2} \quad ; \quad \tau > 0 \tag{4.11.37}$$

The next examples will include the limits of elementary and special functions based on using (4.11.14). From the Laplace transform of the Bessel function [37]

$$\int_0^\infty e^{-tx} J_\nu(\nu x)\, dx = \frac{\nu^\nu}{\sqrt{\nu^2 + t^2}\left[t + \sqrt{\nu^2 + t^2}\right]^\nu} \quad ; \quad t > 0 \tag{4.11.38}$$

the asymptotic limit for the exponential function is

$$e^{-t} = \lim_{\nu \to \infty} \left\{\frac{\nu^{\nu+1}}{\sqrt{\nu^2 + t^2}\left[t + \sqrt{\nu^2 + t^2}\right]^\nu}\right\} \quad ; \quad t > 0 \tag{4.11.39}$$

From the Bessel function integrals [13]

$$\begin{aligned} \int_0^\infty \sin(tx) J_\nu(\nu x)\, dx &= \frac{\sin\left[\nu \sin^{-1}\left(\frac{t}{\nu}\right)\right]}{\sqrt{\nu^2 - t^2}} \quad ; \quad \nu > t > 0 \\ \int_0^\infty \cos(tx) J_\nu(\nu x)\, dx &= \frac{\cos\left[\nu \sin^{-1}\left(\frac{t}{\nu}\right)\right]}{\sqrt{\nu^2 - t^2}} \quad ; \quad \nu > t > 0 \end{aligned} \tag{4.11.40}$$

the asymptotic limits for the trigonometric functions are:

$$\begin{aligned} \lim_{\nu \to \infty} \left\{\frac{\nu \sin\left[\nu \sin^{-1}\left(\frac{t}{\nu}\right)\right]}{\sqrt{\nu^2 - t^2}}\right\} &= \sin t \quad ; \quad \nu > t > 0 \\ \lim_{\nu \to \infty} \left\{\frac{\nu \cos\left[\nu \sin^{-1}\left(\frac{t}{\nu}\right)\right]}{\sqrt{\nu^2 - t^2}}\right\} &= \cos t \quad ; \quad \nu > t > 0 \end{aligned} \tag{4.11.41}$$

and by dividing these limits, we have

$$\begin{aligned} \lim_{\nu \to \infty} \left\{\tan\left[\nu \sin^{-1}\left(\frac{t}{\nu}\right)\right]\right\} &= \tan t \quad ; \quad \nu > t > 0 \\ \lim_{\nu \to \infty} \left\{\cot\left[\nu \sin^{-1}\left(\frac{t}{\nu}\right)\right]\right\} &= \cot t \quad ; \quad \nu > t > 0 \end{aligned} \tag{4.11.42}$$

In the following two examples the limits will include special functions, but not the Bessel functions. Using the following integral

$$\int_0^\infty x^{\lambda-1} e^{-t^2x^2} J_\nu(\nu x)\, dx = \frac{\Gamma\left(\frac{\lambda+\nu}{2}\right) \nu^\nu}{2^{\nu+1}\Gamma(\nu+1)\, t^{\nu+\lambda}}\, {}_1F_1\left(\frac{\lambda+\nu}{2}; \nu+1; -\frac{\nu^2}{4t^2}\right) \tag{4.11.43}$$

$$\mathrm{Re}(\lambda+\nu)>0 \quad ; \quad t>0$$

from (4.11.14) the limit of the confluent hypergeometric function is

$$\lim_{\nu\to\infty}\left\{\frac{\Gamma\left(\frac{\lambda+\nu}{2}\right)}{\Gamma(\nu+1)}\left(\frac{\nu}{2t}\right)^{\nu+1} {}_1F_1\left(\frac{\lambda+\nu}{2}; \nu+1; -\frac{\nu^2}{4t^2}\right)\right\} = t^{\lambda-1}e^{-t^2} \tag{4.11.44}$$

$$\mathrm{Re}\,\lambda>0 \quad ; \quad t>0$$

In the second example the integral includes the Whittaker functions

$$\begin{aligned}&\int_0^\infty \frac{\left[1+\sqrt{1+(tx)^2}\right]^\lambda}{x^\lambda \sqrt{1+(tx)^2}} J_\nu(\nu x)\, dx \\ &= \frac{\Gamma\left(\frac{\nu-\lambda+1}{2}\right) t^\lambda}{\nu\Gamma(\nu+1)} M_{-\lambda/2,\,\nu/2}\left(\frac{\nu}{t}\right) W_{\lambda/2,\,\nu/2}\left(\frac{\nu}{t}\right)\\ &\mathrm{Re}\,(\nu-\lambda)>-1 \quad ; \quad t>0\end{aligned} \tag{4.11.45}$$

and therefore from (4.11.4) it follows that

$$\begin{aligned}&\lim_{\nu\to\infty}\left\{\frac{\Gamma\left(\frac{\nu-\lambda+1}{2}\right) t^\lambda}{\Gamma(\nu+1)} M_{-\lambda/2,\,\nu/2}\left(\frac{\nu}{t}\right) W_{\lambda/2,\,\nu/2}\left(\frac{\nu}{t}\right)\right\}\\ &= \frac{\left[1+\sqrt{1+t^2}\right]^\lambda}{t^\lambda\sqrt{1+t^2}}\end{aligned} \tag{4.11.46}$$

The Lamborn formula for the shifted Dirac delta function given in (4.11.7) can be explored to obtain additional representations of functions by considering its derivatives with respect to the argument x. The first derivative of the shifted Dirac delta function is

$$\begin{aligned}\delta(x-1) &= \lim_{\nu\to\infty}\{\nu J_\nu(\nu x)\}\\ \delta'(x-1) &= \frac{1}{2}\lim_{\nu\to\infty}\{\nu^2[J_{\nu-1}(\nu x) - J_{\nu+1}(\nu x)]\}\end{aligned} \tag{4.11.47}$$

but

$$\delta^{(n)}(x) = (-1)^n n! \frac{\delta(x)}{x^n} \tag{4.11.48}$$

and in similar way as in (4.11.13), we have

$$\begin{aligned} f'(t) &= \int_0^\infty \delta'(x-1) f(tx)\, dx \\ &= -\frac{1}{2t} \lim_{v\to\infty} \int_0^\infty \{v^2 [J_{v-1}(vx) - J_{v+1}(vx)] f(tx)\}\, dx \end{aligned} \tag{4.11.49}$$

Higher derivatives of function, $f^{(n)}(t)$, can be derived in the same manner by using (4.11.48) and by considering that it is known [11].

$$\frac{d^n J_v(vx)}{dx^n} = n! \left(\frac{v}{2}\right)^n \sum_{k=0}^{n} \frac{(-1)^k}{(n-k)!\,k!} J_{v-n+2k}(vx) \tag{4.11.50}$$

An application of the first derivative of the shifted Dirac delta function can be illustrated by using (4.11.34)

$$\begin{aligned} &\int_0^\infty e^{-t^2x^2} J_v(vx)\, dx = \frac{\sqrt{\pi}}{2t} e^{-v^2/8t^2} I_{v/2}\left(\frac{v^2}{8t}\right) \\ &f(t) = e^{-t^2} \\ &f'(t) = -2te^{-t^2} \end{aligned} \tag{4.11.51}$$

and therefore from (4.11.48) we have

$$\lim_{v\to\infty} \left\{ v^2 e^{-v^2/8t^2} \left[I_{(v-1)/2}\left(\frac{v^2}{8t}\right) - I_{(v+1)/2}\left(\frac{v^2}{8t}\right) \right] \right\} = \frac{8t^3 e^{-t^2}}{\sqrt{\pi}} \tag{4.11.52}$$

So far, in order to derive asymptotic limit expressions, the same function $f(tx)$ appeared on both sides of equations (4.11.13) and (4.11.49). This can be changed if integral transforms are combined with the Lamborn formula. If integral transformation of a function $f(t)$ is defined by

$$T\{f(t)\} = \int_0^\infty K(s,t) f(t)\, dx = T(s) \tag{4.11.53}$$

and the function is $f(t) = J_v(vt)$, then we have

$$\int_0^\infty \delta(t-1)K(s,t)\,dt = \lim_{v\to\infty}\left[v\int_0^\infty K(s,t)\,J_v(vt)\,dt\right] = K(s;1) \tag{4.11.54}$$

where $K(s,t)$ denotes the kernel of integral transformation.

Most of integral transformations satisfy the following similarity rule

$$T\{f(at)\} = \frac{1}{a}\,T\left(\frac{s}{a}\right) \quad ; \quad a>0 \tag{4.11.55}$$

and therefore by comparing (4.11.54) with (4.11.55), the equivalent forms of asymptotic limits are [85]

$$T(s) = \lim_{v\to\infty}\left[v\int_0^\infty J_v(v\xi)\,T(s\xi)\,d\xi\right] = \lim_{v\to\infty}\left[v\int_0^\infty J_v(vx)\,\frac{s}{x}\,T\left(\frac{s}{x}\right)\,dx\right] \tag{4.11.56}$$

An application of the asymptotic limit in (4.11.54) can be demonstrated by using the Laplace transform of $f(t) = J_v(vt)$ from (4.11.38)

$$K(s,t) = e^{-st} \quad ; \quad s>0$$

$$L\{J_v(vt),s\} = \int_0^\infty e^{-st}J_v(vt)\,dt = \frac{v^v}{\sqrt{v^2+s^2}\left[s+\sqrt{v^2+s^2}\right]^v} \qquad \operatorname{Re} v > 0 \tag{4.11.57}$$

and therefore,

$$\lim_{v\to\infty}\left\{\frac{v^{v+1}}{\sqrt{v^2+s^2}\left[s+\sqrt{v^2+s^2}\right]^v}\right\} = e^{-s} \quad ; \quad s>0 \tag{4.11.58}$$

For the Stieltjes transform of $J_v(vt)$, we have

$$K(s,t) = \frac{1}{s+t} \quad ; \quad s>0$$

$$S\{J_v(vt)\} = \int_0^\infty \frac{J_v(vt)}{s+t}\,dt = \frac{\pi}{\sin(\pi v)}[\mathbf{J}_v(vs) - J_v(vs)] \qquad \operatorname{Re} v > 0 \tag{4.11.59}$$

which gives immediately from (4.11.54)

$$\lim_{\nu\to\infty}\left\{\frac{\nu\,[\mathbf{J}_\nu(\nu s)-J_\nu(\nu s)]}{\sin(\pi\nu)}\right\}=\frac{1}{\pi(s+1)}\quad;\quad s>0 \tag{4.11.60}$$

The Anger function is represented by

$$\mathbf{J}_\nu(\nu s)-J_\nu(\nu s)=\frac{\sin(\pi\nu)}{\pi}\int_0^\infty e^{-\nu\tau-\nu s\sinh\tau}\,d\tau \tag{4.11.61}$$

and therefore, the limit of this integral is derived as a by-product

$$\lim_{\nu\to\infty}\left\{\nu\int_0^\infty e^{-\nu\tau-\nu s\sinh\tau}\,d\tau\right\}=\frac{1}{s+1}\quad;\quad s>0 \tag{4.11.62}$$

The limit in (11.4.54) contains only the kernel of integral transformation $K(s,t)$, but if also the function $f(\xi;\lambda)$ where λ is a parameter, is introduced, then usefulness of derived expression is considerably enlarged. Multiplication of (4.11.54) by $f(\xi;\lambda;)$ and integration from zero to infinity with respect to variable ξ gives

$$\begin{aligned}
&\lim_{\nu\to\infty}\left\{\nu\int_0^\infty\int_0^\infty K(\xi,x)\,J_\nu(\nu x)\,f(\xi,\lambda)\,dx\,d\xi\right\}\\
&=\lim_{\nu\to\infty}\left\{\nu\int_0^\infty f(\xi,\lambda)\left[\int_0^\infty K(\xi,x)\,J_\nu(\nu x)\,dx\right]d\xi\right\}\\
&=\lim_{\nu\to\infty}\left\{\nu\int_0^\infty J_\nu(\nu x)\left[\int_0^\infty K(\xi,x)\,f(\xi,\lambda)\,d\xi\right]dx\right\}
\end{aligned} \tag{4.11.63}$$

or in the compact form

$$\begin{aligned}
&\lim_{\nu\to\infty}[\nu\int_0^\infty f(\xi,\lambda)\,T\{J_\nu(\nu x),\xi\}\,d\xi]\\
&=\lim_{\nu\to\infty}[\nu\int_0^\infty J_\nu(\nu x)\,T\{f(\xi,\lambda),x\}\,dx]=T(1,\lambda)
\end{aligned} \tag{4.11.64}$$

For example, for the Laplace transformation we have

$$L\{f(\xi,\lambda),x\} = \int_0^\infty e^{-\xi x} J_\nu(\nu\xi)\, d\xi = \frac{\nu^\nu}{\sqrt{\nu^2+\xi^2}\,\left[\xi+\sqrt{\nu^2+\xi^2}\right]^\nu}$$

$$\lim_{\nu\to\infty}\left\{\nu^{\nu+1}\int_0^\infty \frac{f(\xi,\lambda)}{\sqrt{\nu^2+\xi^2}\,\left[\xi+\sqrt{\nu^2+\xi^2}\right]^\nu}\, d\xi\right\} \tag{4.11.65}$$

$$= \lim_{\nu\to\infty}\left\{\nu\int_0^\infty J_\nu(\nu x)\, L(x,\lambda)\, dx\right\} = L(1,\lambda)$$

By suitable choice of the original function transform pair, the above expression permits to represent mathematical constants and special functions as limits of integrals with the integrands being elementary functions from the Laplace transform

$$L\left\{\frac{\sin\xi}{\xi}\right\} = \tan^{-1}\left(\frac{1}{x}\right) \tag{4.11.66}$$

it follows from (4.11.65) that π number is expressed by the following limit

$$\lim_{\nu\to\infty}\left\{\nu^{\nu+1}\int_0^\infty \frac{\sin\xi}{\xi\sqrt{\nu^2+\xi^2}\left[\xi+\sqrt{\nu^2+\xi^2}\right]^\nu}\, d\xi\right\} = \tan^{-1}(1) = \frac{\pi}{4} \tag{4.11.67}$$

The Euler constant $\gamma = 0{\cdot}577215{\cdot}$is represented by starting from the Laplace transform

$$L\{\ln\xi\} = -\frac{1}{x}(\gamma + \ln x) \tag{4.11.68}$$

and therefore,

$$\lim_{\nu\to\infty}\left\{\nu^{\nu+1}\int_0^\infty \frac{\ln\xi}{\sqrt{\nu^2+\xi^2}\left[\xi+\sqrt{\nu^2+\xi^2}\right]^\nu}\, d\xi\right\} = -\gamma \tag{4.11.69}$$

The Bernoulli numbers can be evaluated using

$$L\left\{\frac{\xi^{\mu-1}}{\sinh\xi}\right\} = 2^{1-\mu}\Gamma(\mu)\,\zeta\left(\mu, \frac{x+1}{2}\right) \quad ; \quad \mu > 1 \tag{4.11.70}$$

where the generalized zeta function $\zeta(z,\alpha)$ becomes $\zeta(z,1) = \zeta(z)$ and therefore,

$$\lim_{\nu\to\infty}\left\{\nu^{\nu+1}\int_0^\infty \frac{\xi^{\mu-1}}{\sinh\xi\,\sqrt{\nu^2+\xi^2}\left[\xi+\sqrt{\nu^2+\xi^2}\right]^\nu}\,d\xi\right\} \tag{4.11.71}$$

$$=2^{1-\mu}\,\Gamma(\mu)\,\varsigma(\mu)$$

but

$$\zeta(2n)=\frac{(2\pi)^{2n}}{2\,(2n)!}|B_{2n}| \quad ; \quad n=1,2,3,\ldots \tag{4.11.72}$$

and therefore,

$$\lim_{\nu\to\infty}\left\{\nu^{\nu+1}\int_0^\infty \frac{\xi^{2n-1}}{\sqrt{\nu^2+\xi^2}\left[\xi+\sqrt{\nu^2+\xi^2}\right]^\lambda \sinh\xi}\,d\xi\right\}=\frac{\pi^{2n}}{2n}|B_{2n}| \tag{4.11.73}$$

Limit representations of special functions are given in Appendix B, but those derived for the Bessel and related functions are presented below. Most of them are determined by applying the Laplace transforms, but the analogs of (4.11.65) of the Fourier cosine and sine and the Stieltjes transformations are also used

$$\lim_{\nu\to\infty}\left[\nu\int_0^\infty \frac{\cos\left[\nu\sin^{-1}\left(\frac{\xi}{\nu}\right)\right]\,f(\xi,\lambda)}{\sqrt{\nu^2-\xi^2}}\,d\xi\right]=F_c(1,\lambda)$$

$$\lim_{\nu\to\infty}\left[\nu\int_0^\infty \frac{\sin\left[\nu\sin^{-1}\left(\frac{\xi}{\nu}\right)\right]\,f(\xi,\lambda)}{\sqrt{\nu^2-\xi^2}}\,d\xi\right]=F_s(1,\lambda) \tag{4.11.74}$$

$$\lim_{\nu\to\infty}\left\{\nu\int_0^\infty f(\xi,\lambda)\left[\int_0^\infty e^{-\nu\eta-\nu\xi\sinh\eta}d\eta\right]d\xi\right\}=S(1,\lambda)$$

The asymptotic limits of the Bessel functions of the first kind are:

$$J_\mu(t)$$

$$=\frac{1}{2^{\mu-1}\sqrt{\pi}\,\Gamma\left(\mu+\frac{1}{2}\right)t^\mu}\lim_{\nu\to\infty}\left\{\nu\int_0^t \frac{\cos\left[\nu\sin^{-1}\left(\frac{\xi}{\nu}\right)\right](t^2-\xi^2)^{\mu-1/2}}{\sqrt{\nu^2-\xi^2}}\,d\xi\right\} \tag{4.11.75}$$

$$\operatorname{Re}\mu>-\frac{1}{2}$$

and

$$Y_{-(\mu+1)}(t)$$

$$= \frac{1}{2^{\mu-1}\sqrt{\pi}\,\Gamma\left(\mu+\frac{1}{2}\right)t^{\mu+1}} \lim_{\nu\to\infty}\left\{\nu\int_t^\infty \frac{\sin\left[\nu\sin^{-1}\left(\frac{\xi}{\nu}\right)\right]\xi\,(\xi^2-t^2)^{\mu-1/2}}{\sqrt{\nu^2-\xi^2}}d\xi\right\}$$

$$-\frac{1}{2}<\operatorname{Re}\mu<0$$

(4.11.76)

In the case of modified Bessel functions, we have

$$I_\mu(t)$$

$$= \frac{e^t}{\sqrt{\pi}\,\Gamma\left(\mu+\frac{1}{2}\right)(2t)^\mu} \lim_{\nu\to\infty}\left\{\nu^{\nu+1}\int_0^{2t} \frac{(2t\xi-\xi^2)^{\mu-1/2}}{\sqrt{\nu^2+\xi^2}\left[\xi+\sqrt{\nu^2+\xi^2}\right]^\nu}d\xi\right\} \tag{4.11.77}$$

$$\mu>-\frac{1}{2}$$

$$K_\mu(t) = \frac{\sqrt{\pi}\,e^{-t}}{\Gamma\left(\mu+\frac{1}{2}\right)(2t)^\mu} \lim_{\nu\to\infty}\left\{\nu^{\nu+1}\int_0^\infty \frac{(\xi^2+2t\xi)^{\mu-1/2}}{\sqrt{\nu^2+\xi^2}\left[\xi+\sqrt{\nu^2+\xi^2}\right]^\nu}d\xi\right\} \tag{4.11.78}$$

$$\mu>-\frac{1}{2}$$

Similarly, the Struve functions $\mathrm{H}_\mu(t)$ and $\mathrm{L}_\mu(t)$ are represented by

$$H_\mu(t) = Y_\mu(t)$$

$$+ \frac{1}{2^{\mu-1}\sqrt{\pi}\,\Gamma\left(\mu+\frac{1}{2}\right)t^\mu} \lim_{\nu\to\infty}\left\{\nu^{\nu+1}\int_0^\infty \frac{(t^2+\xi^2)^{\mu-1/2}}{\sqrt{\nu^2+\xi^2}\left[\xi+\sqrt{\nu^2+\xi^2}\right]^\nu}d\xi\right\}$$

$$\mu>-\frac{1}{2}\ ,\quad t>0$$

(4.11.79)

and

$$L_\mu(t) = I_\mu(t)$$

$$-\frac{1}{2^{\mu-1}\sqrt{\pi}\,\Gamma\left(\mu+\frac{1}{2}\right)t^\mu}\lim_{\nu\to\infty}\left\{\nu^{\nu+1}\int_0^t \frac{(t^2-\xi^2)^{\mu-1/2}}{\sqrt{\nu^2+\xi^2}\left[\xi+\sqrt{\nu^2+\xi^2}\right]^\nu}d\xi\right\}$$

$$\operatorname{Re}\mu > -\frac{1}{2} \quad , \quad t>0 \tag{4.11.80}$$

The asymptotic limits for the Anger function $\mathbf{J}_\mu(t)$ and Weber function $\mathbf{E}_\mu(t)$ are

$$\mathbf{J}_\mu(t) = J_\mu(t)$$

$$-\frac{t^\mu\sin(\pi\mu)}{\pi\mu}\lim_{\nu\to\infty}\left\{\nu^{\nu+1}\int_0^\infty \frac{\left[\left(\xi+\sqrt{t^2+\xi^2}\right)^{-\mu}-t^{-\mu}\right]}{\sqrt{\nu^2+\xi^2}\left[\xi+\sqrt{\nu^2+\xi^2}\right]^\lambda}d\xi\right\} \tag{4.11.81}$$

and

$$\boldsymbol{E}_\mu(t) = -Y_\mu(t)$$

$$+t\lim_{\nu\to\infty}\left\{\nu^{\nu+1}\int_0^\infty \frac{I(\xi,t,\mu)}{\sqrt{1+\frac{\xi^2}{t^2}}\sqrt{\nu^2+\xi^2}\left[\xi+\sqrt{\nu^2+\xi^2}\right]^\nu}d\xi\right\} \tag{4.11.82}$$

$$I(\xi,t,\mu) = \left[-\left(\frac{\xi}{t}+\sqrt{1+\frac{\xi^2}{t^2}}\right)^\mu + \cos(\pi\mu)\sqrt{1+\frac{\xi^2}{t^2}}\left(\sqrt{1+\frac{\xi^2}{t^2}}-\frac{\xi}{t}\right)^\mu\right]$$

The Lommel functions limits are presented here in two cases only

$$S_{0,\mu}(t) = \frac{1}{\mu t}\lim_{\nu\to\infty}\left\{\nu^{\nu+1}\int_0^\infty \frac{\sinh\left[\mu\sinh^{-1}\left(\frac{\xi}{t}\right)\right]}{\sqrt{\nu^2+\xi^2}\left[\xi+\sqrt{\nu^2+\xi^2}\right]^\nu}d\xi\right\} \tag{4.11.83}$$

$$S_{1,\mu}(t) = \frac{1}{t}\lim_{\nu\to\infty}\left\{\nu^{\nu+1}\int_0^\infty \frac{\cosh\left[\mu\sinh^{-1}\left(\frac{\xi}{t}\right)\right]}{\sqrt{\nu^2+\xi^2}\left[\xi+\sqrt{\nu^2+\xi^2}\right]^\nu}d\xi\right\} \tag{4.11.84}$$

If the shifted Dirac delta function is expressed by the orthogonality identity of the Bessel functions (4.11.5)

$$\delta(x-1) = \int_0^\infty xuJ_\nu(u)J_\nu(xu)\,du \tag{4.11.85}$$

then by multiplication of both sides of (4.11.85) by function $f(tx)$ and performing integration from zero to infinity with respect to variable x gives

$$f(t) = \int_0^\infty \delta(x-1)f(tx)\,dx = \int_0^\infty \left[\int_0^\infty xuJ_\nu(u)J_\nu(xu)\,du\right] f(tx)\,dx \tag{4.11.86}$$

which can be rearranged to

$$\begin{aligned} g(t) &= t^{1/2}f(t) \\ &= \int_0^\infty u^{1/2}J_\nu(u)\left\{\int_0^\infty (xu)^{1/2}J_\nu(xu)\left[(tx)^{1/2}f(tx)\right]dx\right\}du \end{aligned} \tag{4.11.87}$$

The inner integral can be recognized as the Hankel transform of order ν of the function $g(\xi)$

$$\begin{gathered} G\{g(\xi),\nu\} = \int_0^\infty (s\xi)^{1/2}J_\nu(s\xi)\,g(\xi)\,d\xi = G(s,\nu) \\ s > 0 \quad ; \quad \operatorname{Re}\nu > -\frac{1}{2} \end{gathered} \tag{4.11.88}$$

Taking into account the operational rule of the Hankel transformation

$$G\{g(a\xi),\nu\} = \frac{1}{a}G\left(\frac{s}{a},\nu\right) \quad ; \quad a > 0 \tag{4.11.89}$$

we have

$$t^{3/2}f(t) = tg(t) = \int_0^\infty u^{1/2}J_\nu(u)\,G\left(\frac{u}{t},\nu\right)du \tag{4.11.90}$$

which can be treated as some kind of inversion formula The corresponding asymptotic limit of (4.11.90) is

$$t^{1/2} f(t) = \lim_{\nu\to\infty}\left[\nu \int_0^\infty J_\nu(\nu x)\ [(tx)^{1/2} f(tx)]\, dx\right]$$

$$G\left(\frac{\nu}{t},\nu\right) = (\nu t)^{1/2} \int_0^\infty J_\nu(\nu x)\ [(tx)^{1/2} f(tx)]\, dx \tag{4.11.91}$$

$$f(t) = \lim_{\nu\to\infty}\left\{\nu^{1/2} G\left(\frac{\nu}{t},\nu\right)\right\}$$

If $\nu = \pm 1/2$, the Hankel transforms are reduced to the Fourier sine and cosine transforms, and (4.11.90) becomes

$$t^{3/2} f(t) = t g(t) = \frac{2}{\pi}\int_0^\infty \sin u\, F_s\left(\frac{u}{t}\right) du$$

$$t^{3/2} f(t) = t g(t) = \frac{2}{\pi}\int_0^\infty \sin u\, F_c\left(\frac{u}{t}\right) du \tag{4.11.91}$$

If the integral transform with the Struve function as the kernel is considered

$$H_\nu\{g(\xi)\} = \int_0^\infty (s\xi)^{1/2} H_\nu(s\xi)\, g(\xi) d\xi = H(s,\nu) \tag{4.11.92}$$

the analogs of (4.11.90) and (4.11.91) have similar forms

$$t^{3/2} f(t) = t g(t) = \int_0^\infty u^{1/2} Y_\nu(u) H\left(\frac{u}{t},\nu\right) du$$

$$t f(t) = \lim_{\nu\to\infty}\left\{\nu^{1/2} H\left(\frac{\nu}{t},\nu\right)\right\} \tag{4.11.93}$$

Since tables of the Hankel and Struve transforms, ($G\,(s,\nu)$ and $H(s,\nu)$ – transforms of order ν of functions $g(t) = t^{1/2} f(t)$) are available in the literature, the above formulas are a source of many rather complicated finite and infinite integrals of Bessel functions and corresponding limits of elementary and special functions.

In order to illustrate the asymptotic limit of elementary function, let us start with the following Hankel function transform pair

$$g(t) = t^{n+1/2} e^{-\alpha t} \quad ; \quad n = 0,1,2,3,\dots \quad ; \quad \mathrm{Re}\,\alpha > 0 \quad , \quad \mathrm{Re}\,\nu > 0$$

$$G\{\nu, s\} = (-1)^{n+1} s^{\nu+1/2} \frac{d^{n+1}}{d\alpha^{n+1}}\left\{\frac{1}{\sqrt{\alpha^2+s^2}\left[\alpha+\sqrt{\alpha^2+s^2}\right]^\nu}\right\} \tag{4.11.94}$$

and using (4.11.91) we have

$$\lim_{\nu\to\infty}\left\{\frac{d^{n+1}}{d\alpha^{n+1}}\left[\frac{\nu^{\nu+1}}{\sqrt{(\alpha t)^2+\nu^2}\,[\alpha t+\sqrt{(\alpha t)^2+\nu^2}]^{\nu}}\right]\right\}$$
$$=(-1)^{n+1}t^{n+1}e^{-\alpha t} \tag{4.11.95}$$
$$n=0,1,2,3,\ldots\quad;\quad \mathrm{Re}\,\alpha>0$$

In the case of special function, starting from

$$g(t)=t^{\mu-1/2}e^{-\alpha t^2}.\quad;\quad \mathrm{Re}\,\alpha>0\quad,\quad \mathrm{Re}\,(\mu+\nu)>0$$
$$G\{\nu,s\}=\frac{\Gamma\left(\frac{\mu+\nu+1}{2}\right)e^{-s^2/8\alpha}}{\Gamma(\nu+1)\,\alpha^{\mu/2}s^{1/2}M_{\mu/2,\nu/2}\left(\frac{s^2}{4\alpha}\right)} \tag{4.11.96}$$

the asymptotic limit for the Whittaker function is

$$\lim_{\nu\to\infty}\left\{\frac{\Gamma\left(\frac{\mu+\nu+1}{2}\right)e^{-\nu^2/8\alpha t^2}}{\Gamma(\nu+1)}M_{\mu/2,\nu/2}\left(\frac{\nu^2}{4\alpha t^2}\right)\right\}=\alpha^{\mu}t^{2\mu}e^{-\alpha t^2}. \tag{4.11.97}$$

Lommel functions can be treated by using the following Hankel transform:

$$g(t)=\frac{t^{\mu-1/2}}{\alpha+t}.\quad;\quad \mathrm{Re}\,\mu<\frac{3}{2}\quad,\quad \mathrm{Re}\,(\mu+\nu)>-1$$
$$G\{\nu,s\}$$
$$=(2\alpha)^{\mu}\sqrt{s}\left\{\frac{\Gamma\left(\frac{\mu+\nu+1}{2}\right)}{\Gamma\left(\frac{1-\mu+\nu}{2}\right)}S_{-\mu,\nu}(\alpha s)-\frac{2\Gamma\left(\frac{\mu+\nu+2}{2}\right)}{\Gamma\left(\frac{\nu-\mu}{2}\right)}S_{-\mu-1,\nu}(\alpha s)\right\} \tag{4.11.98}$$

and therefore,

$$\lim_{\nu\to\infty}\nu\left\{\frac{\Gamma\left(\frac{\mu+\nu+1}{2}\right)}{\Gamma\left(\frac{1-\mu+\nu}{2}\right)}S_{-\mu,\nu}\left(\frac{\alpha\nu}{t}\right)-\frac{2\Gamma\left(\frac{\mu+\nu+2}{2}\right)}{\Gamma\left(\frac{\nu-\mu}{2}\right)}S_{-\mu-1,\nu}\left(\frac{\alpha\nu}{t}\right)\right\}$$
$$=\frac{t^{\mu+1}}{(2\alpha)^{\mu}(\alpha+1)} \tag{4.11.99}$$

which for $\mu=0$ reduces to

$$\lim_{\nu\to\infty}\left\{S_{0,\nu}\left(\frac{\alpha\nu}{t}\right)-\nu S_{-1,\nu}\left(\frac{\alpha\nu}{t}\right)\right\}=\frac{t}{\alpha+1} \tag{4.11.100}$$

The procedure presented here is also efficient in evaluation of integrals. For example, from the Hankel transform function pair

$$\begin{aligned} g(t) &= \frac{\xi^{\nu+1/2}}{(\alpha^2+\xi^2)^{\mu}} \\ G(s,\nu) &= \frac{2^{1-\mu}\alpha^{\nu-\mu+1}s^{\mu-1/2}}{\Gamma(\mu)}K_{\nu-\mu+1}(\alpha s) \\ &\operatorname{Re}\nu>-1 \quad , \quad \operatorname{Re}(\mu-\nu)>0 \end{aligned} \tag{4.11.101}$$

by using (4.11.90) we have the following infinite integral of the Bessel functions:

$$\int_0^\infty u^{\mu}\,J_\nu(u)\,K_{\mu-\nu+1}\left(\frac{\alpha u}{\xi}\right)du=\frac{2^{\mu-1}\Gamma(\mu)\,\alpha^{\mu-\lambda-1}\xi^{\mu+\nu+1}}{(\alpha^2+\xi^2)^{\mu}} \tag{4.11.102}$$

By changing variables, this integral can be written in a much simpler form

$$\begin{aligned} \int_0^\infty x^{\nu-\mu+1}J_\nu(\alpha x)K_\mu(\beta x)\,dx &= \frac{2^{\nu-\mu}\Gamma(\nu-\mu+1)\,\alpha^{\nu}}{\beta^{\mu}(\alpha^2+\beta^2)^{\nu-\mu+1}} \\ \operatorname{Re}\nu>-1 \quad &, \quad \operatorname{Re}(\nu-2\mu)>-\frac{3}{2} \end{aligned} \tag{4.11.103}$$

For $\nu=\pm 1/2$, the above integral is reduced to integrals with trigonometric functions

$$\begin{aligned} \int_0^\infty x^{1-\nu}\sin(\alpha x)\,K_\nu(\beta x)\,dx &= \frac{\sqrt{\pi}\,\Gamma(\frac{3}{2}-\nu)\,\alpha}{(2\beta)^{\nu}(\alpha^2+\beta^2)^{3/2-\nu}} \\ \int_0^\infty x^{-\nu}\cos(\alpha x)\,K_\nu(\beta x)\,dx &= \frac{\sqrt{\pi}\,\Gamma(\frac{1}{2}-\nu)}{2^{\nu+1}\beta^{\nu}(\alpha^2+\beta^2)^{1/2-\nu}} \end{aligned} \tag{4.9.104}$$

If the Hankel transform includes the Heaviside step function $u(x)$ or the shifted step function $u(x\text{-}a)$

$$\begin{aligned} g(t) &= t^{1/2}J_{\nu/2-1/4}\left(\frac{\alpha t}{2}\right)J_{\nu/2-1/4}\left(\frac{\alpha t}{2}\right) \quad , \quad \operatorname{Re}\nu>-1 \\ G(s,\nu) &= \frac{1}{\pi\left(\frac{\alpha t}{2}\right)\sqrt{(\alpha-s)}}\,[u(s)-u(s-\alpha)] \end{aligned} \tag{4.11.105}$$

then the Bessel function integral becomes the finite integral

$$\int_0^\alpha \frac{J_\nu(tx)}{\sqrt{\alpha-x}}\,dx=\pi\sqrt{\frac{\alpha}{2}}\,J_{\nu/2-1/4}\left(\frac{\alpha t}{2}\right)J_{\nu/2+1/4}\left(\frac{\alpha t}{2}\right)\quad,\quad \operatorname{Re}\nu>-1 \tag{4.11.106}$$

The above integral for $\nu=\pm 1/2$, reduces to the trigonometric integrals

$$\begin{aligned}\int_0^\alpha \frac{\sin(tx)}{\sqrt{x(\alpha-x)}}\,dx&=\pi\sin\left(\frac{\alpha t}{2}\right)J_0\left(\frac{\alpha t}{2}\right)\\ \int_0^\alpha \frac{\cos(tx)}{\sqrt{x(\alpha-x)}}\,dx&=\pi\cos\left(\frac{\alpha t}{2}\right)J_0\left(\frac{\alpha t}{2}\right)\end{aligned} \tag{4.11.107}$$

If two shifted Heaviside step functions appear in the Hankel function transform pair

$$\begin{aligned}&g(t)=t^{1/2-\nu}J_\nu(at)J_\nu(bt)\\ &G(s,\lambda)=\frac{s^{1/2-\nu}\{[s^2-(a-b)^2][(a+b)^2-s^2]\}^{\nu-1/2}}{2^{3\nu-1}\sqrt{\pi}\,\Gamma\left(\nu+\frac{1}{2}\right)(ab)^\nu}I(s,a,b)\\ &I(s,a,b)=\{u[s-|a-b|]-u[s-(a+b)]\}\quad;\quad \operatorname{Re}\nu>-\frac{1}{2}\end{aligned} \tag{4.11.108}$$

the finite integral has limits of integration which are not at the origin as in (4.11.106) or as in (4.11.107)

$$\begin{aligned}&\int_{|a-b|}^{a+b}\xi^{1-\nu}\{[\xi^2-(a-b)^2][(a+b)^2-\xi^2]\}^{\nu-1/2}J_\nu(t\xi)\,d\xi\\ &=2^{3\nu-1}\sqrt{\pi}\,\Gamma\left(\nu+\frac{1}{2}\right)\left(\frac{ab}{t}\right)^\nu J_\nu(at)J_\nu(bt)\end{aligned} \tag{4.11.109}$$

This integral can be reduced to more simpler forms

$$\begin{aligned}&\int_\alpha^\beta \xi^{1-\nu}\{[\xi^2-\alpha^2][\beta^2-\xi^2]\}^{\nu-1/2}J_\nu(t\xi)\,d\xi\\ &=2^{3\nu-1}\sqrt{\pi}\,\Gamma\left(\nu+\frac{1}{2}\right)\left(\frac{\beta^2-\alpha^2}{t}\right)^\nu J_\lambda\left(\frac{\alpha+\beta}{2}t\right)J_\lambda\left(\frac{\beta-\alpha}{2}t\right)\end{aligned} \tag{4.11.110}$$

and

$$\int_0^{\alpha} \xi^{\nu}(\alpha^2-\xi^2)^{\nu-1/2} J_\nu(t\xi)\, d\xi = 2^{3\nu-1}\sqrt{\pi}\,\Gamma\left(\nu+\frac{1}{2}\right)\left(\frac{\alpha^2}{t}\right)^{\nu}\left[J_\nu\left(\frac{\alpha t}{2}\right)\right]^2 \tag{4.11.111}$$

The Struve transforms $H(s,\nu)$ are less numerous than the Hankel transforms $G(s,\nu)$. However, if used, they give interesting integrals of the Bessel functions of the second kind, as it is illustrated below. From the function transform pair

$$\begin{aligned} g(t) &= J_{2\nu+1}(\alpha t^{1/2}) \quad ; \quad \alpha>0 \quad , \quad -\frac{3}{2}<\operatorname{Re}\nu<\frac{1}{4} \\ H(s;\nu) &= -\frac{\alpha}{2s^{3/2}}\, Y_{\nu+1}\left(\frac{\alpha^2}{4s}\right) \\ H\left(\frac{u}{t};\nu\right) &= -\frac{\alpha t^{3/2}}{2u^{3/2}}\, Y_{\nu+1}\left(\frac{\alpha^2 t}{4u}\right) \end{aligned} \tag{4.11.112}$$

we have

$$\int_0^{\infty} \frac{1}{u} Y_\nu(u)\, Y_{\nu+1}\left(\frac{\alpha^2 t}{4u}\right) du = -\frac{2}{\alpha t^{1/2}}\, J_{2\nu+1}(\alpha t^{1/2}) \tag{4.11.113}$$

For $\nu=-1/2$, the above integral can be reduced to the trigonometric integral

$$\int_0^{\infty} \frac{1}{x}\sin(\alpha x)\cos\left(\frac{\beta}{x}\right) dx = \frac{\pi}{2} J_0(2\sqrt{\alpha\beta}) \tag{4.11.114}$$

A very similar function transform pair

$$\begin{aligned} g(t) &= \frac{1}{t^{1/2}} J_{2\nu}(\alpha t^{1/2}) \quad ; \quad \alpha>0 \quad , \quad -1<\operatorname{Re}\nu<\frac{5}{4} \\ H(s;\nu) &= -\frac{\alpha}{2s^{3/2}}\, Y_{\nu+1}\left(\frac{\alpha^2}{4s}\right) \end{aligned} \tag{4.11.115}$$

gives after few steps the following integral:

$$\int_0^{\infty} Y_\nu(\alpha x)\, Y_\nu\left(\frac{\beta}{x}\right) dx = -\frac{1}{\alpha} J_{2\nu}(2\sqrt{\alpha\beta}) \tag{4.11.116}$$

Introducing $\nu=\pm 1/2$ into (4.11.116), we have

$$\begin{aligned}&\int_0^\infty \cos(\alpha x)\,\cos\left(\frac{\beta}{x}\right)dx = -\frac{\pi}{2}\sqrt{\frac{b}{a}}J_1(2\sqrt{\alpha\beta})\\&\int_0^\infty \sin(\alpha x)\,\sin\left(\frac{\beta}{x}\right)dx = \frac{\pi}{2}\sqrt{\frac{b}{a}}J_1(2\sqrt{\alpha\beta})\end{aligned} \tag{4.11.117}$$

Using the following Struve transform

$$\begin{aligned}&g(t) = t^{1/2}\,Y_\nu(\alpha t^{1/2})\,K_\nu(\alpha t^{1/2}) \quad ; \quad \operatorname{Re}\nu > -\frac{3}{2}\\&H(s;\nu) = \frac{1}{2s^{3/2}}e^{-\alpha^2/2s}\end{aligned} \tag{4.11.118}$$

it follows from (4.11.93) that

$$\int_0^\infty \frac{e^{-\alpha^2 t/x}}{x}Y_\nu(x)\,dx = 2Y_\nu(\alpha t^{1/2})\,K_\nu(\alpha t^{1/2}) \tag{4.11.119}$$

In the final example, the Struve transform has the shifted Heaviside step function

$$\begin{aligned}&g(t) = t^{1/2}\left\{\left[J_{\nu/2}\left(\frac{\alpha t}{2}\right)\right]^2 - \left[Y_{\nu/2}\left(\frac{\alpha t}{2}\right)\right]^2\right\} \quad ; \quad -\frac{3}{2} < \operatorname{Re}\nu < 1\\&H(s;\lambda) = \frac{4u(s-\alpha)}{\pi\sqrt{s(s^2-\alpha^2)}}\end{aligned} \tag{4.11.120}$$

and this leads to the integral with the limit of integration which is not at the origin

$$\int_t^\infty \frac{Y_\nu(x)}{\sqrt{x^2-t^2}}dx = \frac{\pi}{4}\left\{\left[J_{\nu/2}\left(\frac{t}{2}\right)\right]^2 - \left[Y_{\nu/2}\left(\frac{t}{2}\right)\right]^2\right\} \tag{4.11.121}$$

Finally, the Lamborn expression for the shifted Dirac delta function will be applied to the sequence of functions $f_n(t,\lambda)$. This sequence converges to the sum $S(t,\lambda)$

$$S(t,\lambda) = \sum_n^\infty f_n(t,\lambda) \tag{4.11.122}$$

where λ denotes a parameter. In the same way as applying in (4.11.13), we have

$$\begin{aligned}S(t,\lambda) &= \int_0^\infty S(tx,\lambda)\,\delta(x-1)\,dx\\&= \lim_{\nu\to\infty}\left\{\nu\sum_n^\infty\left[\int_0^\infty f_n(tx,\lambda)\,J_\nu(\nu x)\,dx\right]\right\}\end{aligned} \tag{4.11.123}$$

It is assumed that the order of summation, integration and the limit of the series can be reversed and the integrals in (4.11.123) exist. If the sum in (4.11.123) is known then the expression (4.11.123) permits to determine the limit of sum of integrals.

For example, from (4.11.34)

$$\int_0^\infty e^{-\xi^2 x^2} J_\nu(\nu x)\,dx = \frac{\sqrt{\pi}}{2t} e^{-\nu^2/8\xi^2} I_{\nu/2}\left(\frac{\nu^2}{8\xi}\right) \tag{4.11.124}$$

it follows that

$$\lim_{\nu\to\infty}\left\{\frac{\nu e^{-\nu^2/8nt^2}}{\sqrt{n}} I_{\nu/2}\left(\frac{\nu^2}{8nt^2}\right)\right\} = \frac{2t}{\sqrt{\pi}} e^{-nt^2} \tag{4.11.125}$$

$$\xi = t\sqrt{n}$$

but it is known that

$$\sum_{n=1}^\infty e^{-nt^2} = \frac{e^{-t^2}}{1-e^{-t^2}} \tag{4.11.126}$$

and therefore we have limit of the sum

$$\lim_{\nu\to\infty}\left\{\nu \sum_{n=1}^\infty \frac{e^{-\nu^2/8nt^2}}{\sqrt{n}} I_{\nu/2}\left(\frac{\nu^2}{8nt^2}\right)\right\} = \frac{2t}{\sqrt{\pi}}\left(\frac{e^{-t^2}}{1-e^{-t^2}}\right) \tag{4.11.127}$$

Similarly from

$$\begin{aligned} &\sum_{n=0}^\infty J_n(\lambda)\, J_{n+1}(\lambda)\, \sin[(2n+1)x] = \frac{1}{2} J_1(2\lambda \sin x) \\ &\sum_{n=0}^\infty (-1)^n J_n(\lambda)\, J_{n+1}(\lambda)\, \cos[(2n+1)x] = \frac{1}{2} J_1(2\lambda \cos x) \end{aligned} \tag{4.11.128}$$

and considering that

$$\begin{aligned} &\int_0^\infty \sin(ax) J_\lambda(bx)\,dx = \frac{\sin\left[\lambda \sin^{-1}\left(\frac{a}{b}\right)\right]}{\sqrt{b^2-a^2}} \quad ; \quad 0<a<b \\ &\int_0^\infty \cos(ax) J_\lambda(bx)\,dx = \frac{\cos\left[\lambda \sin^{-1}\left(\frac{a}{b}\right)\right]}{\sqrt{b^2-a^2}} \end{aligned} \tag{4.11.129}$$

the following limits of series can be evaluated:

$$\lim_{\nu\to\infty}\left\{\nu\sum_{n=0}^{\infty}J_n(\lambda)J_{n+1}(\lambda)\frac{\sin\{\nu\sin^{-1}[(2n+1)(\frac{t}{\nu})]\}}{\sqrt{\nu^2-[(2n+1)(\frac{t}{\nu})]^2}}\right\} \tag{4.11.130}$$
$$=\frac{1}{2}J_1(2\lambda\sin t)$$

and

$$\lim_{\nu\to\infty}\left\{\nu\sum_{n=0}^{\infty}(-1)^n J_n(\lambda)J_{n+1}(\lambda)\frac{\cos\{\nu\sin^{-1}[(2n+1)(\frac{t}{\nu})]\}}{\sqrt{\nu^2-[(2n+1)(\frac{t}{\nu})]^2}}\right\} \tag{4.11.131}$$
$$=\frac{1}{2}J_1(2\lambda\cos t)$$

As demonstrated in this section, mathematical operations performed with the order ν of the Bessel function $J_\nu(\nu x)$ lead to many interesting results. Established asymptotic limits, representations of mathematical constants and functions, evaluated integrals and series undoubtedly prove that the Lamborn sequence for the shifted Dirac delta function $\delta(x-1)$, is a very efficient tool in the theory of special functions.

References

[1] Apelblat, A., Katzir-Katchalsky, A., Silberberg, A. (1974) A mathematical analysis of capillary – tissue fluid exchange. Biorheology, 11:1–49.

[2] Apelblat, A. (1980) Mass transfer with a chemical reaction of the first order. Analytical solutions. Chem. Eng. J. 19:19–37.

[3] Apelblat, A. (1982) Mass transfer with a chemical reaction of the first order. Effect of axial diffusion. Chem. Eng. J. 23:193–203.

[4] Apelblat, A., Kravitsky, N. (1985) Integral representations of derivatives and integrals with respect to the order of the Bessel functions $J_\nu(t)$, $I_\nu(t)$, the Anger function $\mathbf{J}_\nu(t)$ and the integral Bessel function $Ji_\nu(t)$. IMA J. Appl. Math. 34:187–210.

[5] Apelblat, A. (1989) Derivatives and integrals with respect to the order of the Struve functions $H_\nu(x)$ and $L_\nu(x)$ J. Math. Anal. Appl. 137:13–36.

[6] Apelblat, A. (1991) Integral representation of Kelvin functions and their derivatives with respect to the order. Z. Angew. Math. Phys 42:708–714.

[7] Watson, G.N. (1958) A Treatise on the Theory of Bessel Functions. Sec. Ed. Cambridge University Press, Cambridge.

[8] Hayek Calil, N. (1989) Tablas de Ecuaciones Differenciales Integrables Mediante Funciones de Bessel-Clifford. Univ. de la Laguna, Tenerife, Canarias, Spain.

[9] Abramowitz, M., Stegun, I.A. (1964) Handbook of Mathematical Functions with Formulas, Graphs, and Mathematical Tables. U.S. National Bureau of Standards. Applied Mathematics Series, vol. 55, Washington, D.C.

[10] Andrew, L.C. (1985) Special Functions for Engineers and Applied Mathematicians. MacMillan Publ. Co., New York.

[11] Brychkov, Yu. A. (2008) Handbook of Special Functions. Derivatives, Integrals, Series and Other Formulas. CRC Press, Boca Raton.

[12] Erdélyi, A., Magnus, W., Oberhettinger, F., Tricomi, F.G. (1953) Higher Transcendental Functions. McGraw-Hill, New York.

[13] Gradstein, I., Ryzhik, I. (1981) Tables of Series, Products and Integrals. Verlag Harri Deutsch, Thun-Frankfurt, 1981.

[14] Gray, A., Mathews, G.B., MacRobert, T.M. (1966) A Treatise on Bessel Functions Their Application to Physics. Sec. Ed. Dover. Publ., Inc New York

[15] Hansen, E.R. (1975) A Table of Series and Products. Prentice Hall, Englewood Cliff.

[16] Kamke, E. (1977) Differentialgleichungen: Lösungsmethoden und Lösungen. Tenbner, Leipzig.

[17] Lebedev, N.N. Special Functions and Their Applications. (1965) Prentice-Hall, London.

[18] Magnus, W., Oberhettiner, F., Soni, R.P. (1966) Formulas and Theorems for the Special Functions of Mathematical Physics. 3rd ed. Springer-Verlag, Berlin.

[19] Olver, F.W.J., Lozier, D.W., Boisvet, R.F., Clark, C.W. (2010) NIST Hanbook of Mathematical Functions. Cambridge University Press, New York.

[20] Whittaker, E., Watson, G. A (1963) Course of Modern Analysis. Cambridge University Press, Cambridge.

[21] Wong, Z.X., Guo, D.R. (1989) Special Functions. World Scientific, Singapore.

[22] McLachlan, N.W. (1955) Bessel Functions for Engineers. The Clarendon Press, Oxford.

[23] Petiau, G. (1955) La Thêorie des Fonctions de Bessel. CNRS, Paris.

[24] Tranter, C.J. (1968) Bessel Functions With Some Physical Applications. Hart Publ. Co., Inc. New York.

[25] Vallée, O., Soares, M. (1998) Les Functions D'Airy Pour La Physique. Diderot Multimedia, Paris.

https://doi.org/10.1515/9783110681642-005

[26] Whellon, A.D. (1968) Tables of Summable Series and Integrals of Bessel Functions. Holden-Day, San Francisco.

[27] Apelblat, A. (1983) Table of Definite and Infinite Integrals. Elsevier Scientific Publishing Co., Amsterdam.

[28] Apelblat, A. (1996) Tables of Integrals and Series. Verlag Harri Deutsch, Frankfurt am Main.

[29] Apelblat, A. (2004) The application of the Dirac delta function $\delta(x-1)$ to the evaluation of limits and integrals of elementary and special functions. Int. J. Appl. Math. 16:323–339.

[30] Apelblat, A. (2008) Volterra Functions. Nova Sci. Publ. Inc. New York.

[31] Apelblat, A. (2009) Integral Transforms and Volterra Functions. Nova Sci. Publ. Inc. New York.

[32] Apelblat, A. (2012) Laplace Transforms and Their Applications. Nova Sci. Publ. Inc. New York.

[33] Erdélyi, A., Magnus, W., Oberhettinger, F., Tricomi, F.G. (1954) Tables of Integral Transforms. McGraw-Hill, New York.

[34] Hladik, J. La Transformation de Laplace a Plusieurs Variables. Masson et Cie Éditeurs. Paris, 1969.

[35] Luke, Y. (1962) Integrals of Bessel Functions. McGraw-Hill Book Co., Inc., New York.

[36] Oberhettinger, F., Badii, L. (1970) Laplace Transforms, Springer-Verlag, Berlin.

[37] Oberhettinger, F., Badii, L. (1972) Table of Bessel Transforms, Springer-Verlag, Berlin.

[38] Oberhettinger, F., Badii, L. (1974) Table of Mellin Transforms, Springer-Verlag, Berlin.

[39] Prudnikov, A.P., Brychkov, Y.A., Marichev, D.I. (1986) Special Functions. Vol. 2. Integrals and Series. Gordon and Breach, New York.

[40] Prudnikov, A.P., Brychkov, Y.A., Marichev, D.I. (1990) Integrals and Series. Vol. 3. More Special Functions. Gordon and Breach, New York.

[41] Roberts, G.E., Kaufman, H. (1966) Table of Laplace Transforms. W.B. Saunders Co., Philadelphia.

[42] Airey, J.R. (1928) Tables of the Bessel functions derivatives $\partial J_\nu(x)/\partial\nu$: $\nu = \pm 1/2$ and $\nu = \pm 3/2$. Report of the Math. Tables Comm. British Association.

[43] Airey, J.R. XVIII. (1935) The Bessel function derivatives $\partial J_\nu(x)/\partial\nu$ and $\partial^2 J_\nu(x)/\partial\nu^2$. Phil. Mag. 19:236–243.

[44] Brychkov, Yu. A., Geddes, K.O. (2005) On the derivatives of the Bessel and Struve functions with respect to the order. Integral Transform Special Functions 16:187–198.

[45] Brychkov, Yu. A. (2007) On some formulas for Weber $E_\nu(z)$ and $J_\nu(z)$ functions. ITSF 18: 187–198.

[46] Cooke, J.C. (1954) Note on some integrals of Bessel functions with respect to their order. Monath. Math. 58:1–4.

[47] Coulomb, J. (1936) Sur les zeros des functions de Bessel considérées comme function de l'ordre. Bull. Sci. Math., 60:297–302.

[48] Dunster, T.M. (2017) On the order derivatives of Bessel functions. Constr. Approx. 46:47–68.

[49] Erber, T., Gordon, A. (1963) Tabulation of the function $\partial I_\nu/\partial\nu$ for $\nu = \pm 1/3$. Comp. Math. 17:162–169.

[50] Fényes, T. (1993) On the Fourier transform of the modified Bessel function with respect to the order. Studia Sci. Math. Hungar. 28:189–196.

[51] Fényes, T. (1993) On the Fourier transform of the Bessel function with respect to the order. Studia Sci. Math. Hungar. 28:197–204.

[52] Fényes, T. (1997) Integral representations of the Laplace transform and moments of the Bessel function with respect to the order. Periodica Math. Hung. 35:1–8.

[53] González-Santander, J.L. (2018) Closed-form expressions for derivatives of Bessel functions with respect to the order. J. Math. Anal. Appl. 466:1060–1081.

[54] González-Santander, J.L. (2018) On the n-th derivative and the fracional integration of Bessel functions with respect to the order. arXiv:1808.05608v1.

[55] Lee, K., Radosevich, L.D. (1960) Evaluation of $\partial J_\nu/\partial\nu$. J. Math. Phys.39:293–299.
[56] Oberhettinger, F. (1958) On the derivative of Bessel functions with respect to the order. J. Math. and Phys. 37:75–78.
[57] Wienke, B.R. (1977) Order derivatives of Bessel functions. Bull. Calcutta Math. Soc. 69:389–392.
[58] Jahnke, E., Emde, F. (1945) Tables of Functions with Formulae and Curves. Dover Publ. New York.
[59] Oldham, K.B., Myland, J., Spanier, J. (1987) An Atlas of Functions. Hemisphere, Washington, DC.
[60] Ahmed, S., Calogero, F. (1978) On the zeros of Bessel functions III. Lett. Nuovo Cimento. 21:311–314.
[61] Beckmann, P., Franz, W. (1957) Über die Greenschen Funktionen transparenter Zylinder. Z. Naturf. 12a:257–267.
[62] Calogero, F. (1977) On the zeros of Bessel functions. Lett. Nuovo Cimento. 20:254–256.
[63] Calogero, F. (1977) On the zeros of Bessel functions II. Lett. Nuovo Cimento. 20:476–478.
[64] Cohen, D.S. (1964) Zeroes of Bessel functions and eigenvalues of non-self- adjointvalue problems. J. Math. and Phys. 43:133–139.
[65] Cochran, J.A. (1965) The zeros of Henkel functions as function of their order. Num. Math. 7:238–250.
[66] Keller, J.B., Rubinow, S.I., Goldstein, M. (1963) Zeros of Hankel functions and poles of scattering amplitudes. J. Math. Phys. 4:829–832.
[67] Levy, B.R., Keller, J.B. (1959) Diffraction by a smooth object. Comm. Pure Appl. Math. 12:159–209.
[68] Magnus, W., Kotin, L. (1960) The zeros of the Hankel function as a function of its order. Num. Math. 2:228–224.
[69] Muldoon, M.E. (1981) The variation with respect to order of zeros of Bessel functions. Rend. Sem. Mat. Univers. Politecn. Torino 39:15–25.
[70] Streifer, W., Kodis, R.D. (1964) On the solution of a transcendental equation arising in the theory of scattering by a dielectric cylinder. Quart. Appl. Math. 21:285–298.
[71] Streifer, W., Kodis, R.D. (1964) On the solution of a transcendental equation scattering theory. Quart. Appl. Math. 23:27–38.
[72] Streifer, W. (1965) On the zeros of Bessel functions as a function of order. J. Math. Phys. 47:400–405.
[73] Müller, R. (1940) Uber die partielle Ableitung der Besselschen Funktionen nach Ihrem Parameter. ZAMM 20:61–62.
[74] Mitra, S.C. (1936) On certain new connections between Legendre and Bessel functions. Math. Zeitsch. 41:680–685.
[75] Mitra, S.C. (1938) On parabolic cylinder functions which are self-reciprocical in Hankel-transform. Math. Zeitsch. 43:205–211.
[76] Fényes, T. (1997) Integral representations of the Laplace transform and moments of the modified Bessel function with respect to the order. Periodica Math. Hung. 35:9–14.
[77] Watson, G.N. (1916) XXV. Bessel functions of equal order and argument. Phil. Mag. 32: 232–237.
[78] Lorch, L. (1992) On Bessel functions of equal order and argument. Rend. Sem. Math. Univ. Pol. Torino. 50:209–216.
[79] Franz, W. (1954) Über die Greenschen Funktionen des Zylinders und der Kugel. Z. Naturf. 9a:705–716.
[80] Franz, W., Galle, R. (1955) Semiasymptotische Reihen für die Beugung einer ebenen Welle am Zylinder. Z. Naturf. 10a:374–378.

[81] van der Pol, B. (1929) On the operational solution of linear differential equations and an investigation of the properties of these solutions. Phil. Mag. 80:861–898.
[82] van der Pol, B., Bremmer, H. (1964) Operational Calculus Based on the Two- Sized Laplace Integral. Cambridge University Press, Cambridge.
[83] Lamborn, B.N. A. (1969) An expression of the Dirac delta function. SIAM Rev. 11:603.
[84] Apelblat, A. (1999) The integral representation of functions as the asymptotic limit of infinite integral of the Bessel function $J_\nu(\nu x)$. Int. J. Appl. Math. 1:19–27.
[85] Apelblat, A. (2000) The asymptotic limit of infinite integral of the Bessel function $J_\nu(\nu x)$ as integral representation of elementary and special functions. Int. J. Appl. Math. 2:743–762.
[86] González-Santander, J.L. (2018) A note on the order derivatives of Kelvin functions. Results Math. 74:31.
[87] González-Santander, J.L. (2018) Reflection formulas for order derivatives of Bessel functions. arXiv:1809.08124v1.
[88] Becker, P.A. (2009) Infinite integrals of Whittaker and Bessel functions with respect their indices. J. Math. Phys. 50:123515.
[89] Humbert, P. (1933) Bessel-integral functions. Proc. Edinburgh. Math. Soc. 3:276–285.
[90] Conde, S., Kalla, S.L. (1979) The ν-Zeros of $J_{-\nu}(x)$. Math. Comp. 33:423–426.
[91] Cruz, A., Esparza, J., Sesma, J. (1991) Zeros of the Hankel function of real order of the principal Riemann sheet. J. Comp. Appl. Math. 37:89–99.
[92] Cruz, A., Sesma, J. (1982) Zeros of the Hankel function of real order and its derivative. Math. Comp. 39:639–345.
[93] Hethcote, H.M. (1970) Error bounds for asymptotic approximations of zeros of transcendental functions. SIAM J. Math. Anal. 1:147–152.
[94] Hethcote, H.M. (1970) Bounds for zeros of some special functions. Proc. Amer. Math. Soc. 25:72–74.
[95] Hethcote, H.W. (1970) Error bounds for approximations of zeros of Hankel functions occurring in diffraction problems. J. Math. Phys. 11:2501–2504.
[96] Nagase, M. (1954) On the zeros of certain transcendental functions related to Hankel function. Part I. J. Phys. Soc. Japan, 9:826–841.
[97] Nagase, M. (1954) On the zeros of certain transcendental functions related to Hankel function. Part II. J. Phys. Soc. Japan, 9:842–853.
[98] Aktaş, I., Baricz, A., Orhan, H. (2018) Bounds for radii of starlikeness and convexity of some special functions. arXiv:1610.03233v. Tirkish J. Math. 42:211–226.
[99] Baricz, Á. (2015) Bounds for Turánians of modified Bessel functions. arXiv:1202.4853. Expo. Math. 33:223–251.
[100] Baricz, Á., Kokologiannaki, C.G., Pogány, T.K. (2018) Zeros of Bessel function derivatives. Proc. Amer. Math. Soc. 146:209–222.
[101] Boyer, T.H. (1969) Concerning the zeros of some functions related to Bessel functions. J. Math Phys. 10:1729–1744.
[102] Breen, S. (1995) Uniform upper and lower bounds on the zeros of Bessel functions of the first kind. J. Math. Anal. Appl. 196:1–17.
[103] Budzinkiy, S.S., Kharitonov, D.M. (2017) On inflection points of Bessel functions of the second kind of positive order. ITSF 28:909–914.
[104] Cho, Y.K., Chung, S.Y. (2019). On the positivity and zeros of Lommel functions. Hyperbolic extension and interlacing. J. Math. Anal. Appl. 470:898–910.
[105] Davis, H.T., Kirkham, W.J. (1927) A new table of the zeros of the Bessel functions $J_0(x)$ and $J_1(x)$ and corresponding values of $J_1(x)$ and $J_0(x)$. Bull. Amer. Math. Soc. 33:760–772.
[106] Deniz, E., Topkaya, S., Çačlar, M. (2018) Geometric properties of Bessel function derivatives, arXiv:1802.05462v2.

[107] Elbert, Á., Laforgia, A. (1983) On the zeros of derivatives of Bessel functions. J. Appl. Math. Phys (ZAMP) 34:774–786.

[108] Elbert, Á., Laforgia, A. (1997) An upper bound for the zeros of the derivative of Bessel functions. Rend. Circ. Matem. Palermo 46:123–130.

[109] Elbert, Á. (2001) Some recent results on the zeros of Bessel functions and orthogonal polynomials. J. Comp. Appl. Math. 133:65–83.

[110] Elbert, Á., Siafarikas, P. (1999) On the square of the first zero of the Bessel function $J_\nu(z)$. Can. Math. Bull. 42:56–67.

[111] Elbert, Á., Laforgia, A. (1987) Some consequences of a lower bound of the second derivative of the zeros of Bessel functions. J. Math. Anal. Appl. 125:1–5.

[112] Elbert, Á., Laforgia, A. (1984) On the square of the zeros of Bessel functions. SIAM J. Math. Anal. 14:206–212.

[113] Elizalde, E., Leseduarte, S., Romeo, A. (1993) Sum rules for zeros of Bessel functions and application to spherical Aharonov-Bohm quantum bags. J. Phys. A. Math. Gen. 26:2409–2419.

[114] Gatteschi, L. (1955) Sulla reppresentacione asintotica della funzioni di Bessel di uguale ordine ed argomento. Ann. Mat. Pure Appl. 38:267–280.

[115] Gatteschi, L. (1955) Sulla reppresentacione asintotica della funzioni di Bessel di uguale ordine ed argomento. Boll. Union Mat. Ital. 10:531–532.

[116] Gatteschi, L. (1957) Sul comportamento asintotica della funzioni di Bessel di prima specie di ordine ed argomento quasi uguale. Ann. Mat. Pure Appl. 43:97–117.

[117] Gateschi, L., Giordano, C. (2000) Error bounds for McMahon's asymptotic approximations for the zeros of the Bessel functions. ITSF 10:41–66.

[118] Giusti, A., Mainardi, F. (2016) On infinite series concerning zeros of Bessel functions of the first kind. Eur. Phys. J. Plus 131:206–218.

[119] Gordano, C., Laforgia, A. (1983) Elementary approximations for zeros of Bessel functions. J. Comp. Appl. Math. 9:221–228.

[120] Ismail, M.E.H., Muldoon, M.E. (1988) On the variation with respect to the parameter of zeros of Bessel and q-Bessel functions. J. Math. Anal. Appl. 135:187–201.

[121] Ismail. M.E.H., Muldoon. M.E. (1995) Bounds for the small real and purely imaginary zeros of Bessel and related functions. Methods Appl. Math. 2:1–21.

[122] Ifantis, E.K., Siafarikas, P.D. (1990) Differential inequalities for positive zeros of Bessel functions. J. Comp. Appl. Math. 30:139–143.

[123] Ifantis, E.K., Siafarikas, P.D. (1992) A differential inequality for the positive zeros of Bessel functions. J. Comp. Appl. Math. 42:115–120.

[124] Ifantis, E.K., Kokologiannaki, C.G., Kouris, C.B. (1991) On the positive zeros of the second derivative of Bessel functions. J. Comp. Appl. Math. 34:21–31.

[125] Kerimov, M.K. (2014) Studies on the zeros of Bessel functions and methods for their computation. Comp. Math. Math. Phys. 54:3–41.

[126] Kerimov, M.K. (2016) Studies on the zeros of Bessel functions and methods for their computation. 2. Monotonicity, convexity, concavity, and other properties. Comp. Math. Math. Phys. 56:1175–1208.

[127] Kerimov, M.K. (2016) Studies on the zeros of Bessel functions and methods for their computation. 3. Some new works on monotonicity, convexity, and other properties. Comp. Math. Math. Phys. 56:1949–1208.

[128] Kerimov, M.K. (2018) Studies on the zeros of Bessel functions and methods for their computation. IV. Inequalities, estimations, expansions, etc. for zeros of Bessel functions. Comp. Math. Math. Phys. 58:1337–1388.

[129] Kokologiannaki, C. (2002) Convexity of the square of the first zero of the derivative of Bessel functions. ITSF 13:471–481.

[130] Koumandos, S., Lamprecht, M. (2012) The zeros of certain Lommel functions. Proc. Amer. Math. Soc. 140:3091–3100.
[131] Laforgia, A. (1980) Sugli zeri delle funzione di Bessel. Calcolo 17:211–220.
[132] Laforgia, A., Muldoon, M.E. (1983) Inequalities and approximations for zeros of Bessel functions of small order. SIAM J. Math. Anal. 14:383–388.
[133] Laforgia, A., Muldoon, M.E. (1984) Monotonicity and concavity properties of zeros of Bessel functions. J. Math. Anal. Appl. 98:470–477.
[134] Laforgia, A. (1984) Monotonicity properties of the zeros of orthogonal polynomials and Bessel functions. In Polynômes Orthogonaux er Applications. 267–277. Eds. Brezinski, C., Draux, A., Magnus, A.P., Maroni, P. Bar-le-Duc, France, Springer-Verlag, Berlin.
[135] Laforgia, A., Natalini, P. (2007) Zeros of Bessel functions: monotonicity, concavity, inequalities. Matematiche (Catania) 62:255–270.
[136] Lang, T., Wong, R. (1996) "best possible" upper bounds for the first two positive zeros of the Bessel function $J_\nu(x)$. The infinite case. J. Comp. Appl. Math. 71:311–329.
[137] Lewis, J.T., Muldoon, M.E. (1977) Monotonicity and convexity properties of zeros of Bessel functions. SIAM J. Math. Anal. 8:171–178.
[138] Lorch, L. (1990) Monotonicity in terms of order of the zeros of the derivatives of Bessel functions. Proc. Amer. Math. Soc. 108:387–389.
[139] Lorch, L., Szego, P. (1990) On the points of inflection of Bessel functions of positive order. I. Can. J. Math. 42:933–948.
[140] Lorch, L. (1993) Some inequalities for the first positive zeros of Bessel functions. SIAM J. Math. Anal. 24:814–823.
[141] Lorch, L. (1995) The zeros of the third derivatives of Bessel functions of order less than one. Methods Appl. Anal. 2:147–159.
[142] Lorch, L., Uberti, R. (1996) "Best possible" upper bounds for the first positive zeros of Bessek functions –the finite part J. Comp. Appl. Math. 75:249–258.
[143] McCann, R.101–103. (1977) Lower bounds for the zeros of Bessel functions. Proc. Amer. Math. Soc. 64:101–103.
[144] McMahon, J. (1894–1895) On the roots of the Bessel and certain related functions. Ann. Math. 9:23–30.
[145] Nalesso, G.F. (1989) On the zeros of a class of Bessel functions whose argument and order are function of a complex variable. IMA J. Appl. Math. 43:195–217.
[146] Paris, R.B. (1984) An inequality for the Bessel function $J_\nu(\nu x)$. SIAM J. Math. Anal. 15:203–205.
[147] Pálmai, T., Apagyi, B. (2011) Interlacing of positive real zeros of Bessel functions. J. Math. Anal. Appl. 375:320–322.
[148] Piessens, R. (1984) A series expansion for the first positive zero of the Bessel functions. Math. Comp. 42:195–197.
[149] Qu, C.K., Wong, R. "Best possible" upper and lower bounds for the zeros of the Bessel function $J_\nu(x)$. Trans. Amer. Math. Soc. 351:2833–2850.
[150] Sneddon, I.N. (1960) On some infinite series involving the zeros of Bessel functions of the first kind. Proc. Glasgow Math. Assoc. 7:144–156.
[151] Steining, J. (1970) The real zeros of Struve's function. SIAM J. Math. Anal. 1:365–375.
[152] Steining, J. (1972) The sign of Lommel's function. Trans. Amer. Math. Soc. 163:123–129.
[153] Colombaro, I., Giusti, A., Mainardi, F. (2016) A class of linear viscoelastic models based on Bessel functions. Meccanica DOI.10.1007/s11012-016-0456-5.
[154] Buschman, R.G. (1974) Finite sums representations for partial derivatives of special functions with respect to parameters. Math. Comp. 28:817–824.
[155] Grosjean, C.C. (1984) The orthogonality property of the Lommel polynomials and a twofold infinity of relations between Rayleigh's σ – series. J. Comp. Appl. Math. 10:355–382.

[156] Hanson, M.T., Puja, I.W. (1997) The evaluation of certain infinite integrals involving products of Bessel functions. A correlation formula. Quart. Appl. Math. 55:505–524.
[157] Siegel, K.M. (1953) An inequality involving Bessel functions of argument nearly equal to their order. Proc. Amer. Math. Soc. 4:858–859.
[158] Siegel, K.M., Sleator, F.B. (1954) Inequalities involving cylindrical functions of nearly equal argument and order. Proc. Amer. Math. Soc. 5:337–344.
[159] Hoskins, R.F. (1979) Generalized Functions. Ellis Horwood Ltd., Chichester.
[160] Katz, M.G., Tall, D. (2013) A Cauchy-Dirac delta function. Found. Sci. 18:107–123.
[161] Li, Y.T., Wong, R. (2008) Integral and series representations of the Dirac delta function. arXiv:1303.1943v1. Commun. Pure Appl. Anal. 7:229–247.
[162] Laforgia, A., Natalini, P. (2017) Asymptotic limits in infinite integrals from an expression of Dirac function. J. Inequalities Spec. Funct. 8: 203–206.62:255–270.
[163] Abramochkin, E.G., Razueva, E.V. (2016) Mellin transform of quartic products of shifted Airy functions. ITSF 26:454–467.
[164] Askari, H., Ansari, A. (2018) Airy functions and Riesz fractional trigonometric operators. ITSF = Integral Transforms Spec. Func. 29:585–604.
[165] Bailey, D.H., Borwein, Crandall, R. (2006) Integrals of the Ising class. J. Phys. A. Math. Gen. 39:12271–12302.
[166] Bailey, D.H., Borwein, J.M., Broadhurst, D., Glasser, M.L. (2008) Elliptic integral evaluation of Bessel moments and applications. J. Phys. A. Math. Theor. 41:205203–205249.
[167] Chaundhry, A.S. (1994) On a integral of Lommel and Bessel functions. J. Austral. Math. Soc. B 35:439–444.
[168] Cohl, H.S., Nair, S.J., Palmer, R. M. (2016) Some dual definite integrals for Bessel functions of the first kind. Scientia, A. Math. Sci. 27:15–30.
[169] Conway, J.T. (2015) Infinite integrals of some special functions from a new method. ITSF 26:845–858.
[170] Desbois, J., Ouvry, S. (2012) Bessel integrals, periods and zeta numbers. arXiv:1209.1036v2.
[171] Fabrikant, V.I., Dôme, G. (2001) Elementary evaluation of certain infinite integrals involving Bessel functions. Quart. Appl. Math. 54:1–21.
[172] Fabrikant, V.I. (2003) Computation of infinite integrals involving three Bessel functions by introduction of new formalism. J. Angew. Math. Mech. (ZAMM) 83:363–374.
[173] Furtlehmer, C., Ouvry, S. (2004) Integrals involving four Macdonalds functions and their relations. arXiv: math-ph/0306004v2.
[174] Glasser, M.L., Montaldi, E. (1994) Some integrals involving Bessel functions. J. Math. Anal. Appl. 183:377–390.
[175] Glasser, M.L., Kowalenko, V. (1997) A method for evaluating Laplace transforms and other integrals. ITSF 5:161–184.
[176] Glasser, M.L. (2010) Integral representations for the exceptional univariable Lommel functions. J. Phys. A Math. Theor. 43:1–4.
[177] Glasser, L., Kohl, K.T., Koutschan, C., Moll, V.H., Straub, A. (2012) The integrals of Gradshteyn and Ryzhik. Part 22. Bessel K-functions. Scientia A.
[178] González-Santander, J.L. (2015) Calculation of some integrals arising in the Samara-Valencia solution for dry flat grinding. Math. Probl. Eng. ID 928461.
[179] González-Santander, J.L. (2017) New integrals arising in heat transfer the Samara-Valencia model in grinding. J. Appl. Math. ID 35911713.
[180] González-Santander, J.L. (2016) Calculation of some integrals involving the Macdonald function by using Fourier transform. J. Math. Anal. Appl. 441:349–363.
[181] Grandits, P. (2019) Some notes on Sonine-Gegenbauer integrals. ITSP 30:128–137.

[182] Koumandos, S. (2017) Positive trigonometric integrals associated with some Lommel functions of the first kind. Mediterr. J. Math. 14:15.
[183] Kölbig, K.S. (1995) Two infinite integrals of products of modified of Bessel functions and powers of logarithms. J. Comp. Appl. Math. 62:41–65.
[184] Kölbig, K.S. (1996) An infinite integral of Bessel function. J. Comp. Appl. Math. 64:161–183.
[185] Laurenzi, B.J. (1993) Moment integrals of powers of Airy functions. J. Angew. Math. Phys. (ZAMP) 44:891–907.
[186] Laurenzi, B.J. (2017). Integrals containing the logarithm of the Airy function Ai'(x). arXiv: 1712,10332v.
[187] Lin, Q.G. (2014) Infinite integrals involving Bessel functions by an improved approach of contour integration and the residue theorem. Ramanujan J. 35:443–466.
[188] Mashkevich, S., Ouvry, S. (2008) Random Aharonov-Bohm vortices and some exact families of integrals. II. J. Stat. Mech. P03018.
[189] Mavromatis, H.A. (1994) Two integrals arising in inverse scattering theory. J. Math. Anal. Appl. 188:458–464.
[190] McPhedran, R.C., Dawes, D.H., Scott, T.C. (1992) On a Bessel function integral. Appl. Alg. Eng. Comm. Comp. (AAECC) 2:207–216.
[191] McPhedran, R.C., Stout, B. (2018) "Killing Mie Softly" Analytic integrals for resonant scattering. arXiv:1811.07132v1.
[192] Ouvry, S. (2005) Random Aharonov-Bohm vortices and some exactly solvable families of integrals. J. Stat. Mech. P09004.
[193] Reid, W.H. (1995) Integral representations for products of Airy functions. Z. Angew. Math. Phys. (ZAMP) 46:159–170.
[194] Reid, W.H. (1997) Integral representations for products of Airy functions. Part 2. Cubic products. Z. Angew. Math. Phys. (ZAMP) 48:646–655.
[195] Reid, W.H. (1997) Integral representations for products of Airy functions. Part 3. Quartic products. Z. Angew. Math. Phys. (ZAMP) 48:656–664.
[196] Schulz-DuBois, E.O. (1969) Integral relations among Bessel functions. Math. Comp. 23:848–847.
[197] Vallee, O., Soares, M., de Izarra, C. (1997) An integral representation of the product of Airy functions. Z. Angew. Math. Phys. (ZAMP) 48:156–160.
[198] Vallée, O. (2002) Some integrals involving Airy functions and Volterra functions. ITSF 13:403–408.
[199] Varlamov, V. (2010) Integrals involving products of Airy functions, their derivatives and Bessel functions. J. Math. Anal. Appl. 270:26–30. Math. Sci. 22:129–151.

Appendix A The Bessel and Related Functions Integrals

Miscellaneous integrals recently reported in the literature [158, 165–201], mainly from the period 1990–2017, are presented here and therefore they are not available in major tabulations of integrals of the Bessel and related functions.

A1 Integrals Containing One Bessel or Related Function, and Elementary and Special Functions

$$\int_0^a \frac{J_\nu(bx)}{\sqrt{x(a-x)}}\,dx = \pi\sqrt{\frac{a}{2}}J_{\nu/2-1/4}\left(\frac{ab}{2}\right)J_{\nu/2+1/4}\left(\frac{ab}{2}\right) \quad ; \quad \mathrm{Re}\,\nu > -1 \tag{A.1.1}$$

$$\int_a^\infty \frac{J_\nu(bx)}{\sqrt{x^2-a^2}}\,dx = -\frac{\pi}{2}J_{\nu/2}\left(\frac{ab}{2}\right)Y_{\nu/2}\left(\frac{ab}{2}\right) \quad ; \quad a>0 \quad ; \quad \nu \geq -\frac{1}{2} \tag{A.1.2}$$

$$\int_0^a x^\nu (a^2-x^2)^{\nu-1/2} J_\nu(bx)\,dx = 2^{3\nu-1}\sqrt{\pi}\,\Gamma\left(\nu+\frac{1}{2}\right)\frac{a^{2\nu}}{b^\nu}\left[J_\nu\left(\frac{ab}{2}\right)\right]^2 \tag{A.1.3}$$

$\mathrm{Re}\,\nu > -\frac{1}{2}$

$$\int_a^b x^{1-\nu}\left[(x^2-a^2)(b^2-x^2)\right]^{\nu-1/2} J_\nu(cx)\,dx$$

$$= 2^{3\nu-1}\sqrt{\pi}\,\Gamma\left(\nu+\frac{1}{2}\right)\frac{(b^2-a^2)^\nu}{c^\nu}J_\nu\left[\frac{(a+b)c}{2}\right]J_\nu\left[\frac{(b-c)c}{2}\right] \quad ; \quad \mathrm{Re}\,\nu > -\frac{1}{2} \tag{A.1.4}$$

$$\int_0^\infty x^{\alpha/2-1}(1+x)^{1-\alpha-\beta} J_\nu(\lambda\sqrt{1+x})\,dx$$

$$= \frac{\Gamma(\frac{\nu}{2}-\beta+1)}{\Gamma(\frac{\nu}{2}+\beta)}\left(\frac{\lambda^2}{4}\right)^{\beta-1} {}_1F_2\left(1-\alpha;\beta-\frac{\nu}{2},\beta+\frac{\nu}{2};-\frac{\lambda^2}{4}\right) \tag{A.1.5}$$

$$+ \frac{\Gamma(\alpha)\,\Gamma(\beta-\frac{\nu}{2}-1)}{\Gamma(\nu+1)\,\Gamma(\alpha+\beta-\frac{\nu}{2}-1)}\,{}_1F_2\left(\frac{\nu}{2}-\alpha-\beta+2;\frac{\nu}{2}-\beta+2,1+\nu;-\frac{\lambda^2}{4}\right)$$

$$\int_0^\infty x^{-1/4}(1+x)^{-1/2} J_\nu(\lambda\sqrt{1+x})\,dx = \pi J_{\nu/2}\left(\frac{\lambda}{2}\right)Y_{\nu/2}\left(\frac{\lambda}{2}\right) \tag{A.1.6}$$

https://doi.org/10.1515/9783110681642-006

$$\int_0^\infty x^{-3/4} J_\nu(\lambda\sqrt{1+x})\,dx = -\frac{\pi}{2}\left[J_{(\nu-1)/2}\left(\frac{\lambda}{2}\right) Y_{(\nu+1)/2}\left(\frac{\lambda}{2}\right) + J_{(\nu+1)/2}\left(\frac{\lambda}{2}\right) Y_{(\nu-1)/2}\left(\frac{\lambda}{2}\right)\right] \tag{A.1.7}$$

$$\int_0^\infty [x(1+x)]^{-1/2} J_0(\lambda\sqrt{1+x})\,dx = 2\left[\frac{1}{\lambda} - J_0(\lambda)\right] + \pi\,[J_0(\lambda)\,H_1(\lambda) - J_1(\lambda)\,H_0(\lambda)] \tag{A.1.8}$$

$$\int_0^\infty (1+x)^{-1/2} J_1\left(\lambda\sqrt{1+x}\right)dx = 2\,\frac{J_0(\lambda)}{\lambda} \tag{A.1.9}$$

$$\int_0^\infty x^{-3/4}(1+x)^{-1/4} J_1\left(\lambda\sqrt{1+x}\right)dx = 2\,\frac{\sin\lambda}{\lambda} \tag{A.1.10}$$

$$\int_0^\infty x^{-1/2}(1+x)^{-1/4} J_\nu(\lambda\sqrt{1+x})\,dx = \frac{2}{\sqrt{\lambda}}\left[J_{\nu-1}(\lambda)\,S_{\nu/2}(\lambda) + \left(\frac{1}{2} - \nu\right) J_\nu(\lambda)\,S_{-\nu/2}(\lambda)\right] \tag{A.1.11}$$

$$\int_0^\infty x^{-3/4} J_0\left(\lambda\sqrt{1+x}\right)dx = 2\,\frac{\cos\lambda}{\lambda} \tag{A.1.12}$$

$$\int_0^a x(1-2x^2)\,e^{-x^2}\,[J_{1/2}(ax)]^2 dx = -\frac{a\,e^{-a^2}}{\sqrt{\pi}} \tag{A.1.13}$$

$$\int_0^\infty \frac{x^\lambda}{(x^2-a^2)^2} J_\nu(bx)\,dx = \frac{\pi a^{\lambda-2} b}{4}\tan\left[\frac{\pi(\lambda+\nu)}{2}\right] J_{\nu+1}(ab)$$
$$+\frac{\Gamma\left(\frac{\lambda+\nu-3}{2}\right)}{2\Gamma\left(\frac{5-\lambda+\nu}{2}\right)}\left(\frac{b}{2}\right)^{3-\lambda} {}_1F_2\left(2;\frac{5-\lambda+\nu}{2},\frac{5-\lambda-\nu}{2};\frac{a^2b^2}{4}\right)$$
$$-(\operatorname{Re}\nu+1) < \operatorname{Re}\lambda < \frac{9}{2} \tag{A.1.14}$$

$$\int_0^\infty \frac{e^{-2/x}}{x} J_\nu(2\sqrt{ax})\,dx = 2J_\nu(2\sqrt{a})\,K_\nu(2\sqrt{a}) \quad ; \quad \nu \geq -\frac{1}{2} \tag{A.1.15}$$

$$\int_0^a \ln\left(1-\frac{x^2}{a^2}\right) J_1(bx)\,dx = -\frac{\pi}{a} Y_0(ab) \quad ; \quad a>0 \tag{A.1.16}$$

$$\int_0^\infty \ln(1+x^2)\, J_1(ax)\, dx = \frac{2}{a} K_0(a) \tag{A.1.17}$$

$$\int_0^\infty \frac{\sin(bx)}{\sqrt{1+x^2}} \sinh\left(\frac{ax}{1+x^2}\right) J_0\left(\frac{a}{1+x^2}\right) dx = K_0(b)\, \mathrm{bei}(2\sqrt{ab}) \tag{A.1.18}$$

$$\int_0^\infty \frac{\cos(bx)}{\sqrt{1+x^2}} \cosh\left(\frac{ax}{1+x^2}\right) J_0\left(\frac{a}{1+x^2}\right) dx = K_0(b)\, \mathrm{ber}(2\sqrt{ab}) \tag{A.1.19}$$

$$\int_0^a \sin^{-1}\left(\frac{x}{a}\right) J_1(bx)\, dx = \frac{\pi}{2b}\left\{\left[J_0\left(\frac{a}{2}b\right)\right]^2 - J_0(ab)\right\} \tag{A.1.20}$$

$$\int_0^\infty e^{-x} J_0(bx)\, \mathrm{ber}\left[a\sqrt{(1+b^2)x}\right] dx = \frac{\cos\left(\frac{a^2}{4}\right)}{\sqrt{1+b^2}} I_0\left(\frac{a^2 b}{4}\right) \tag{A.1.21}$$

$$\int_0^\infty e^{-x} J_0(bx)\, \mathrm{bei}\left[a\sqrt{(1+b^2)x}\right] dx = \frac{\sin\left(\frac{a^2}{4}\right)()}{\sqrt{1+b^2}} I_0\left(\frac{a^2 b}{4}\right) \tag{A.1.22}$$

$$\int_0^\infty \frac{1}{\sqrt{x}} J_1(2\sqrt{ax}) E_\alpha(x^\alpha)\, dx = \frac{1 - E_\alpha(a^\alpha)}{\sqrt{a}} \tag{A.1.23}$$

$$\int_0^\infty \frac{1}{\sqrt{x}} J_{2\beta-1}(2\sqrt{ax}) E_{\alpha,\beta}(x^\alpha)\, dx = \frac{1}{\sqrt{a}}\left[\frac{1}{\Gamma(\beta)} - E_{\alpha,\beta}(a^\alpha)\right] \tag{A.1.24}$$

$$\int_0^1 x J_0(ax) P_n(1-2x^2)\, dx = \frac{1}{a} J_{2n+1}(a) \tag{A.1.25}$$

$$\int_0^\infty x^n e^{-x^2/4} J_n(ax)\, D_{2n-1}(x)\, dx = (-1)^n a^{n-1} e^{-a^2/4} D_{2n-1}(a) \tag{A.1.26}$$

$$\int_0^\infty x^{1-\nu} J_\nu(ax)\, \gamma(\nu, bx^2)\, dx = 2^{1-\nu} a^{\nu-2} e^{-a^2/4b} \tag{A.1.27}$$

$$\int_0^\infty x^{-(\alpha+1)/2} J_{\alpha+1}(2\sqrt{ax})\, \nu(x)\, dx = -\, a^{-(\alpha+1)/2}\, \nu(a,\alpha) \quad ; \quad \alpha > -1 \tag{A.1.28}$$

$$\int_0^\infty \frac{e^{-a^2/x}}{x} Y_\nu(x)\, dx = 2\, Y_\nu(a) K_\nu(a) \quad ; \quad \mathrm{Re}\,\nu > -\frac{3}{2} \tag{A.1.29}$$

$$\int_a^\infty \frac{Y_\nu(x)}{\sqrt{x^2-a^2}}\,dx = \frac{\pi}{4}\left\{\left[J_{\nu/2}\left(\frac{a}{2}\right)\right]^2 - \left[Y_{\nu/2}\left(\frac{a}{2}\right)\right]^2\right\} \quad ; \quad -\frac{3}{2} < \operatorname{Re}\nu < 1 \tag{A.1.30}$$

$$\int_0^1 x\,Y_0(ax)\,P_n(1-2x^2)\,dx = \frac{1}{\pi a}\left[S_{n+1}(a) + \pi\,Y_{n+1}(a)\right] \tag{A.1.31}$$

$$\int_0^a x^{\nu+2}\,Y_\nu(x)\,{}_2F_1\left(1, 2\nu+\frac{3}{2}; \nu+2; \frac{x^2}{a^2}\right)dx = \frac{\pi^{3/2}\Gamma\left(\nu+\frac{1}{2}\right)a^{2\nu+3}}{2^{\nu+1}\Gamma\left(2\nu+\frac{3}{2}\right)}\,J_{\nu/2}\left(\frac{a}{2}\right)Y_{\nu/2}\left(\frac{a}{2}\right) \quad ; \quad -\frac{3}{4} < \operatorname{Re}\nu < 0 \tag{A.1.32}$$

$$\int_0^a \frac{x\cosh\left(b\sqrt{a^2-x^2}\right)}{\sqrt{a^2-x^2}}\,I_0(x)\,dx = \frac{\sinh\left(a\sqrt{1+b^2}\right)}{\sqrt{1+b^2}} \tag{A.1.33}$$

$$\int_0^\infty K_0(x)\,dx = \frac{\pi}{2} \tag{A.1.34}$$

$$\int_0^\infty x\,K_0(x)\,dx = 1 \tag{A.1.35}$$

$$\int_0^\infty x^n\,K_0(x)\,dx = 2^{n-1}\left[\Gamma\left(\frac{n+1}{2}\right)\right]^2 \quad ; \quad n = 0, 1, 2, 3, \ldots \tag{A.1.36}$$

$$\int_0^\infty \frac{x}{\sqrt{1+x^2}}\,K_0(ax)\,dx = \frac{1}{a}\left[\sin a\,\operatorname{Ci}(a) - \cos a\,\operatorname{Si}(a)\right] \tag{A.1.37}$$

$$\int_0^\infty \ln x\,K_0(x)\,dx = -\frac{\pi}{2}(\ln 2 + \gamma) \tag{A.1.38}$$

$$\int_0^\infty e^{-x}\ln x\,K_0(x)\,dx = -(\gamma - \ln 2) \tag{A.1.39}$$

$$\int_0^\infty \frac{\sinh x}{x^{3/2}}\ln x\,K_0(x)\,dx = 2^{3/2}\sqrt{\pi}\,(4 - 5\ln 2 - \gamma - \pi) \tag{A.1.40}$$

$$\int_0^\infty \cosh(ax)\,K_0\left(b\sqrt{c^2+x^2}\right)dx = \frac{\pi\,e^{-c\sqrt{b^2-a^2}}}{2\sqrt{b^2-a^2}} \quad ; \quad b > a > 0 \quad ; \quad c > 0 \tag{A.1.41}$$

$$\int_0^\infty \sin(bx)\,K_0(x)\,\mathrm{bei}\left(a\sqrt{(1+b^2)x}\right)dx = \frac{\pi}{2\sqrt{1+b^2}}\sinh\left(\frac{a^2b}{4}\right)J_0\left(\frac{a^2}{4}\right) \tag{A.1.42}$$

$$\int_0^\infty \cos(bx)\,K_0(x)\,\mathrm{ber}\left(a\sqrt{(1+b^2)x}\right)dx = \frac{\pi}{2\sqrt{1+b^2}}\cosh\left(\frac{a^2b}{4}\right)J_0\left(\frac{a^2}{4}\right) \tag{A.1.43}$$

$$\int_0^\infty e^{-x}J_0(bx)\,\mathrm{ber}\left(a\sqrt{(1+b^2)x}\right)dx = \frac{1}{\sqrt{1+b^2}}\cos\left(\frac{a^2}{4}\right)I_0\left(\frac{a^2b}{4}\right) \tag{A.1.44}$$

$$\int_0^\infty e^{-x}J_0(bx)\,\mathrm{bei}\left(a\sqrt{(1+b^2)x}\right)dx = \frac{1}{\sqrt{1+b^2}}\sin\left(\frac{a^2}{4}\right)I_0\left(\frac{a^2b}{4}\right) \tag{A.1.45}$$

$$\int_0^\infty e^{-x}I_{2n}(x\sin\theta)\,\mathrm{bei}_{4n}\left(2\cos\theta\sqrt{\alpha x}\right)dx = (-1)^n\sec\theta\sin\alpha\,I_{2n}(\alpha\sin\theta) \tag{A.1.46}$$

$$\int_0^\infty e^{-x}I_{2n+1}(x\sin\theta)\,\mathrm{ber}_{4n+2}(2\cos\theta\sqrt{\alpha x})\,dx = (-1)^{n+1}\sec\theta\sin\alpha\,I_{2n+1}(\alpha\sin\theta) \tag{A.1.47}$$

$$\int_0^\infty \sqrt{x}\,e^{-x^2/a}K_{1/4}\left(\frac{x^2}{a}\right)\mu(x,\beta,\alpha)\,dx = 2^{\beta}\pi\sqrt{a}\,\mu\left(\frac{a}{8},\beta.\frac{2\alpha+1}{4}\right) \tag{A.1.48}$$

$\alpha,\beta > -1$

$$\int_0^\infty \sqrt{x}\,K_{1/3}\left(\frac{2x^{3/2}}{\sqrt{27a}}\right)\mu(x,\beta,\alpha)\,dx = 3^{\beta+1}\pi\sqrt{a}\,\mu\left(a,\beta.\frac{\alpha}{3}\right) \tag{A.1.49}$$

$$\int_0^\infty x^{3/2}K_{1/3}\left(\frac{2x^{3/2}}{\sqrt{27a}}\right)\mu(x,\beta,\alpha)\,dx = 3^{\beta+2}\pi\,a^{3/2}\mu\left(a,\beta.\frac{\alpha-2}{3}\right) \tag{A.1.50}$$

$$\int_0^\infty x\,K_{2/3}\left(\frac{2x^{3/2}}{\sqrt{27a}}\right)\mu(x,\beta,\alpha)\,dx = 3^{\beta+3/2}\pi\,a\,\mu\left(a,\beta.\frac{\alpha-1}{3}\right) \tag{A.1.51}$$

$$\int_0^\infty x^{3/2}e^{-x^2/a}K_{3/4}\left(\frac{x^2}{a}\right)\mu(x,\beta,\alpha)\,dx = 2^{\beta-1}\pi\sqrt{a}\left[\mu\left(\frac{a}{8},\beta.\frac{2\alpha+3}{4}\right)+4\mu\left(\frac{a}{8},\beta.\frac{2\alpha-1}{4}\right)\right]\quad;\quad \alpha,\beta>-1 \tag{A.1.52}$$

$$\int_a^\infty \frac{K_\nu(\sqrt{x})}{(a-x)^{\mu+1}}\,dx = \frac{\Gamma(-a)}{2^\mu a^{(\mu+\nu)/2}}K_{\mu+\nu}(\sqrt{a}) \tag{A.1.53}$$

$$\int_0^\infty x^{n-1/2}e^{-x^2/4}K_{n+1/2}(ax)\,D_{2n}(x)\,dx=(-1)^n\sqrt{\frac{\pi}{2}}\Gamma(2n)\,a^{n-1/2}\,e^{a^2/4}D_{-2n}(a) \tag{A.1.54}$$

$$\int_0^\infty x^{n+3/2}e^{-x^2/4}K_{n+1/2}(ax)\,D_{2n+1}(x)\,dx=(-1)^n\sqrt{\frac{\pi}{2}}\Gamma(2n+3)\,a^{n-1/2}\,e^{a^2/4}D_{-(2n+3)}(a) \tag{A.1.55}$$

$$\int_0^\infty \left(\frac{x}{a^2+x^2}\right)^{\nu+1/2}H_{\nu-1/2}(bx)\,dx=\frac{\pi b^{\nu-1/2}}{2^{\nu+1/2}\Gamma(\nu+1)}[I_0(ab)-L_0(ab)] \tag{A.1.56}$$

$$\int_0^\infty \frac{x^\nu}{x^2+a^2}H_\nu(bx)\,dx=\frac{\pi a^{\nu-1}}{2\sin(\pi\nu)}[L_{-\nu}(ab)-L_\nu(ab)]\quad;\quad|\operatorname{Re}\nu|<1 \tag{A.1.57}$$

$$\int_0^\infty \frac{x^{\nu+2k+1}}{x^2+a^2}H_\nu(bx)\,dx=(-1)^k\frac{\pi a^{\nu+2k}\sec(\pi\nu)}{2}[I_{-\nu}(ab)-L_\nu(ab)]$$

$$-\left(k+\frac{3}{2}\right)<\operatorname{Re}\nu<\min\left(\frac{3}{2}-2k,\frac{1}{2}-k\right)\quad;\quad k=0,1,2 \tag{A.1.58}$$

$$\int_0^\infty \frac{x^\nu}{x^2-a^2}H_\nu(bx)\,dx=\frac{\pi a^{\nu-1}}{2}[\csc(\pi\nu)\,H_{-\nu}(ab)-\cot(\pi\nu)\,H_\nu(ab)] \tag{A.1.59}$$

$|\operatorname{Re}\nu|<1$

$$\int_0^\infty \frac{x^{\nu+2k+1}}{x^2-a^2}H_\nu(bx)\,dx=\frac{\pi a^{\nu+2k}}{2}[\sec(\pi\nu)\,J_{-\nu}(ab)+\tan(\pi\nu)\,H_\nu(ab)]$$

$$-\left(k+\frac{3}{2}\right)<\operatorname{Re}\nu<\min\left(\frac{3}{2}-2k,\frac{1}{2}-k\right)\quad;\quad k=0,1,2 \tag{A.1.60}$$

$$\int_0^\infty \frac{x^{\nu+1}}{(x^2+a^2)^{\lambda+1}}H_\nu(bx)\,dx=\frac{\pi a^{\nu-\lambda}b^\lambda\sec[\pi(\lambda-\nu)]}{2^{\lambda+1}\Gamma(\lambda+1)}[I_{\lambda-\nu}(ab)-L_{\nu-\lambda}(ab)] \tag{A.1.61}$$

$$-\frac{3}{2}<\operatorname{Re}\nu<\min\left(\operatorname{Re}\lambda+\frac{1}{2},2\operatorname{Re}\lambda+\frac{3}{2}\right)$$

$$\int_0^1 x\,E_0(ax)\,P_n(1-2x^2)\,dx=\frac{1}{a}E_{2n+1}(a)-\frac{2}{(2n+1)\pi a} \tag{A.1.62}$$

$$\int_0^\infty \frac{\operatorname{ber}(2\sqrt{ax})}{b^2+x^2}dx=\frac{\pi}{2b}J_0\left(2\sqrt{ab}\right) \tag{A.1.63}$$

$$\int_0^\infty e^{-ax}\,\mathrm{Ji}_0(bx)\,dx = \frac{1}{2a}\ln\left[\frac{\sqrt{a^2+b^2}-a}{\sqrt{a^2+b^2}+a}\right] \tag{A.1.64}$$

$$\int_0^\infty si(x)\,\mathrm{Ji}_0\left(2\sqrt{ax}\right)\,dx = \frac{\sin a}{2a} - \frac{ci(a)}{2} \tag{A.1.65}$$

$$\int_0^\infty ci(x)\,\mathrm{Ji}_0\left(2\sqrt{ax}\right)\,dx = \frac{1-\cos a}{2a} - \frac{si(a)}{2} \tag{A.1.66}$$

$$\int_0^\infty \frac{1}{\sqrt{x}} si(x)\,\mathrm{Ji}_1\left(2\sqrt{ax}\right)\,dx = \frac{\pi}{2\sqrt{a}} + \frac{2\sin a + si(a)}{\sqrt{a}} + 4C\left(\sqrt{a}\right) \tag{A.1.67}$$

$$\int_0^\infty \frac{1}{\sqrt{x}} ci(x)\,\mathrm{Ji}_1\left(2\sqrt{ax}\right)\,dx = \frac{2+\gamma+\ln a + 2\cos a - ci(a)}{\sqrt{a}} + 4S\left(\sqrt{a}\right) \tag{A.1.68}$$

$$\int_0^\infty e^{-ax}\mathrm{Ji}_0(x)\,dx = -\frac{1}{a}\ln\left(a+\sqrt{a^2+1}\right) \tag{A.1.69}$$

$$\int_0^\infty e^{-ax}\mathrm{Ji}_\nu(x)\,dx = \frac{1}{a\nu}\left[\frac{1}{\left(a+\sqrt{a^2+1}\right)^\nu} - 1\right] \quad ; \quad \mathrm{Re}\,\nu > 0 \tag{A.1.70}$$

$$\int_0^\infty e^{-ax}\mathrm{Yi}_0(x)\,dx = \frac{\left[\ln\left(a+\sqrt{a^2+1}\right)\right]^2}{\pi a} \tag{A.1.71}$$

$$\int_0^\infty e^{-ax}\mathrm{Yi}_\nu(x)\,dx = \frac{\csc(\pi\nu)}{a\nu}\left[\left(a+\sqrt{a^2+1}\right)^\nu - 1\right] - \frac{\cot(\pi\nu)}{a\nu}\left[1 - \frac{1}{\left(a+\sqrt{a^2+1}\right)^\nu}\right]$$

$|\mathrm{Re}\,\nu| < 1$

(A.1.72)

$$\int_0^\infty e^{-ax}\mathrm{Ki}_0(x)\,dx = -\frac{1}{2a}\left[\ln\left(a+\sqrt{a^2+1}\right)\right]^2 - \frac{\pi^2}{8a} \tag{A.1.73}$$

$$\int_0^\infty e^{-ax}\mathrm{Ki}_\nu(x)\,dx = -\frac{\pi\csc(\pi\nu)}{2a\nu}\left[\left(a+\sqrt{a^2+1}\right)^\nu + \frac{1}{\left(a+\sqrt{a^2+1}\right)^\nu}\right] - \frac{\cos\left(\frac{\pi\nu}{2}\right)}{a\nu}$$

$|\mathrm{Re}\,\nu < 1|$

(A.1.74)

A2 Integrals Containing Product of Two Bessel or Related Functions, and Elementary and Special Functions

$$\int_0^\tau J_0\left[2\sqrt{a(\tau-x)}\right] J_0\left[2\sqrt{bx}\right] dx = \sqrt{\frac{ab\tau}{a+b}} J_1\left[2\sqrt{(a+b)\tau}\right] \tag{A.2.1}$$

$$\int_0^\infty e^{-cx} J_0(ax)\, J_0(bx)\, dx = \frac{2}{\pi\beta} F(k)$$

$$F(k) = \int_0^1 \frac{dx}{\sqrt{(1-x^2)(1-k^2x^2)}} \quad ; \quad k = \frac{\alpha}{\beta} \quad ; \quad 0 < k^2 < \gamma < 1 \quad ; \quad \gamma = \frac{\alpha^2}{b^2} \tag{A.2.2}$$

$$\alpha = \frac{1}{2}\left\{\sqrt{(a+b)^2 + c^2} - \sqrt{(a-b)^2 + c^2}\right\}$$

$$\beta = \frac{1}{2}\left\{\sqrt{(a+b)^2 + c^2} + \sqrt{(a-b)^2 + c^2}\right\}$$

$$\int_0^\infty x\, e^{-cx} J_0(ax)\, J_0(bx)\, dx = \frac{2c}{\pi a \beta^3 (1-k^2)} \left\{\frac{2}{1-k^2} E(k) - F(k)\right\}$$

$$E(k) = \int_0^1 \sqrt{\frac{1-k^2x^2}{1-x^2}}\, dx \tag{A.2.3}$$

$$\int_0^{\pi/2} J_0(a\cos\theta)\, J_1(a\sin\theta)\, \cos\theta\, d\theta = \frac{1}{a}\left[J_0(a) - J_0(\sqrt{2}a)\right] \tag{A.2.4}$$

$$\int_0^\infty e^{-cx} J_1(ax)\, J_0(bx)\, dx = -\frac{2c}{\pi a \beta} \{F(k) - \Pi(\gamma, k)\}$$

$$\Pi(\gamma, k) = \int_0^1 \frac{dx}{(1-\gamma x^2)\sqrt{(1-x^2)(1-k^2x^2)}} \tag{A.2.5}$$

$$\int_0^\infty x\, e^{-cx} J_0(ax)\, J_1(bx)\, dx = \frac{2}{\pi b \beta^3 (1-k^2)} \left\{|\beta^2 - b^2| F(k) - \frac{a^2 + c^2 - b^2}{1-k^2} E(k)\right\} \tag{A.2.6}$$

$$\int_0^\infty x^2 e^{-cx} J_0(ax)\, J_1(bx)\, dx = \frac{2c(7\alpha^2 + \beta^2 - 5b^2 - 5b^2k^2)}{\pi b \beta^5 (1-k^2)^3} F(k) + \frac{2c\left[8b^2(1+k^2) - \alpha^2k^2 - 14\alpha^2 - \beta^2\right]}{\pi b \beta^5 (1-k^2)^4} E(k) \tag{A.2.7}$$

$$\int_0^\infty \frac{e^{-cx}}{x} J_0(ax)\, J_1(bx)\, dx = \frac{2}{\pi b\beta}\left\{-\frac{\pi c\beta}{2} + \beta^2 E(k) - \left|\beta^2 - b^2\right| F(k) + c^2 \Pi(\gamma, k)\right\} \tag{A.2.8}$$

$$\int_0^\infty e^{-cx} J_0(ax)\, J_2(bx)\, dx = -\frac{2c}{b^2} + \frac{2(b^2 - 2\beta^2)}{\pi b^2 \beta} F(k) + \frac{4\beta}{\pi b^2} E(k) + \frac{4c^2}{\pi b^2 \beta} \Pi(\gamma, k) \tag{A.2.9}$$

$$\int_0^\infty x e^{-cx} J_0(ax)\, J_2(bx)\, dx = \frac{2}{b^2} + \frac{2c}{\pi \beta^3 (1-k^2)}\left\{F(k) - \frac{2}{1-k^2} E(k) - \frac{4c}{\pi b^2 \beta} \Pi(\gamma, k)\right\} \tag{A.2.10}$$

$$\int_0^\infty x^2 e^{-cx} J_0(ax)\, J_0(bx)\, dx = \frac{2\left[\beta^2(1-k^2)^2 - 5c^2 - 3c^2k^2\right]}{\pi \beta^5 (1-k^2)^3} F(k) - \frac{4\left[\beta^2(1-k^2)^2 - 4c^2(1+k^2)\right]}{\pi \beta^5 (1-k^2)^4} E(k) \tag{A.2.11}$$

$$\int_0^\infty \frac{e^{-cx}}{x} J_2(ax)\, J_0(bx)\, dx = \frac{c}{\pi a^2 \beta}\left[(3\beta^2 + b^2 - 2a^2 - 2c^2) F(k) - 3\beta^2 E(k) + (a^2 - 2b^2 + 2c^2)\Pi(\gamma, k)\right] \tag{A.2.12}$$

$$\begin{aligned}\int_0^\infty \frac{e^{-cx}}{x^2} J_2(ax)\, J_0(bx)\, dx = \frac{1}{9\pi a^2 \beta}\{&\beta^2(8a^2 + 11c^2 - 6b^2) E(k)\\ &- [\beta^2(2\beta^2 + 3c^2) - 6\alpha^2(a^2 + c^2)\\ &+ b^2(6\alpha^2 + 4a^2 + 9c^2 - 6b^2)] F(k)\\ &+ c^2(9b^2 - 9a^2 - cb^2)\Pi(\gamma, k)\}\end{aligned} \tag{A.2.13}$$

$$\int_0^\infty x e^{-ax^2} J_0(x)\, Y_0(x)\, dx = \frac{e^{-1/2a}}{2\pi a} K_0\left(\frac{1}{2a}\right) \tag{A.2.14}$$

$$\int_0^\infty \frac{[J_0(ax) J_0(bx) - Y_0(ax) Y_0(bx)]}{c^2 + x^2} dx = -\frac{2}{\pi c} K_0(ac) K_0(bc) \tag{A.2.15}$$

$a, b, c > 0$

$$\int_0^\infty \frac{[J_0(ax) J_0(bx) - Y_0(ax) Y_0(bx)]}{x^2 - c^2} dx = -\frac{\pi}{2c}\left[J_0(ac) Y_0(bc) + J_0(bc) Y_0(ac)\right] \quad ; \quad a, b, c > 0 \tag{A.2.16}$$

$$\int_0^{\pi/2} J_0(a\cos\theta)\, I_1(a\sin\theta)\,\cos\theta\, d\theta = \frac{1}{a}\left[1 - J_0(\sqrt{2}a)\right] \tag{A.2.17}$$

$$\int_0^{\infty} \frac{x}{\sqrt{1+x^2}} \sin\left(\frac{a}{1+x^2}\right) J_0(bx) I_0\left(\frac{ax}{1+x^2}\right) dx = \frac{e^{-b}}{b}\,\mathrm{bei}\left(2\sqrt{ab}\right) \tag{A.2.18}$$

$$\int_0^{\infty} \frac{x}{\sqrt{1+x^2}} \cos\left(\frac{a}{1+x^2}\right) J_0(bx) I_0\left(\frac{ax}{1+x^2}\right) dx = \frac{e^{-b}}{b}\,\mathrm{ber}\left(2\sqrt{ab}\right) \tag{A.2.19}$$

$$\int_0^{\infty} x\, e^{-2x^2} J_0(ax)\, K_0(ax)\, dx = \frac{\pi}{16}\left[H_0\left(\frac{a^2}{4}\right) - Y_0\left(\frac{a^2}{4}\right)\right] \tag{A.2.20}$$

$$\int_0^{1} x\,[J_0(ax)]^2 P_n(1-2x^2)\, dx = \frac{1}{2(2n+1)}\left\{[J_n(a)]^2 + [J_{n+1}(a)]^2\right\} \tag{A.2.21}$$

$$\int_0^{1} x J_0(2\sqrt{ax})\, K_0(2\sqrt{ax})\, P_n(1-2x^2)\, dx$$
$$= \frac{1}{2(2n+1)}\left[J_{2n}(2\sqrt{a})\, K_{2n}(2\sqrt{a}) + J_{2n+2}(2\sqrt{a})\, K_{2n+2}(2\sqrt{a})\right] \tag{A.2.22}$$

$$\int_0^{1} x\,[J_0(ax)]^2 P_n(1-2x^2)\, dx = \frac{1}{2(2n+1)}\left\{[J_n(a)]^2 + [J_{n+1}(a)]^2\right\} \tag{A.2.23}$$

$$\int_0^{\infty} \frac{e^{-cx}}{x^{3/2}} J_1(ax)\, J_{1/2}(bx)\, dx = \frac{1}{a\sqrt{\pi b}}\left[\alpha\sqrt{a^2-\alpha^2} + a^2 \sin^{-1}\left(\frac{\alpha}{a}\right) + 2c\left(\sqrt{b^2-\alpha^2} - b\right)\right]$$

$\mathrm{Re}\, c > |\mathrm{Im}\,(a \pm b)|$

(A.2.24)

$$\int_0^{\infty} \frac{e^{-cx}}{\sqrt{x}} J_1(ax)\, J_{1/2}(bx)\, dx = \sqrt{\frac{2}{\pi a^2 b}}\left[b - \sqrt{b^2-\alpha^2}\right] \tag{A.2.25}$$

$$\int_0^{\infty} \sqrt{x}\, e^{-cx} J_1(ax)\, J_{1/2}(bx)\, dx = \sqrt{\frac{2}{\pi a^2 b}}\left[\frac{\alpha\sqrt{a^2-\alpha^2}}{\beta^2-\alpha^2}\right] \tag{A.2.26}$$

$$\int_0^{\infty} \sqrt{x}\, e^{-cx} J_1(ax)\, J_{3/2}(bx)\, dx = \sqrt{\frac{2}{\pi a^2 b^3}}\left[\frac{\alpha^2\sqrt{b^2-\alpha^2}}{\beta^2-\alpha^2}\right] \tag{A.2.27}$$

$$\int_0^{\infty} \frac{e^{-cx}}{\sqrt{x}} J_1(ax)\, J_{3/2}(bx)\, dx = \sqrt{\frac{1}{2\pi a^2 b^3}}\left[a^2 \sin^{-1}\left(\frac{\varepsilon}{a}\right) - \alpha\sqrt{a^2-\alpha^2}\right] \tag{A.2.28}$$

$$\int_0^\infty \frac{e^{-cx}}{\sqrt{x}} J_1(ax)\,J_{5/2}(bx)\,dx = \sqrt{\frac{c^2}{2\pi a^2 b^5}}\left[\alpha\sqrt{a^2-\alpha^2} + \frac{2a^2\alpha}{\sqrt{a^2-\alpha^2}} - 3\sin^{-1}\left(\frac{\alpha}{a}\right)\right] \tag{A.2.29}$$

$$\int_0^{\pi/2} J_1(a\sin\theta)\,J_1(a\cos\theta)\;d\theta = \frac{\sqrt{2}}{a}\left[\sqrt{2}J_1(a) - J_1(\sqrt{2}a)\right] \tag{A.2.30}$$

$$\int_0^\infty e^{-cx} J_1(ax)\,J_1(bx)\,dx = -\frac{2\beta}{\pi a b}\{F(k) - E(k)\} \tag{A.2.31}$$

$$\int_0^\infty x\,e^{-cx} J_1(ax)\,J_1(bx)\,dx = \frac{2c}{\pi a b \beta(1-k^2)}\left\{\frac{1+k^2}{1-k^2}E(k) - F(k)\right\}$$

$$\int_0^\infty x^2 e^{-cx} J_1(ax)\,J_1(bx)\,dx = \frac{2\left[c^2 + 7c^2k^2 - \beta^2(1-k^2)^2\right]}{\pi a b \beta^3 (1-k^2)^3}F(k)$$

$$- \frac{2\left[\beta^2(1+k^2)(1-k^2)^2 - c^2(1+14k^2+k^4)\right]}{\pi a b \beta^3(1-k^2)^4}E(k) \tag{A.2.32}$$

$$\int_0^\infty \frac{e^{-cx}}{x} J_1(ax)\,J_1(bx)\,dx = \frac{c}{\pi a b \beta}\{\beta^2 E(k) - (\beta^2+b^2)F(k) - (a^2-b^2)\,\Pi(\gamma,k)\} \tag{A.2.33}$$

$$\int_0^\infty \frac{e^{-cx}}{x^2} J_1(ax)\,J_1(bx)\,dx = \frac{1}{3\pi a b \beta}\{[\beta^2(4\alpha^2 + c^2 - 2a^2 - 2b^2) + 3b^2c^2]\mathrm{F}(k)$$

$$+ [\beta^2(2a^2+2b^2-c^2)]\,E(k)$$

$$+ 3c^2(a^2-b^2)\Pi(\gamma,k)\} - \frac{ac}{2b} \tag{A.2.34)}$$

$$\int_0^\infty x e^{-ax^2} J_1(x)\,Y_1(x)\,dx = -\frac{1}{\pi} + \frac{e^{-1/2a}}{2\pi a}K_1\left(\frac{1}{2a}\right)$$

$$\int_0^\infty e^{-cx} J_1(ax)\,J_2(bx)\,dx = \frac{a}{b^2} + \frac{2c\beta}{\pi a b^2}\left\{E(k) - F(k) - \frac{a^2}{\beta^2}\Pi(\gamma,k)\right\} \tag{A.2.35}$$

$$\int_0^\infty x\,e^{-cx} J_2(ax)\,J_1(bx)\,dx = \left\{\frac{(2\beta - \alpha^2 - b^2)}{\pi a b^2 \beta(1-k^2)}F(k) - \frac{[b^2(b^2+c^2-a^2) - 2(\beta^2-\alpha^2)]}{\pi a b^2 \beta^3(1-k^2)}E(k)\right\} \tag{A.2.36}$$

$$\int_0^\infty x^2 e^{-cx} J_1(ax) J_2(bx)\,dx = \frac{2c\,[2\beta^2(1-k^2)+7b^2k^2+b^2-5\alpha^2-3\alpha^2k^2]}{\pi\, a b^2 \beta^3 (1-k^2)^3} F(k) + \frac{2c\,[2(1+k^2)(\beta^2-6\alpha^2+\alpha^2k^2)+b^2(1+14k^2+k^4)]}{\pi\, a b^2\beta^3(1-k^2)^4} E(k) \tag{A.2.37}$$

$$\int_0^\infty \frac{e^{-cx}}{x} J_1(ax) J_2(bx)\,dx = -\frac{ac}{b^2} + \frac{2ac^2}{\pi b^2\beta}\Pi(\gamma,k) + \frac{2\beta}{3\pi a b^2}\{(2a^2-b^2-c^2)E(k)+(\alpha^2-2a^2+b^2+c^2)F(k)\} \tag{A.2.38}$$

$$\int_0^\infty \sqrt{x}\, e^{-cx} J_2(ax) J_{3/2}(bx)\,dx = \sqrt{\frac{2}{\pi}}\frac{a^2 b^{3/2}\sqrt{\beta^2-b^2}}{\beta^4(\beta^2-\alpha^2)} \tag{A.2.39}$$

$$\int_0^\infty \frac{e^{-cx}}{\sqrt{x}} J_2(ax) J_{3/2}(bx)\,dx = \sqrt{\frac{2b^3}{\pi a^4}}\left[\frac{2}{3} - \frac{\sqrt{b^2-\alpha^2}}{b} + \frac{(b^2-\alpha^2)^{3/2}}{3b^3}\right] \tag{A.2.40}$$

$$\int_0^\infty \frac{e^{-cx}}{x^2} J_1(ax) J_2(bx)\,dx = \frac{a(2b^2+4c^2-a^2)}{8b^2} + \frac{c(5b^2+2c^2-13a^2)}{12\pi b \alpha}E(k) - \frac{c[\alpha^2(13a^2-5\alpha^2-5b^2-2c^2)-3b^4]}{4\pi a b^2\beta}F(k) + \frac{c[a^2(2b^2+4c^2-a^2)-b^4]}{4\pi a^2\beta}\Pi(\gamma,k) \tag{A.2.41}$$

$$\int_0^\infty e^{-cx} J_2(ax) J_2(bx)\,dx = \frac{2\beta}{3\pi a^2 b^2}\{(2\beta^2-\alpha^2)F(k) - 2(a^2+b^2+c^2)E(k)\} \tag{A.2.42}$$

$$\int_0^\infty x\, e^{-cx} J_2(ax) J_2(bx)\,dx = \frac{2c\beta}{\pi a^2 b^2(1-k^2)}\left\{\frac{2(1-k^2+k^4)}{(1-k^2)}E(k) - (2-k^2)F(k)\right\} \tag{A.2.43}$$

$$\int_0^\infty x^2 e^{-cx} J_2(ax) J_2(bx)\,dx = \frac{2\left[(1-k^2)^2(2c^2+2\beta^2-\alpha^2)-c^2k^2(5+3k^2)\right]}{\pi a^2 b^2\beta(1-k^2)^3}F(k) + \frac{4\left[4c^2k^2(1+k^2)-(1-k^2)^2(\beta^2-\alpha^2+\alpha^2k^2+c^2+c^2k^2)\right]}{\pi a^2 b^2 \beta(1-k^2)^4}E(k) \tag{A.2.44}$$

$$\int_0^\infty \frac{e^{-cx}}{x} J_2(ax) J_2(bx)\, dx = \frac{a^2}{4b^2} + \frac{c}{6\pi a^2 b^2 \beta}\{\beta^2(5a^2+5b^2+2c^2)E(k) - [\beta^2(4a^2+4b^2+c^2+\beta^2)+3b^4]F(k) + 3(b^4-a^4)\Pi(\gamma,k)\} \tag{A.2.45}$$

$$\int_0^\infty \frac{e^{-cx}}{x^2} J_2(ax) J_2(bx)\, dx$$

$$= -\frac{a^2 c}{4b^2} + \frac{1}{30\pi a^2 b^2 \beta}\{[c^2\beta^2(c^2+\beta^2) + 3b^4(c^2+4\alpha^2-4b^2) + 4\beta^2(\alpha^2 a^2 + 2a^2c^2 - 2b^2\beta^2 - 2a^4 - b^4)]\, F(k) + 15c^2(a^4-b^4)\Pi(\gamma,k) + [\beta^2(\alpha^2+\beta^2)(8a^2+8b^2-2c^2) - 3\beta(8a^2b^2+5b^2c^2+5a^2c^2)]\, E(k)\} \tag{A.2.46}$$

$$\int_0^\infty x e^{-ax^2} J_2(x) Y_2(x)\, dx = -\frac{2}{\pi}(1-2a) - \frac{e^{-1/2a}}{2\pi a} K_2\left(\frac{1}{2a}\right) \tag{A.2.47}$$

$$\int_0^\infty x^3 e^{-x^2/a} J_2(x) Y_2(x)\, dx = -\frac{4}{\pi} + \frac{a^2(a+2)\, e^{-a/2}}{4\pi} K_0\left(\frac{a}{2}\right) + \frac{a(8+4a+a^2)\, e^{-a/2}}{4\pi} K_1\left(\frac{a}{2}\right) \tag{A.2.48}$$

$$\int_0^\infty \frac{e^{-cx}}{\sqrt{x}} J_3(ax) J_{1/2}(bx)\, dx = \sqrt{\frac{2}{9\pi a^6 b}}\left[3a^2b - 4b^3 + 12bc^2 - \sqrt{b^2-\alpha^2}\,(12\beta^2 - 16b^2 + 4\alpha^2 - 3\alpha^2)\right] \tag{A.2.49}$$

$$\int_0^\infty \sqrt{x} e^{-cx} J_3(ax) J_{3/2}(bx) dx = \sqrt{\frac{2b^3}{\pi}}\left[\frac{4}{a^3}\left(\frac{2}{3} - \frac{\sqrt{b^2-\alpha^2}}{b} + \frac{(b^2-\alpha^2)^{3/2}}{3b^3}\right) - \frac{a\sqrt{\beta^2-a^2}}{\beta^3(\beta^2-\alpha^2)}\right] \tag{A.2.50}$$

$$\int_0^\infty \frac{e^{-cx}}{x^{3/2}} J_3(ax) J_{3/2}(bx)\, dx = \sqrt{\frac{2b^3}{9\pi a^6}}\left\{\frac{\sqrt{\beta^2-b^2}\,[4b^4(2b^2-\alpha^2)-\alpha^4] - 8b^4c}{b^4}\right\} \tag{A.2.51}$$

$$\int_0^\infty x e^{-ax^2} J_3(x) Y_3(x)\, dx = -\frac{3}{\pi}\left(1 - \frac{16a}{3} + \frac{32a^2}{3}\right) + \frac{e^{-1/2a}}{2\pi a} K_3\left(\frac{1}{2a}\right) \tag{A.2.52}$$

$$\int_0^\infty J_\nu(ax)\,J_{2\nu}(2\sqrt{x})\,dx = \frac{1}{a}\,J_\nu\left(\frac{1}{a}\right) \quad ; \quad \nu > 0 \tag{A.2.53}$$

$$\int_0^\infty \frac{1}{x^\lambda} J_\mu(ax)\,J_\nu(bx)\,dx = \frac{\Gamma\left(\frac{\mu+\nu-\lambda+1}{2}\right) b^\nu}{2^\lambda\,\Gamma(\nu+1)\,\Gamma\left(\frac{\mu-\nu+\lambda+1}{2}\right) a^{\nu-\lambda+1}}\,I$$

$$I = {}_2F_1\left(\frac{\nu-\mu-\lambda+1}{2}, \frac{\nu+\mu-\lambda+1}{2}; \nu+1; \frac{b^2}{a^2}\right) \tag{A.2.54}$$

$$\mathrm{Re}(\nu+\mu-\lambda) > -1 \quad ; \quad 0 < b < a$$

$$\int_0^\infty x J_\nu\left(\frac{x^2}{4}\right) J_{2\nu}(ax)\,dx = 2J_\nu(a^2) \quad ; \quad \nu \geq -\frac{1}{4} \tag{A.2.55}$$

$$\int_0^\infty J_\nu(ax)\,J_\nu\left(\frac{1}{4x}\right) dx = \frac{1}{a}\,J_{2\nu}(\sqrt{a}) \quad ; \quad \nu > -\frac{1}{2} \tag{A.2.56}$$

$$\int_0^\infty x^2 J_{2\nu}(ax)\,J_{\nu+1/2}(x^2)\,dx = \frac{a}{4}\,J_{\nu-1/2}\left(\frac{a^2}{4}\right) \quad ; \quad \nu \geq -\frac{1}{4} \tag{A.2.57}$$

$$\int_0^\infty x^2 J_{2\nu}(ax)\,J_{\nu-1/2}(x^2)\,dx = \frac{a}{4}\,J_{\nu+1/2}\left(\frac{a^2}{4}\right) \quad ; \quad \nu \geq -\frac{1}{4} \tag{A.2.58}$$

$$\int_0^\tau (\tau-x)^{\mu/2} x^{\nu/2} J_\mu\left[2\sqrt{a(\tau-x)}\right] J_\nu\left[2\sqrt{bx}\right] dx =$$

$$a^{\mu/2} b^{\nu/2} \left(\frac{\tau}{a+b}\right)^{(\mu+\nu)/2} J_{\mu+\nu+1}\left[2\sqrt{(a+b)\tau}\right] \tag{A.2.59}$$

$$\int_0^\infty x e^{-bx^2} [J_\nu(ax)]^2 dx = \frac{e^{-a^2/2b}}{2b}\,I_\nu\left(\frac{a^2}{2b}\right) \tag{A.2.60}$$

$$\int_0^\infty x^2 e^{-bx^2} J_\nu(ax)\,J_\nu(bx)\,dx = \frac{a^{2\nu-1}}{2^{2\nu}\Gamma(\nu)\,b^{\nu+1}}\,{}_1F_1\left(\nu+\frac{1}{2}; 2\nu; -\frac{a^2}{b}\right) \tag{A.2.61}$$

$$\int_0^\infty \frac{x^{\lambda-\nu+2k+1}}{x^2-a^2} J_\lambda(bx)\,J_\nu(cx)\,dx = \frac{\pi a^{\lambda-\nu+2k}}{2}\,J_\nu(bc)\,Y_\lambda(bc)$$

$$-(k+1) < \mathrm{Re}\,\lambda < \mathrm{Re}\,\nu - 2k + 2 \quad ; \quad b > c \quad ; \quad k = 0, 1, 2, \ldots$$

$$-(k+1) < \mathrm{Re}\,\lambda < \mathrm{Re}\,\nu - 2k + 1 \quad ; \quad b = c$$

(A.2.62)

$$\int_0^\infty x^{-\mu} J_\lambda(ax)\, J_\mu(ax)\, dx =$$

$$\frac{a^{\mu-\nu-1} b^\nu \Gamma\left(\frac{\lambda+\nu-\mu+1}{2}\right)}{2^\mu \Gamma(\nu+1)\,\Gamma\left(\frac{\lambda-\nu+\mu+1}{2}\right)}\, {}_2F_1\left(\frac{\nu-\lambda-\mu+1}{2}, \frac{\lambda+\nu-\mu+1}{2}; \nu+1; \frac{b^2}{a^2}\right) \tag{A.2.63}$$

$\mathrm{Re}(\lambda+\nu-\mu) > -1 \quad ; \quad 0 < b < a$

$$\int_0^{\pi/2} J_1(a\sin\theta)\, I_0(a\cos\theta)\; d\theta = \frac{1}{a}\, I_0(a) \tag{A.2.64}$$

$$\int_0^{\pi/2} J_1(a\sin\theta)\, I_0(a\cos\theta)\; \cos\theta\, d\theta = \frac{1}{a}\left[I_0(\sqrt{2}a) - I_0(a)\right] \tag{A.2.65}$$

$$\int_0^{\pi/2} J_1(a\sin\theta)\, I_1(a\cos\theta)\; d\theta = \frac{1}{a}\left[I_1(a) - J_1(a)\right] \tag{A.2.66}$$

$$\int_0^\infty J_\nu(ax)\, H_\nu(ax)\, dx = \frac{1}{2a} \quad ; \quad \mathrm{Re}\, a > 0 \tag{A.2.67}$$

$$\int_0^\infty x J_\nu(ax)\, H_{\nu/2}\left(\frac{x^2}{4}\right) dx = -2Y_{\nu/2}(a^2) \quad ; \quad \nu \geq -\frac{1}{2} \tag{A.2.68}$$

$$\int_0^\infty x^{\nu-\lambda} J_{\nu+\lambda}(2ax)\, H_{2\lambda}(2ax)\, dx = \frac{\Gamma\left(\lambda+\nu+\frac{1}{2}\right)\, \Gamma\left(\nu-\lambda+\frac{1}{2}\right)\, a^{\lambda-\nu-1}}{4\sqrt{\pi}\,\Gamma(\nu-\lambda+1)\, \Gamma\left(2\lambda+\frac{1}{2}\right)} \tag{A.2.69}$$

$\mathrm{Re}(\nu\pm\lambda) > -1) \quad ; \quad \mathrm{Re}\,\lambda > -\frac{1}{4} \quad ; \quad \mathrm{Re}\, a > 0$

$$\int_0^\infty x^\mu J_\nu(ax)\, K_{\nu-\mu+1}(bx)\, dx = \frac{2^{\mu-1}\Gamma(\mu)\, a^\nu b^{\mu-\nu-1}}{(a^2+b^2)^\mu} \tag{A.2.70}$$

$\mathrm{Re}\,\nu > -1 \quad ; \quad \mathrm{Re}(2\mu-\nu) > \frac{1}{2}$

$$\int_0^\infty x J_\nu\left(\frac{x}{\sqrt{2}}\right) K_\nu\left(\frac{x}{\sqrt{2}}\right) J_{2\nu}(ax)\, dx = \frac{\left(\sqrt{a^4+1}-1\right)^\nu}{a^{2\nu}\sqrt{a^4+1}} \tag{A.2.71}$$

$$\int_0^\infty e^{-ax} J_\nu(b\sqrt{x})\, K_\nu(b\sqrt{x})\, dx = \frac{1}{2a}\left[S_{0,\nu}\left(\frac{b^2}{2a}\right) - \nu S_{-1,\nu}\left(\frac{b^2}{2a}\right)\right] \tag{A.2.72}$$

$$\int_0^a x\,{}_3F_2\left(\frac{3}{2},\frac{\lambda}{2},\frac{\lambda+1}{2};1-\nu,1+\nu;-4x^2\right)[J_\nu(ax)]^2dx=-\frac{\nu a^{\lambda-2}e^{-a}}{2\Gamma(\lambda)} \tag{A.2.73}$$

$\nu\geq\frac{1}{2}\quad;\quad\lambda>1$

$$\int_0^a x\,{}_2F_1\left(\frac{3}{2},\nu+\frac{1}{2};1-\nu;-4x^2\right)[J_\nu(ax)]^2dx=-\frac{a^{2\nu-1}e^{-a}}{4\Gamma(2\nu)} \tag{A.2.74}$$

$$\int_0^a x\,{}_2F_1\left(\frac{3}{2},\nu+\frac{3}{2};1-\nu;-4x^2\right)[J_\nu(ax)]^2dx=-\frac{\nu a^{2\nu}e^{-a}}{2\Gamma(2\nu+2)} \tag{A.2.75}$$

$$\int_0^a x\,{}_2F_2\left(\frac{3}{2},\frac{\lambda}{2};\frac{\lambda+1}{2};1-\nu,1+\nu;-x^2\right)[J_\nu(ax)]^2dx=-\frac{\nu a^{\lambda-2}e^{-a^2}}{\Gamma\left(\frac{\lambda}{2}\right)} \tag{A.2.76}$$

$\nu\geq\frac{1}{2}\quad;\quad\lambda>1$

$$\int_0^a x\,{}_1F_1\left(\frac{3}{2};1-\nu;-x^2\right)[J_\nu(ax)]^2dx=-\frac{a^{2\nu}e^{-a^2}}{\Gamma(\nu)} \tag{A.2.77}$$

$$\int_0^a x\,e^{-x^2}{}_1F_1\left(-\frac{1}{2}-\nu;1-\nu;-x^2\right)[J_\nu(ax)]^2dx=-\frac{a^{2\nu}e^{-a^2}}{\Gamma(\nu)} \tag{A.2.78}$$

$$\int_0^a x\,{}_2F_1\left(\frac{3}{2},1;\frac{1}{2};-4x^2\right)[J_{1/2}(ax)]^2dx=-\frac{e^{-a^2}}{4} \tag{A.2.79}$$

$$\int_0^\infty x^{\nu-k}J_{\nu+k}(2bx)\,s_{2k,2\lambda}(2ax)\,dx=\frac{2^{2k-3}\Gamma\left(\lambda+\nu+\frac{1}{2}\right)\Gamma\left(\nu-\lambda+\frac{1}{2}\right)a^{2\lambda}}{\Gamma(\nu-k+1)\,b^{2\lambda+\nu-k+1}}I$$

$$I={}_2F_1\left(\lambda+\nu+\frac{1}{2},\lambda-k+\frac{1}{2};\nu-k+1;1-\frac{a^2}{b^2}\right) \tag{A.2.80}$$

$\mathrm{Re}\left(\nu\pm\lambda;k\pm\lambda\right)>-\frac{1}{2}\quad;\quad\mathrm{Re}(\nu+k)>-1\quad;\quad\mathrm{Re}\,b>0$

$$\int_0^\infty x^{\nu-k}J_{\nu+k}(2ax)\,s_{2k,2\lambda}(2ax)\,dx=\frac{2^{2\lambda-3}\Gamma\left(\lambda+\nu+\frac{1}{2}\right)\Gamma\left(\nu-\lambda+\frac{1}{2}\right)}{\Gamma(\nu-k+1)\,a^{\nu-k+1}} \tag{A.2.81}$$

$\mathrm{Re}\left(\nu\pm\lambda;k\pm\lambda\right)>-\frac{1}{2}\quad;\quad\mathrm{Re}(\nu+k)>-1\quad;\quad\mathrm{Re}\,a>0$

$$\int_0^\infty x^{\nu-\lambda} J_{\nu+\lambda}(2bx)\, H_{2\lambda}(2ax)\, dx = \frac{\Gamma\left(\lambda+\nu+\frac{1}{2}\right)\Gamma\left(\nu-\lambda+\frac{1}{2}\right) a^{2\lambda}}{4\sqrt{\pi}\,\Gamma(\nu-\lambda+1)\,\Gamma\left(2\lambda+\frac{1}{2}\right) b^{\lambda+\nu+1}}\, {}_2F_1\left(\lambda+\nu+\frac{1}{2}, \frac{1}{2}; \nu-\lambda+1; 1-\frac{a^2}{b^2}\right) \tag{A.2.82}$$

$\mathrm{Re}(\nu\pm\lambda) > -1$; $\mathrm{Re}\lambda > -\frac{1}{4}$; $\mathrm{Re}\, b > 0$; $\arg a < \pi$

$$\int_0^\infty Y_\nu(ax)\, Y_\nu\left(\frac{b}{x}\right) dx = -\frac{1}{a} J_{2\nu}\left(2\sqrt{ab}\right) \quad -1 < \mathrm{Re}\,\nu < \frac{5}{4} \quad ; \quad a, b > 0 \tag{A.2.83}$$

$$\int_0^\infty \frac{1}{x} Y_\nu\left(\frac{a}{x}\right) Y_{\nu+1}\left(\frac{b}{x}\right) dx = -\frac{1}{\sqrt{ab}} J_{2\nu+1}\left(2\sqrt{ab}\right) \quad -\frac{3}{2} < \mathrm{Re}\,\nu < \frac{1}{4} \quad ; \quad a, b > 0 \tag{A.2.84}$$

$$\int_0^\tau I_0\left[2\sqrt{a(\tau-x)}\right] I_0\left[2\sqrt{bx}\right] dx = \sqrt{\frac{ab\tau}{a+b}}\, I_1\left[2\sqrt{(a+b)\tau}\right] \tag{A.2.85}$$

$$\int_0^{\pi/2} I_1(a\sin\theta)\, I_1(a\cos\theta)\, d\theta = \frac{\sqrt{2}}{a}\left[I_1(\sqrt{2}a) - \sqrt{2} I_1(a)\right] \tag{A.2.86}$$

$$\int_0^\infty \frac{(\ln x)^2}{x} I_1(x) K_0(x)\, dx = 6 + \frac{\pi^2}{6} + \frac{1}{14}(\gamma + 3\ln 2)^2 + (\gamma + 3\ln 2) \tag{A.2.87}$$

$$\int_0^\infty \frac{(\ln x)^2}{x} I_2(x) K_0(x)\, dx = \frac{3}{8} + \frac{\pi^2}{24} + \frac{1}{4}(\gamma - \ln 2)^2 - \frac{1}{2}(\gamma - \ln 2) \tag{A.2.88}$$

$$\int_0^\infty \frac{(\ln x)^3}{x} I_2(x) K_0(x)\, dx = \frac{3}{4} + \frac{\pi^2}{8} + \frac{1}{4}\zeta(3) - \frac{1}{4}(\gamma - \ln 2)^3 + \frac{3}{4}(\gamma - \ln 2)^2 - \left(\frac{\pi^2}{8} + \frac{9}{8}\right)(\gamma - \ln 2) \tag{A.2.89}$$

$$\int_0^\infty x^5 e^{-x^2/a} I_3(x) K_3(x)\, dx = -32 + \frac{4}{\pi} + \frac{e^{a/2} a^2 (32 - 16a + 5a^2 - a^3)}{8} K_0\left(\frac{a}{2}\right) + \frac{e^{a/2} a\,(128 - 64a + 25a^2 - 6a^3 + a^4)}{8} K_1\left(\frac{a}{2}\right) \tag{A.2.90}$$

$$\int_0^\infty x e^{-x^2/a} I_3(x) K_3(x)\, dx = -\frac{(32 + 16a + 3a^2)}{2a^2} + \frac{a e^{a/2}}{4} K_3\left(\frac{a}{2}\right) \tag{A.2.91}$$

$$\int_0^\infty \frac{\ln x}{x} I_4(x)K_0(x)\,dx = \frac{5}{32} - \frac{1}{16}(\gamma - \ln 2) \tag{A.2.92}$$

$$\int_0^\infty \frac{\ln x}{x} I_5(x)K_2(x)\,dx = \frac{194}{735} - \frac{1}{21}(\gamma + 3\ln 2) \tag{A.2.93}$$

$$\int_0^\tau x^\alpha(\tau - x)^\beta I_\alpha(ax) I_\beta[a(\tau - x)]\,dx = \frac{\Gamma(\alpha + \frac{1}{2})\,\Gamma(\alpha + \frac{1}{2})\,\tau^{\alpha+\beta+1/2}}{\sqrt{2\pi a}\,\Gamma(\alpha+\beta+1)} I\alpha + \beta + 1/2(\tau a)$$

$$\operatorname{Re}\alpha, \beta > -\frac{1}{2} \quad ; \quad \tau > 0 \tag{A.2.94}$$

$$\int_0^\tau x^\alpha(\tau - x)^{\beta+1} I_\alpha(ax) I_\beta[a(\tau - x)]\,dx = \frac{\Gamma(\alpha + \frac{1}{2})\,\Gamma(\alpha + \frac{3}{2})\,\tau^{\alpha+\beta+3/2}}{\sqrt{2\pi a}\,\Gamma(\alpha+\beta+2)} I_{\alpha+\beta+1/2}(\tau a)$$

$$\operatorname{Re}\alpha > -\frac{1}{2} \quad ; \quad \operatorname{Re}\beta > -\frac{3}{2} \quad ; \quad \tau > 0 \tag{A.2.95}$$

$$\int_0^\tau (\tau - x)^{\mu/2} x^{\nu/2} I_\mu\left[2\sqrt{a(\tau - x)}\right] I_\nu\left[2\sqrt{bx}\right] dx = a^{\mu/2} b^{\nu/2} \left(\frac{\tau}{a+b}\right)^{(\mu+\nu)/2} I_{\mu+\nu+1}\left[2\sqrt{(a+b)\tau}\right] \tag{A.2.96}$$

$$\int_0^\infty [K_0(x)]^2 dx = \frac{3\zeta(2)}{2} = \frac{\pi^2}{4} \tag{A.2.97}$$

$$\int_{-\infty}^\infty K_0(|x|)\,K_0(|a - x|)\,dx = \frac{\pi^2}{2} e^{-a} \quad ; \quad a > 0 \tag{A.2.98}$$

$$\int_0^\infty K_0(x)\,K_0\left(\sqrt{a^2 + x^2}\right) dx = \frac{\pi^2}{4}[1 - a K_0(a) L_{-1}(a) + K_1(a) L_0(a)] \tag{A.2.99}$$

$$a > 0$$

$$\int_{-\infty}^\infty K_0(|x - a|)\,\frac{K_1(b\sqrt{b^2 + x^2})}{\sqrt{b^2 + x^2}}\,dx = \frac{\pi}{b} K_0\left(\sqrt{b^2 + a^2}\right) \quad ; \quad b, a \text{ real} \tag{A.2.100}$$

$$\int_0^\infty [K_\nu(ax)]^2 dx = \frac{\pi^2}{4a\cos(\pi\nu)} \quad ; \quad |\operatorname{Re}\nu| < \frac{1}{2} \quad ; \quad a > 0 \tag{A.2.101}$$

$$\int_0^\infty x\,[K_0(x)]^2 dx = \frac{1}{2} \tag{A.2.102}$$

$$\int_0^\infty x^n\,[K_0(x)]^2\,dx = 2^{n-1}\frac{\sqrt{\pi}\,\left[\Gamma\left(\frac{n+1}{2}\right)\right]^3}{4\Gamma\left(\frac{n}{2}+1\right)} \quad ; \quad n = 0, 1, 2, 3, \ldots \tag{A.2.103}$$

$$\int_0^\infty \sqrt{x}\,\ln x\,[K_0(x)]^2\,dx = -\frac{\left[\Gamma\left(\frac{3}{4}\right)\right]^4}{\sqrt{2\pi}}(2+3\ln 2+\gamma+\pi) \tag{A.2.104}$$

$$\int_0^\infty x\,\ln x\,[K_0(x)]^2\,dx = -(1-\ln 2+\gamma) \tag{A.2.105}$$

$$\int_0^\infty x\,(\ln x)^2\,[K_0(x)]^2\,dx = 1+(\gamma-\ln 2)+\frac{1}{2}(\gamma-\ln 2)^2 \tag{A.2.106}$$

$$\int_0^\infty \frac{\ln x}{\sqrt{x}}[K_0(x)]^2\,dx = -\frac{\left[\Gamma\left(\frac{1}{4}\right)\right]^4}{2^{3/2}\sqrt{\pi}}(3\ln 2+\gamma+\pi) \tag{A.2.107}$$

$$\int_0^\infty x\,(\ln x)^3\,[K_0(x)]^2\,dx = -\frac{1}{2}[6+(\gamma-\ln 2)^3+3(\gamma-\ln 2)^2+6(\gamma-\ln 2)-\zeta(3)] \tag{A.2.108}$$

$$\int_0^\infty x\,\ln x\,K_0(x)K_1(x)\,dx = -\frac{\pi^2}{8}(-1+3\ln 2+\gamma) \tag{A.2.109}$$

$$\int_0^\infty x\,(\ln x)^2\,K_0(x)K_1(x)\,dx = \frac{\pi^2}{4}\left[\frac{1}{2}(\gamma+3\ln 2)^2-(\gamma+3\ln 2)+\frac{\pi^2}{6}\right] \tag{A.2.110}$$

$$\int_{-\infty}^\infty |x|^\alpha|b-x^\beta|K_\nu(|ax|)\,K_\mu(a|b-x|)\,dx = \sqrt{\frac{\pi}{2a}}B\left(\alpha+\frac{1}{2},\beta\right)\left|b^{\alpha+\beta+1/2}\right|K_{\alpha+\beta+1/2}(a|b|)$$

$$\operatorname{Re}\alpha > -\frac{1}{2} \quad ; \quad \operatorname{Re}\beta > \frac{1}{2} \quad ; \quad b - \text{real} \tag{A.2.111}$$

$$\int_0^\infty \cos(bx)K_0(x)\,\mathrm{ber}\left[a\sqrt{(1+b^2)x}\right]dx = \frac{\pi\cosh\left(\frac{a^2b}{4}\right)}{2\sqrt{1+b^2}}J_0\left(\frac{a^2}{4}\right) \tag{A.2.112}$$

$$\int_0^\infty \sin(bx)K_0(x)\,\mathrm{bei}\left[a\sqrt{(1+b^2)x}\right]dt = \frac{\pi\sinh\left(\frac{a^2b}{4}\right)}{2\sqrt{1+b^2}}J_0\left(\frac{a^2}{4}\right) \tag{A.2.113}$$

$$\int_0^1 x I_0(ax) K_0(ax) P_n(1-2x^2)\,dx = \frac{1}{2(2n+1)}[I_n(a)K_n(a) + I_{n+1}(a)K_{n+1}(a)] \tag{A.2.114}$$

$$\int_0^\tau \operatorname{ber}(\tau - x)\operatorname{bei}(x)\,dx = \int_0^\tau \operatorname{bei}(\tau - x)\operatorname{ber}(x)\,dx$$
$$= \frac{1}{\sqrt{2}}\left[\sin\left(\frac{\tau}{\sqrt{2}}\right)\cosh\left(\frac{\tau}{\sqrt{2}}\right) - \cos\left(\frac{\tau}{\sqrt{2}}\right)\sinh\left(\frac{\tau}{\sqrt{2}}\right)\right] \tag{A.2.115}$$

A3 Integrals Containing Product of Three Bessel Functions and Elementary Functions

$$\int_0^\infty x J_\lambda(ax) J_\nu(bx) K_{\nu-\lambda}(cx)\,dx = \frac{\alpha^{2\lambda}(b^2-\alpha^2)^{\nu-\lambda}}{a^\lambda b^\nu c^{\nu-\lambda}(\beta^2-\alpha^2)}$$
$$\alpha = \frac{1}{2}\left[\sqrt{(a+b)^2+c^2} - \sqrt{(a-b)^2+c^2}\right] \tag{A.3.1}$$
$$\beta = \frac{1}{2}\left[\sqrt{(a+b)^2+c^2} + \sqrt{(a-b)^2+c^2}\right]$$

$$\int_0^\infty x^2 J_{\lambda-1}(ax) J_\nu(bx) K_{\nu-\lambda}(cx)\,dx =$$
$$\frac{2\alpha^{2\lambda-2}(b^2-\alpha^2)^{\nu-\lambda}}{a^{\lambda-1}b^\nu c^{\nu-\lambda}(\beta^2-\alpha^2)^2}\left[\lambda b^2 - \nu\alpha^2 - \frac{\alpha^2(a^2+c^2-b^2)}{\beta^2-\alpha^2}\right] \tag{A.3.2}$$

$$\int_0^\infty x^2 J_{\lambda+1}(ax) J_\nu(bx) K_{\nu-\lambda}(cx)\,dx =$$
$$\frac{2\alpha^{2\lambda-2}(b^2-\alpha^2)^{\nu-\lambda}}{a^{\lambda-1}b^\nu c^{\nu-\lambda}(\beta^2-\alpha^2)}\left[\frac{b^2(\nu b^2-\lambda\alpha^2)}{\beta^2(\beta^2-\alpha^2)} + \frac{\alpha^2(a^2+c^2-b^2)}{(\beta^2-\alpha^2)^2}\right] \tag{A.3.3}$$

$$\int_0^\infty x J_{2\nu}(ax) I_\nu(bx) K_\nu(bx)\,dx = \frac{1}{a(a^2+4b^2)^{1/2}} \quad ; \quad \operatorname{Re}\nu > -\frac{1}{2} \tag{A.3.4}$$

$$\int_0^\infty x^{\nu+1} J_\nu(ax) J_\nu(bx) K_\nu(bx)\,dx = \frac{2^{3\nu}\Gamma\left(\nu+\frac{1}{2}\right)(ab^2)^\nu}{\sqrt{\pi}(a^4+4b^4)^{1/2}} \quad ; \quad \operatorname{Re}\nu > -\frac{1}{2} \tag{A.3.5}$$

$$\int_0^\infty x I_0(x)\,[K_0(x)]^2 dx = \frac{\pi}{3^{3/2}} \tag{A.3.6}$$

$$\int_0^\infty x^3 I_0(x)\,[K_0(x)]^2 dx = \frac{4\pi}{3^{5/2}} \tag{A.3.7}$$

$$\int_0^\infty [K_0(x)]^3\,dx = \frac{3\left[\Gamma\left(\frac{1}{3}\right)\right]^6}{2^{2/3}\cdot 32\,\pi} \tag{A.3.9}$$

$$\int_0^\infty x^2\,[K_0(x)]^3\,dx = \frac{\left[\Gamma\left(\frac{1}{3}\right)\right]^6}{2^{2/3}\cdot 96\,\pi} - \frac{2^{8/3}\,\pi^5}{9\left[\Gamma\left(\frac{1}{3}\right)\right]^6} \tag{A.3.10}$$

$$\int_0^\infty x^2\,[K_0(x)]^3 dx = \left\{\frac{\left[\Gamma\left(\frac{1}{3}\right)\right]^6}{2^{2/3}\cdot 96\,\pi} - \frac{4\cdot 2^{2/3}\pi^5}{9\left[\Gamma\left(\frac{1}{3}\right)\right]^6}\right\} \tag{A.3.11}$$

$$\int_0^\infty x^3\,[K_0(x)]^3\,dx = L_{-3}(2) - \frac{2}{3} \tag{A.3.12}$$

$$\int_0^\infty x^6\,[K_0(x)]^2 K_1(x)\,dx = \frac{-162 + 147\,\zeta(3)}{256} \tag{A.3.13}$$

$$\int_0^\infty x^8\,[K_0(x)]^2 K_1(x)\,dx = \frac{-37 + 63\,\zeta(3)}{8} \tag{A.3.14}$$

A4 Integrals Containing Product of Four and Five Bessel Functions and Elementary Functions

$$\int_0^\infty x\,Y_0(x)\,K_0(x)\,J_1(\alpha x)\,I_1(\alpha x)\,dx = -\frac{\ln(1-\alpha^2)}{2\pi\alpha^2} \quad ; \quad 0<\alpha<1 \tag{A.4.1}$$

$$\int_0^\infty x I_0(x)\,[K_0(x)]^3 dx = \frac{\pi^2}{16} \tag{A.4.2}$$

$$\int_0^\infty x^3 I_0(x)\,[K_0(x)]^3 dx = \frac{\pi^2}{64} \tag{A.4.3}$$

$$\int_0^\infty x^5 I_0(x)\,[K_0(x)]^3 dx = \frac{7\,\pi^2}{16} \tag{A.4.4}$$

$$\int_0^\infty x\, I_1(x)\, [K_0(x)]^2 K_1(x)\, dx = \frac{\zeta(2)}{8} \tag{A.4.5}$$

$$\int_0^\infty x^5\, I_1(x)\, [K_1(x)]^3 dx = \frac{27\,\zeta(2)}{128} \tag{A.4.6}$$

$$\int_0^\infty [K_0(x)]^4 dx = \frac{\pi^4}{4}\, {}_4F_3\left(\frac{1}{2}, \frac{1}{2}, \frac{1}{2}, \frac{1}{2};\, 1, 1, 1{:}1\right) \tag{A.4.7}$$

$$\begin{aligned}\int_0^\infty x^2 [K_0(x)]^4 dx = &- \frac{3\pi^2}{16} + \frac{\pi^4}{64}\left\{4\,{}_4F_3\left(\frac{1}{2}, \frac{1}{2}, \frac{1}{2}, \frac{1}{2};\, 1, 1, 1{:}1\right)\right.\\ &\left. - 3\,{}_4F_3\left(\frac{1}{2}, \frac{1}{2}, \frac{1}{2}, \frac{1}{2};\, 2, 1, 1{:}1\right)\right\}\end{aligned} \tag{A.4.8}$$

$$\int_0^\infty x\, [K_0(x)]^3 K_1(x)\, dx = \frac{7\zeta(3)}{8} \tag{A.4.9}$$

$$\int_0^\infty x^4 [K_0(x)]^3 K_1(x)\, dx = \frac{-\,6 + 7\zeta(3)}{32} \tag{A.4.10}$$

$$\int_0^\infty x^5 [K_0(x)]^2 [K_1(x)]^2 dx = \frac{-\,28 + 42\,\zeta(3)}{256} \tag{A.4.11}$$

$$\int_0^\infty x^6 K_0(x)\, [K_1(x)]^3\, dx = \frac{106 - 63\,\zeta(3)}{256} \tag{A.4.12}$$

$$\int_0^\infty x^8 K_0(x)\, [K_1(x)]^3\, dx = \frac{402 - 315\,\zeta(3)}{128} \tag{A.4.13}$$

$$\int_0^\infty x^4 [K_0(x)]^3 K_1(x)\, dx = \frac{-\,6 + 7\zeta(3)}{32} \tag{A.4.14}$$

$$\int_0^\infty x^5 [K_1(x)]^4 dx = \frac{106 - 153\,\zeta(3)}{128} \tag{A.4.15}$$

$$\int_0^\infty x^5\, I_0(x)\, [K_0(x)]^4 dx = \frac{960\pi}{243\sqrt{3}} \tag{A.4.16}$$

$$\int_0^\infty x^3\, I_0(x)\, K_0(x)\, [K_1(x)]^2\, dx = \frac{8 + 3\,\zeta(2)}{32} \tag{A.4.17}$$

$$\int_0^\infty x^3 I_1(x)\,[K_0(x)]^2 K_1(x)\,dx = \frac{8-3\zeta(2)}{32} \tag{A.4.18}$$

A5 Integrals Containing Airy Functions and Elementary and Special Functions

$$\int_{-\infty}^{+\infty} \mathrm{Ai}\,[2^{2/3}(a+x^2)]\,dx = 2^{1/3}\pi\,[\mathrm{Ai}(a)]^2 \tag{A.5.1}$$

$$\int_0^\infty \frac{\mathrm{Ai}(a+x)}{\sqrt{x}}\,dx = 2^{2/3}\pi\left[\mathrm{Ai}\left(\frac{a}{2^{2/3}}\right)\right]^2 \tag{A.5.2}$$

$$\int_0^\infty \frac{\mathrm{Ai}(a-x)}{\sqrt{x}}\,dx = 2^{2/3}\pi\,\mathrm{Ai}\left(\frac{a}{2^{2/3}}\right)\mathrm{Bi}\left(\frac{a}{2^{2/3}}\right) \tag{A.5.3}$$

$$\int_0^\infty \frac{\mathrm{Ai}'(a+x)}{\sqrt{x}}\,dx = 2\pi\,\mathrm{Ai}\left(\frac{a}{2^{2/3}}\right)\mathrm{Ai}'\left(\frac{a}{2^{2/3}}\right) \tag{A.5.4}$$

$$\int_0^\infty x^n \mathrm{Ai}(x)\,dx = \frac{\Gamma(n+1)}{3^{(n+3)/3}\Gamma\left(\frac{n+3}{3}\right)} \tag{A.5.5}$$

$$\int_\alpha^\infty \frac{1}{x^6}\mathrm{Ai}(x)\,dx = \frac{1}{40}\Big\{\pi\,[\mathrm{Ai}(\alpha)\,\mathrm{Gi}'(\alpha) - \mathrm{Ai}'(\alpha)\,\mathrm{Gi}(\alpha)] + \left(\frac{1}{\alpha}+\frac{2}{\alpha^4}\right)\mathrm{Ai}'(\alpha) + \left(\frac{1}{\alpha^2}+\frac{8}{\alpha^5}\right)\mathrm{Ai}(\alpha)\Big\} \tag{A.5.6}$$

$$\int_\alpha^\infty \frac{1}{x^5}\mathrm{Ai}(x)\,dx = \frac{1}{3^{2/3}12\Gamma\left(\frac{2}{3}\right)\alpha}\,{}_1F_2\left(-\frac{1}{3};\frac{2}{3},\frac{2}{3};\frac{\alpha^3}{9}\right) + \frac{\mathrm{Ai}'(\alpha)}{12\alpha^3} + \frac{\mathrm{Ai}(\alpha)}{4\alpha^4} + \frac{3^{1/6}\Gamma\left(\frac{2}{3}\right)\alpha^3}{864\pi}\,{}_2F_3\left(1,1;2,2,\frac{7}{3};\frac{\alpha^3}{9}\right) + \frac{3^{1/6}\Gamma\left(\frac{2}{3}\right)}{144\pi}\left[\ln\left(\frac{\alpha^6}{3}\right)+4\gamma-6-\frac{\pi}{\sqrt{3}}\right] \tag{A.5.7}$$

$$\int_\alpha^\infty \frac{1}{x^4}\mathrm{Ai}(x)\,dx = \frac{3^{1/6}\Gamma\left(\frac{2}{3}\right)\alpha}{12\pi}\,{}_1F_2\left(\frac{1}{3};\frac{4}{3},\frac{4}{3};\frac{\alpha^3}{9}\right) + \frac{\mathrm{Ai}'(\alpha)}{6\alpha^2} + \frac{\mathrm{Ai}(\alpha)}{3\alpha^3} + \frac{\sqrt{3}\,\alpha^3}{324\Gamma\left(\frac{2}{3}\right)}\,{}_2F_3\left(1,1;2,2,\frac{5}{6};\frac{\alpha^3}{9}\right) - \frac{3^{1/3}}{108\Gamma\left(\frac{2}{3}\right)}\left[\ln\left(\frac{\alpha^6}{3}\right)+4\gamma+\frac{\pi}{\sqrt{3}}\right] \tag{A.5.8}$$

$$\int_{\alpha}^{\infty} \frac{1}{x^3} \operatorname{Ai}(x)\, dx = \frac{\pi}{2}\{\operatorname{Ai}(\alpha)\operatorname{Gi}'(\alpha) - \operatorname{Ai}'(\alpha)\operatorname{Gi}(\alpha)\} + \frac{\operatorname{Ai}'(\alpha)}{2\alpha} + \frac{\operatorname{Ai}(\alpha)}{2\alpha^2} \tag{A.5.9}$$

$$\int_{\alpha}^{\infty} \frac{1}{x^2} \operatorname{Ai}(x)\, dx = \frac{1}{3^{2/3}\Gamma\left(\frac{2}{3}\right)\alpha}\, {}_1F_2\left(-\frac{1}{3}; \frac{2}{3}, \frac{2}{3}; \frac{\alpha^3}{9}\right) + \frac{3^{1/6}\Gamma\left(\frac{2}{3}\right)\alpha^3}{72\pi}\, {}_2F_3\left(1, 1; 2, 2, \frac{7}{3}; \frac{\alpha^3}{9}\right)$$
$$+ \frac{3^{1/6}\Gamma\left(\frac{2}{3}\right)}{12\pi}\left[\ln\left(\frac{\alpha^6}{3}\right) + 4\gamma - 6 - \frac{\pi}{\sqrt{3}}\right] \tag{A.5.10}$$

$$\int_{\alpha}^{\infty} \frac{1}{x} \operatorname{Ai}(x)\, dx = \frac{3^{1/6}\Gamma\left(\frac{2}{3}\right)\alpha}{2\pi}\, {}_1F_2\left(\frac{1}{3}; \frac{4}{3}, \frac{4}{3}; \frac{\alpha^3}{9}\right) - \frac{\sqrt{3}\,\alpha^3}{54\Gamma\left(\frac{2}{3}\right)}\, {}_2F_3\left(1, 1; 2, 2, \frac{5}{6}; \frac{\alpha^3}{9}\right)$$
$$- \frac{3^{1/3}}{18\Gamma\left(\frac{2}{3}\right)}\left[\ln\left(\frac{\alpha^6}{3}\right) + 4\gamma + \frac{\pi}{\sqrt{3}}\right] \tag{A.5.11}$$

$$\int_{\alpha}^{\infty} \operatorname{Ai}(x)\, dx = \pi\{\operatorname{Ai}(\alpha)\operatorname{Gi}'(\alpha) - \operatorname{Ai}'(\alpha)\operatorname{Gi}(\alpha)\} \tag{A.5.12}$$

$$\int_{\alpha}^{\infty} x \operatorname{Ai}(x)\, dx = -\operatorname{Ai}'(\alpha) \tag{A.5.13}$$

$$\int_{\alpha}^{\infty} x^2 \operatorname{Ai}(x)\, dx = -\alpha \operatorname{Ai}'(\alpha) + \operatorname{Ai}(\alpha) \tag{A.5.14}$$

$$\int_{\alpha}^{\infty} x^3 \operatorname{Ai}(x)\, dx = 2\pi\{\operatorname{Ai}(\alpha)\operatorname{Gi}'(\alpha) - \operatorname{Ai}'(\alpha)\operatorname{Gi}(\alpha)\} - \alpha^2 \operatorname{Ai}'(\alpha) + 2\alpha \operatorname{Ai}(\alpha) \tag{A.5.15}$$

$$\int_{\alpha}^{\infty} x^4 \operatorname{Ai}(x)\, dx = -(6 + \alpha^3)\operatorname{Ai}'(\alpha) + 3\alpha^2 \operatorname{Ai}(\alpha) \tag{A.5.16}$$

$$\int_{\alpha}^{\infty} x^5 \operatorname{Ai}(x)\, dx = -(12\alpha + \alpha^4)\operatorname{Ai}'(\alpha) + (12\alpha + 4\alpha^3)\operatorname{Ai}(\alpha) \tag{A.5.17}$$

$$\int_{\alpha}^{\infty} x^6 \operatorname{Ai}(x)\, dx = 40\pi\{\operatorname{Ai}(\alpha)\operatorname{Gi}'(\alpha) - \operatorname{Ai}'(\alpha)\operatorname{Gi}(\alpha)\}$$
$$- (20\alpha^2 + \alpha^5)\operatorname{Ai}'(\alpha) + (40\alpha + 5\alpha^4)\operatorname{Ai}(\alpha) \tag{A.5.18}$$

$$\int_{\alpha}^{\infty} \frac{1}{x^4}\,\mathrm{Ai}'(x)\,dx = \frac{1}{3^{5/3}\Gamma\left(\frac{2}{3}\right)\alpha}\;{}_1F_2\left(-\frac{1}{3};\frac{2}{3},\frac{2}{3};\frac{\alpha^3}{9}\right) - \frac{\mathrm{Ai}'(\alpha)}{3\alpha^3} + \frac{3^{1/6}\,\Gamma\left(\frac{2}{3}\right)\alpha^3}{216\pi}\;{}_2F_3\left(1,1;2,2,\frac{7}{3};\frac{\alpha^3}{9}\right) + \frac{3^{1/6}\Gamma\left(\frac{2}{3}\right)}{36\pi}\left[\ln\left(\frac{\alpha^6}{3}\right) + 4\gamma - 6 - \frac{\pi}{\sqrt{3}}\right] \tag{A.5.19}$$

$$\int_{\alpha}^{\infty} \frac{1}{x^3}\,\mathrm{Ai}'(x)\,dx = \frac{3^{1/6}\Gamma\left(\frac{2}{3}\right)\alpha}{4\pi}\;{}_1F_2\left(\frac{1}{3};\frac{4}{3},\frac{4}{3};\frac{\alpha^3}{9}\right) - \frac{\mathrm{Ai}'(\alpha)}{2\alpha^2} - \frac{\sqrt{3}\,\alpha^3}{108\Gamma\left(\frac{2}{3}\right)}\;{}_2F_3\left(1,1;2,2,\frac{5}{6};\frac{\alpha^3}{9}\right) - \frac{3^{1/3}}{36\Gamma\left(\frac{2}{3}\right)}\left[\ln\left(\frac{\alpha^6}{3}\right) + 4\gamma + \frac{\pi}{\sqrt{3}}\right] \tag{A.5.20}$$

$$\int_{\alpha}^{\infty} \frac{1}{x^2}\mathrm{Ai}'(x)\,dx = \pi\{\mathrm{Ai}(\alpha)\,\mathrm{Gi}'(\alpha) - \mathrm{Ai}'(\alpha)\,\mathrm{Gi}(\alpha)\} + \frac{\mathrm{Ai}'(\alpha)}{\alpha} \tag{A.5.21}$$

$$\int_{\alpha}^{\infty} \frac{1}{x}\,\mathrm{Ai}'(x)\,dx = \frac{1}{3^{2/3}\Gamma\left(\frac{2}{3}\right)\alpha}\;{}_1F_2\left(-\frac{1}{3};\frac{2}{3},\frac{2}{3};\frac{\alpha^3}{9}\right) - \frac{\mathrm{Ai}(\alpha)}{\alpha} + \frac{3^{1/6}\,\Gamma\left(\frac{2}{3}\right)\alpha^3}{72\pi}\;{}_2F_3\left(1,1;2,2,\frac{7}{3};\frac{\alpha^3}{9}\right) + \frac{3^{1/6}\Gamma\left(\frac{2}{3}\right)}{12\pi}\left[\ln\left(\frac{\alpha^6}{3}\right) + 4\gamma - 6 - \frac{\pi}{\sqrt{3}}\right] \tag{A.5.22}$$

$$\int_{\alpha}^{\infty} \mathrm{Ai}'(x)\,dx = -\,\mathrm{Ai}(\alpha) \tag{A.5.23}$$

$$\int_{\alpha}^{\infty} x\,\mathrm{Ai}'(x)\,dx = \pi\{\mathrm{Ai}'(\alpha)\,\mathrm{Gi}(\alpha) - \mathrm{Ai}(\alpha)\,\mathrm{Gi}'(\alpha)\} - \alpha\,\mathrm{Ai}(\alpha) \tag{A.5.24}$$

$$\int_{\alpha}^{\infty} x^2\mathrm{Ai}'(x)\,dx = -\,\alpha^2\,\mathrm{Ai}(\alpha) + 2\mathrm{Ai}'(\alpha) \tag{A.5.25}$$

$$\int_{\alpha}^{\infty} x^3\mathrm{Ai}'(x)\,dx = -(3+\alpha^3)\,\mathrm{Ai}(\alpha) + 2\alpha\,\mathrm{Ai}'(\alpha) \tag{A.5.26}$$

$$\int_{\alpha}^{\infty} x^4\mathrm{Ai}'(x)\,dx = 8\pi\{\mathrm{Ai}'(\alpha)\,\mathrm{Gi}(\alpha) - \mathrm{Ai}(\alpha)\,\mathrm{Gi}'(\alpha)\} - (8\alpha+\alpha^4)\,\mathrm{Ai}(\alpha) + 4\alpha^4\,\mathrm{Ai}'(\alpha) \tag{A.5.27}$$

$$\int_{\alpha}^{\infty} x^5 \mathrm{Ai}'(x)\, dx = -(15\alpha^2 + \alpha^5)\,\mathrm{Ai}(\alpha) + (30 + 5\alpha^3)\,\mathrm{Ai}'(\alpha) \tag{A.5.28}$$

$$\int_{\alpha}^{\infty} x^6 \mathrm{Ai}'(x)\, dx = -(72 + 24\alpha^3 + \alpha^6)\,\mathrm{Ai}(\alpha) + 6\alpha(12 + \alpha^3)\,\mathrm{Ai}'(\alpha) \tag{A.5.29}$$

$$\int_{0}^{\infty} x^n \mathrm{Ai}'(x)\, dx = -\frac{\Gamma(n+1)}{3^{(n+2)/3}\Gamma\left(\frac{n+2}{3}\right)} \tag{A.5.30}$$

$$\int_{-\infty}^{+\infty} e^{ax} \mathrm{Ai}(x)\, dx = e^{a^3/3} \tag{A.5.31}$$

$$\int_{-\infty}^{+\infty} e^{-x^2/4a} \mathrm{Ai}(x)\, dx = 2\sqrt{\pi a}\, e^{2a^3/3} \mathrm{Ai}(a^2) \tag{A.5.32}$$

$$\int \mathrm{Ai}(x-a)\mathrm{Ai}(x-b)\, dx = \frac{1}{a-b}[\mathrm{Ai}(x-a)\mathrm{Ai}'(x-b) - \mathrm{Ai}(x-b)\mathrm{Ai}'(x-a)] \tag{A.5.33}$$

$$\int \mathrm{Bi}(x-a)\mathrm{Ai}(x-b)\, dx = \frac{1}{a-b}[\mathrm{Bi}(x-a)\mathrm{Ai}'(x-b) - \mathrm{Ai}(x-b)\mathrm{Bi}'(x-a)] \tag{A.5.34}$$

$$\int \mathrm{Bi}(x-a)\mathrm{Bi}(x-b)\, dx = \frac{1}{a-b}[\mathrm{Bi}(x-a)\mathrm{Bi}'(x-b) - \mathrm{Bi}(x-b)\mathrm{Bi}'(x-a)] \tag{A.5.35}$$

$$\int_{0}^{\infty} [\mathrm{Ai}(x)]^2 dx = \frac{1}{3^{2/3}\left[\Gamma\left(\frac{1}{3}\right)\right]^2} \tag{A.5.36}$$

$$\int_{\alpha}^{\infty} [\mathrm{Ai}(x)]^2\, dx = -\alpha\,[\mathrm{Ai}(\alpha)]^2 + [\mathrm{Ai}'(\alpha)]^2 \tag{A.5.37}$$

$$\int_{\alpha}^{\infty} x\,[\mathrm{Ai}(x)]^2\, dx = -\frac{\alpha^2}{3}[\mathrm{Ai}(\alpha)]^2 + \frac{\alpha}{3}[\mathrm{Ai}'(\alpha)]^2 - \frac{1}{3}\mathrm{Ai}(\alpha)\,\mathrm{Ai}'(\alpha) \tag{A.5.38}$$

$$\int_{0}^{\infty} \mathrm{Ai}(a-x)\mathrm{Ai}(a+x) dx = 2^{-4/3}\mathrm{Ai}(2^{2/3}a) \tag{A.5.39}$$

$$\int_{-\infty}^{\infty} \mathrm{Ai}(a-\alpha x)\mathrm{Ai}(b-\beta x)\, dx = \frac{1}{(\beta^3-\alpha^3)^{1/3}}\mathrm{Ai}\left[\frac{a\beta - b\alpha}{(\beta^3-\alpha^3)^{1/3}}\right] \tag{A.5.40}$$

$\alpha \neq \beta$

$$\int_0^\infty x^{\alpha-1}\mathrm{Ai}(x)\,\mathrm{Ai}(-x)\,dx = \frac{12^{\beta-1}\sin(\pi\beta)\Gamma\left(\frac{\alpha}{2}\right)\Gamma(\beta)}{\pi^{3/2}} \tag{A.5.41}$$

$$\beta = \left(\frac{\alpha+1}{6}\right) \quad ; \quad \mathrm{Re}\,\alpha > 0$$

$$\int_0^\infty \frac{[\mathrm{Ai}(x)]^2}{\sqrt{x}}\,dx = \frac{1}{6} \tag{A.5.42}$$

$$\int_\alpha^\infty \mathrm{Ai}(x)\,\mathrm{Ai}'(x)\,dx = -\frac{1}{2}[\mathrm{Ai}(\alpha)]^2 \tag{A.5.43}$$

$$\int_\alpha^\infty x\,\mathrm{Ai}(x)\,\mathrm{Ai}'(x)\,dx = -\frac{1}{2}[\mathrm{Ai}'(\alpha)]^2 \tag{A.5.44}$$

$$\int_\alpha^\infty [\mathrm{Ai}'(x)]^2\,dx = \frac{\alpha^2}{3}[\mathrm{Ai}(\alpha)]^2 - \frac{2\alpha}{3}[\mathrm{Ai}'(\alpha)]^2 - \frac{2}{3}\,\mathrm{Ai}(\alpha)\,\mathrm{Ai}'(\alpha) \tag{A.5.45}$$

$$\int_\alpha^\infty x^2\mathrm{Ai}(x)\,\mathrm{Ai}'(x)\,dx = -\frac{\alpha^2}{6}[\mathrm{Ai}(\alpha)]^2 - \frac{\alpha}{3}[\mathrm{Ai}'(\alpha)]^2 + \frac{1}{3}\,\mathrm{Ai}(\alpha)\,\mathrm{Ai}'(\alpha) \tag{A.5.46}$$

$$\int_0^\infty \frac{\mathrm{Ai}(a+x)\mathrm{Ai}'(a+x)}{\sqrt{x}}\,dx = -\frac{2^{2/3}}{4}\,\mathrm{Ai}(2^{2/3}a) \tag{A.5.47}$$

$$\int_0^\infty x\,[\mathrm{Ai}(x)]^2\,dx = \frac{1}{3^2\Gamma\left(\frac{1}{3}\right)\Gamma\left(\frac{2}{3}\right)} \tag{A.5.48}$$

$$\int_0^\infty x^2\,[\mathrm{Ai}(x)]^2\,dx = \frac{1}{3^{4/3}5\,\left[\Gamma\left(\frac{2}{3}\right)\right]^2} \tag{A.5.49}$$

$$\int_0^\infty x^n\,[\mathrm{Ai}(x)]^2\,dx = \frac{\Gamma(n+1)\left(\frac{2}{3}\right)^{2/3}}{4\sqrt{3\pi}\,12^{n/3}\,\Gamma\left(\frac{2n+7}{6}\right)} \tag{A.5.50}$$

$$\int_0^\infty x^n\,[\mathrm{Ai}'(x)]^2\,dx = \frac{(n+2)\Gamma(n+1)\left(\frac{2}{3}\right)^{2/3}}{12^{(n+3)/3}\sqrt{3\pi}\,\Gamma\left(\frac{2n+9}{6}\right)} \tag{A.5.51}$$

$$\int_0^1 \frac{\mathrm{Ai}\left(\frac{a}{x^{1/3}}\right)\mathrm{Ai}\left[\frac{b}{(1-x)^{1/3}}\right]}{[x(1-x)]^{4/3}}\,dx = \frac{(a+b)}{ab}\,\mathrm{Ai}(a+b) \tag{A.5.52}$$

$$\int_0^1 \frac{\mathrm{Ai}'\left(\frac{a}{x^{1/3}}\right)\mathrm{Ai}'\left[\frac{b}{(1-x)^{1/3}}\right]}{[x(1-x)]^{2/3}}\,dx = \mathrm{Ai}(a+b) \tag{A.5.53}$$

$$\int_{-\infty}^{\infty} [\mathrm{Ai}(x)]^3\,dx = \frac{\left[\Gamma\left(\frac{1}{3}\right)\right]^2}{4\pi^2} \tag{A.5.54}$$

$$\int_0^{\infty} [\mathrm{Ai}(x)\,\mathrm{Ai}(-x)]^2\,dt = \frac{1}{24\pi} \tag{A.5.55}$$

$$\int_{-\infty}^{\infty} [\mathrm{Ai}(x)]^2\,\mathrm{Bi}(x)\,dx = \frac{\left[\Gamma\left(\frac{1}{3}\right)\right]^2}{4\sqrt{3}\pi^2} \tag{A.5.56}$$

$$\int_0^{\infty} [\mathrm{Bi}(-x)]^3\,dx = \frac{\left[\Gamma\left(\frac{1}{3}\right)\right]^2}{2\sqrt{3}\pi^2} + \frac{3^{3/2}\Gamma\left(\frac{2}{3}\right)}{2^{5/3}\pi^2}\,{}_2F_1\left(\frac{1}{6},\frac{1}{3};\frac{7}{6};\frac{1}{4}\right) \tag{A.5.57}$$

$$\int_0^{\infty} x\,[\mathrm{Ai}(x)]^3\,dx = \frac{1}{2\cdot 3^{2+1/6}\pi\,\Gamma\left(\frac{2}{3}\right)}\left[{}_2F_1\left(1,\frac{4}{3};\frac{3}{2};\frac{1}{4}\right) - \frac{2}{5}\,{}_2F_1\left(1,\frac{5}{3};\frac{11}{6};\frac{1}{4}\right)\right] \tag{A.5.58}$$

$$\int_0^{\infty} x^2\,[\mathrm{Ai}(x)]^3\,dx = \frac{1}{3^2}\left[\frac{1}{3\left[\Gamma\left(\frac{2}{3}\right)\right]^3} - \frac{2}{\left[\Gamma\left(\frac{1}{3}\right)\right]^3}\right] \tag{A.5.59}$$

$$\int_0^{\infty} [\mathrm{Ai}(x)]^2\,\mathrm{Bi}(x)\,dt = \frac{\left[\Gamma\left(\frac{1}{3}\right)\right]^2}{12\sqrt{3}\pi^2} + \frac{\Gamma\left(\frac{2}{3}\right)}{2^{5/3}\sqrt{3}\pi^2}\,{}_2F_1\left(\frac{1}{6},\frac{1}{3};\frac{7}{6};\frac{1}{4}\right) \tag{A.5.60}$$

$$\int_{-\infty}^{\infty} [\mathrm{Ai}(x)]^2\,\mathrm{Bi}(x)\,dx = \frac{\left[\Gamma\left(\frac{1}{3}\right)\right]^2}{4\sqrt{3}\pi^2} \tag{A.5.61}$$

$$\int_0^{\infty} \mathrm{Ai}(-x)\,[\mathrm{Bi}(-x)]^3\,dt = \frac{1}{12\pi} \tag{A.5.62}$$

$$\int_0^{\infty} x^3\,[\mathrm{Ai}(x)]^4\,dx = \frac{\left(\frac{5}{4}\right)\ln 3 - 1}{96\,\pi^2} \tag{A.5.63}$$

$$\int_0^{\infty} x^{\alpha-1}[\mathrm{Ai}(x)]^4\,dx = \frac{\Gamma(\alpha)}{48^{(\alpha+2)/3}\pi^{3/2}\Gamma\left(\frac{2\alpha+7}{6}\right)}\,{}_2F_1\left(\frac{\alpha+2}{3},\frac{1}{2};\frac{2\alpha+7}{6};\frac{1}{4}\right) \tag{A.5.64}$$

$$\int_0^{\infty} x^{3/2}\,[\mathrm{Ai}(x)]^4\,dx = \frac{1}{2^5\sqrt{3}\,\pi^2}\left[K\left(\frac{1}{2}\right) - E\left(\frac{1}{2}\right)\right] \tag{A.5.65}$$

$$\int_0^\infty x^{\alpha-1}\,[\mathrm{Ai}(x)]^3\,\mathrm{Bi}(x)\,dx = \frac{\Gamma(\alpha)}{2\cdot 12^{(\alpha+2)/3}\pi\,\Gamma\left(\frac{\alpha+2}{3}\right)}\,{}_2F_1\left(\frac{1-\alpha}{2},\frac{1}{2};1;\frac{3}{4}\right) \tag{A.5.66}$$

$$\int_0^\infty x^{3/2}\,[\mathrm{Ai}(x)]^3\,\mathrm{Bi}(x)\,dx = \frac{1}{2^4\sqrt{3}\,\pi^2}E\left(\frac{1}{2}\right) \tag{A.5.67}$$

$$\int_0^\infty x^{\alpha-1}[\mathrm{Ai}(x)]^2[\mathrm{Bi}(x)]^2 dx = \frac{\Gamma(\alpha)\Gamma\left(\frac{1-\alpha}{3}\right)\cos\left(\frac{\pi(1-\alpha)}{3}\right)}{12^{(\alpha+2)/3}\pi^2}\,{}_2F_1\left(\frac{1-\alpha}{3},\frac{1}{2};1;\frac{3}{4}\right) + \frac{\Gamma(\alpha)}{48^{(\alpha+2)/3}\pi^{3/2}\Gamma\left(\frac{2\alpha+7}{6}\right)}\,{}_2F_1\left(\frac{\alpha+2}{3},\frac{1}{2};\frac{2\alpha+7}{6};\frac{1}{4}\right) \tag{A.5.68}$$

$$\int_0^\infty \frac{\mathrm{Gi}(a+x)}{\sqrt{x}}\,dx = 2^{2/3}\pi\,\mathrm{Ai}\left(\frac{a}{2^{2/3}}\right)\mathrm{Bi}\left(\frac{a}{2^{2/3}}\right) \tag{A.5.69}$$

$$\int_0^\infty \frac{\mathrm{Gi}(a-x)}{\sqrt{x}}\,dx = -2^{2/3}\pi\left[\mathrm{Ai}\left(\frac{a}{2^{2/3}}\right)\right]^2 \tag{A.5.70}$$

$$\int_0^\infty \mathrm{Ai}(x)\mathrm{Ai}'(x)J_0(2a\sqrt{x})\,dx = -\frac{1}{2}\mathrm{Ai}(a)\mathrm{Ai}(-a) \tag{A.5.71}$$

$$\int_0^\infty x\,\mathrm{Ai}(a-x)\mathrm{Ai}(a+x)J_0(bx)\,dx = -\mathrm{Ai}\left(a+\frac{b^2}{4}\right)\mathrm{Ai}'\left(a+\frac{b^2}{4}\right) \tag{A.5.72}$$

$$\int_0^\infty x\,\mathrm{Ai}(x)\mathrm{Ai}'(x)\left[J_0(2a\sqrt{x})-J_2(2a\sqrt{x})\right]dx = -\mathrm{Ai}'(a)\mathrm{Ai}'(-a) \tag{A.5.73}$$

$$\int_0^\infty \mathrm{Ai}(x)\,\mu(\alpha x,\beta,\gamma)\,dx = 3^\beta\mu\left(\frac{\alpha^3}{3},\beta,\frac{\gamma}{3}\right) \tag{A.5.74}$$

$$\int_0^\infty \mathrm{Ai}'(x)\,\mu(\alpha x,\beta,\gamma)\,dx = -3^\beta\alpha\,\mu\left(\frac{\alpha^3}{3},\beta,\frac{\gamma-1}{3}\right) \tag{A.5.75}$$

$$\int_0^\infty \mathrm{Ai}^{(n)}(x)\,\mu(\alpha x,\beta,\gamma)\,dx = (-1)^n 3^\beta\alpha^n\,\mu\left(\frac{\alpha^3}{3},\beta,\frac{\gamma-n}{3}\right) \tag{A.5.76}$$

$$\int_0^\infty x\,\mathrm{Ai}(x)\,\mu(\alpha x,\beta,\gamma)\,dx = 3^\beta\alpha^2\mu\left(\frac{\alpha^3}{3},\beta,\frac{\gamma-1}{3}\right) \tag{A.5.77}$$

$$\int_0^\infty x\,\mathrm{Ai}(x)\,\mu\left[(3a)^{1/3}x,\beta,\alpha\right]dx = 3^{\beta+2/3}a^{2/3}\,\mu\left(a,\beta.\frac{\alpha-2}{3}\right) \tag{A.5.78}$$

$$\int_0^\infty \mathrm{Ai}(x)\,\mu\,\left[(3a)^{1/3}x,\beta,\alpha\right]dx = 3^{\beta}a^{1/3}\,\mu\left(a,\beta.\frac{\alpha}{3}\right) \tag{A.5.79}$$

$$\int_0^\infty \mathrm{Ai}'(x)\,\mu\left[(3a)^{1/3}x,\beta,\alpha\right]dx = -\,3^{\beta+1/3}a^{1/3}\,\mu\left(a,\beta.\frac{\alpha-1}{3}\right) \tag{A.5.80}$$

$$\int_0^\infty [\mathrm{Ai}(x)]^2\,\mu(\alpha x,\beta,\gamma)\,dx = \frac{3^{\beta}}{2\sqrt{\pi\alpha}}\alpha^{n}\,\mu\left(\frac{\alpha^3}{12},\beta,\frac{2\gamma+1}{6}\right) \tag{A.5.81}$$

$$\int_0^\infty [\mathrm{Ai}(x)]^2\,\mu\left[(3a)^{1/3}x,\beta,\alpha\right]dx = \frac{3^{\beta-1/6}}{2\sqrt{\pi}}a^{-1/6}\,\mu\left(\frac{a}{4},\beta.\frac{2\alpha+1}{6}\right) \tag{A.5.82}$$

Appendix B Limits Representing Special Functions

Limits of the special function representations are based on the integral transforms (Laplace, Fourier sine, Fourier cosine and Stieltjes) [33, 34, 36].

B.1 Exponential Integral and Related Functions

$$-\mathrm{Ei}(-x) = \lim_{\nu \to \infty} \left\{ \nu^{\nu+1} \int_x^\infty \frac{1}{\xi \sqrt{\nu^2+\xi^2} \left[\xi + \sqrt{\nu^2+\xi^2}\right]^\nu} d\xi \right\} \tag{B.1.1}$$

$$-\mathrm{Ei}(-x) = e^{-x} \lim_{\nu \to \infty} \left\{ \nu^{\nu+1} \int_0^\infty \frac{1}{(\xi + x) \sqrt{\nu^2+\xi^2} \left[\xi + \sqrt{\nu^2+\xi^2}\right]^\nu} d\xi \right\} \tag{B.1.2}$$

$$\mathrm{Ei}(-x) = -\frac{e^{-x}}{x} + e^{-x} \lim_{\nu \to \infty} \left\{ \nu^{\nu+1} \int_0^\infty \frac{1}{(\xi + x)^2 \sqrt{\nu^2+\xi^2} \left[\xi + \sqrt{\nu^2+\xi^2}\right]^\nu} d\xi \right\} \tag{B.1.3}$$

$$-\mathrm{Ei}(-x) = -e^{-x} \ln x + e^{-x} \lim_{\nu \to \infty} \left\{ \nu^{\nu+1} \int_0^\infty \frac{(x+\xi)}{\sqrt{\nu^2+\xi^2} \left[\xi + \sqrt{\nu^2+\xi^2}\right]^\nu} d\xi \right\} \tag{B.1.4}$$

$$-\mathrm{Ei}(-x) = e^{-x} \lim_{\nu \to \infty} \left\{ \nu^{\nu+1} \int_0^\infty \frac{1}{(x+\xi) \sqrt{\nu^2+\xi^2} \left[\xi + \sqrt{\nu^2+\xi^2}\right]^\nu} d\xi \right\} \tag{B.1.5}$$

$$-\mathrm{Ei}(-x) = \lim_{\nu \to \infty} \left\{ \nu^{\nu+1} \int_x^\infty \frac{\ln\left(\frac{\xi}{x}\right)}{\sqrt{\nu^2+\xi^2} \left[\xi + \sqrt{\nu^2+\xi^2}\right]^\nu} d\xi \right\} \tag{B.1.6}$$

$$-\mathrm{Ei}(-x) = -\ln(\gamma x) - \lim_{\nu \to \infty} \left\{ \nu^{\nu+1} \int_0^x \frac{\ln\left(\frac{\xi}{x}\right)}{\sqrt{\nu^2+\xi^2} \left[\xi + \sqrt{\nu^2+\xi^2}\right]^\nu} d\xi \right\} \tag{B.1.7}$$

https://doi.org/10.1515/9783110681642-007

$$-\operatorname{Ei}(-x) = -e^{-x}\ln x + \lim_{\nu\to\infty}\left\{\nu^{\nu+1}\int_x^{\infty}\frac{\ln\xi}{\sqrt{\nu^2+\xi^2}\left[\xi+\sqrt{\nu^2+\xi^2}\right]^{\nu}}\,d\xi\right\} \tag{B.1.8}$$

$$-\operatorname{Ei}(-x) = \frac{2}{\pi}\lim_{\nu\to\infty}\left\{\nu\int_0^{\infty}\frac{\tan^{-1}\left(\frac{\xi}{x}\right)\cos\left[\nu\sin^{-1}\left(\frac{\xi}{\nu}\right)\right]}{\xi\sqrt{\nu^2-\xi^2}}\,d\xi\right\} \tag{B.1.9}$$

$$-\operatorname{Ei}[-(a+x)] = e^{-a}\lim_{\nu\to\infty}\left\{\nu^{\nu+1}\int_x^{\infty}\frac{\ln\left(\frac{\xi}{x}\right)}{(\xi+a)\sqrt{\nu^2+\xi^2}\left[\xi+\sqrt{\nu^2+\xi^2}\right]^{\nu}}\,d\xi\right\} \tag{B.1.10}$$

$$-\operatorname{Ei}(-x) = 2e^{-x}\lim_{\nu\to\infty}\left\{\nu^{\nu+1}\int_x^{\infty}\frac{K_0(2\sqrt{x\xi})}{\sqrt{\nu^2+\xi^2}\left[\xi+\sqrt{\nu^2+\xi^2}\right]^{\nu}}\,d\xi\right\} \tag{B.1.11}$$

$$-\operatorname{Ei}(-x) = \frac{1}{\pi}\lim_{\nu\to\infty}\left\{\nu\int_0^{\infty}\frac{\ln\left[1+\left(\frac{\xi}{x}\right)\right]\sin\left[\nu\sin^{-1}\left(\frac{\xi}{\nu}\right)\right]}{\xi\sqrt{\nu^2-\xi^2}}\,d\xi\right\};\quad x>0 \tag{B.1.12}$$

$$-\operatorname{Ei}(-x) = e^{-x}\lim_{\nu\to\infty}\left\{\nu\int_0^{\infty}\frac{e^{-\nu\xi}}{x+\nu\sinh\xi}\,d\xi\right\} \tag{B.1.13}$$

$$\operatorname{Ei}[-(x+1)] + \operatorname{Ei}(x-1) = -2\lim_{\nu\to\infty}\left\{\nu^{\nu+1}\int_1^{\infty}\frac{\cosh(x\xi)}{\xi\sqrt{\nu^2+\xi^2}\left[\xi+\sqrt{\nu^2+\xi^2}\right]^{\nu}}\,d\xi\right\} \tag{B.1.14}$$

$$\operatorname{Ei}[-(x+1)] - \operatorname{Ei}(x-1) = -2\lim_{\nu\to\infty}\left\{\nu^{\nu+1}\int_1^{\infty}\frac{\sinh(x\xi)}{\xi\sqrt{\nu^2+\xi^2}\left[\xi+\sqrt{\nu^2+\xi^2}\right]^{\nu}}\,d\xi\right\} \tag{B.1.15}$$

$$-\operatorname{Ei}[-(a+x)] = e^{-a}\lim_{\nu\to\infty}\left\{\nu^{\nu+1}\int_x^{\infty}\frac{\ln\left(\frac{\xi}{x}\right)}{(\xi+a)\sqrt{\nu^2+\xi^2}\left[\xi+\sqrt{\nu^2+\xi^2}\right]^{\nu}}\,d\xi\right\} \tag{B.1.16}$$

$$[\operatorname{Ei}(-x)]^2 = 2e^{-x}\lim_{\nu\to\infty}\left\{\nu^{\nu+1}\int_x^{\infty}\frac{\ln\left(\frac{\xi}{x}\right)}{(x+\xi)\sqrt{\nu^2+\xi^2}\left[\xi+\sqrt{\nu^2+\xi^2}\right]^{\nu}}\,d\xi\right\} \tag{B.1.17}$$

$$\frac{1}{2}\left[e^{-x}\operatorname{Ei}(x) - e^{x}\operatorname{Ei}(-x)\right] = \lim_{\nu\to\infty}\left\{\nu\int_0^{\infty}\frac{\tan^{-1}\left(\frac{\xi}{x}\right)\cos\left[\nu\sin^{-1}\left(\frac{\xi}{\nu}\right)\right]}{\sqrt{\nu^2-\xi^2}}\,d\xi\right\} \tag{B.1.18}$$

$x > 0$

B.2 Sine and Cosine Integrals

$$Si(x) = \frac{\pi}{2} - \lim_{\nu\to\infty}\left\{\nu^{\nu+1}\int_0^{\infty}\frac{[x\cos x + \xi\sin x]}{\sqrt{\nu^2+\xi^2}\left[\xi+\sqrt{\nu^2+\xi^2}\right]^{\nu}(x^2+\xi^2)}\,d\xi\right\} \tag{B.2.1}$$

$$Si(x) = \lim_{\nu\to\infty}\left\{\nu\int_0^{x}\frac{\sin\left[\nu\sin^{-1}\left(\frac{\xi}{\nu}\right)\right]}{\xi\sqrt{\nu^2-\xi^2}}\,d\xi\right\} \tag{B.2.2}$$

$$si(x) = -\lim_{\nu\to\infty}\left\{\nu\int_x^{\infty}\frac{\sin\left[\nu\sin^{-1}\left(\frac{\xi}{\nu}\right)\right]}{\xi\sqrt{\nu^2-\xi^2}}\,d\xi\right\} \tag{B.2.3}$$

$$si(x) = -\frac{1}{2\pi}\lim_{\nu\to\infty}\left\{\nu\int_0^{\infty}\frac{\left[\ln\left(\frac{x+\xi}{x-\xi}\right)\right]^2\cos\left[\nu\sin^{-1}\left(\frac{\xi}{\nu}\right)\right]}{\xi\sqrt{\nu^2-\xi^2}}\,d\xi\right\} \tag{B.2.4}$$

$$Ci(x) = \lim_{\nu\to\infty}\left\{\nu^{\nu+1}\int_0^{\infty}\frac{[x\sin x - \xi\cos x]}{\sqrt{\nu^2+\xi^2}\left[\xi+\sqrt{\nu^2+\xi^2}\right]^{\nu}(x^2+\xi^2)}\,d\xi\right\} \tag{B.2.5}$$

$$Ci(x) = -\lim_{\nu\to\infty}\left\{\nu\int_x^{\infty}\frac{\cos\left[\nu\sin^{-1}\left(\frac{\xi}{\nu}\right)\right]}{\xi\sqrt{\nu^2-\xi^2}}\,d\xi\right\} \tag{B.2.6}$$

$$\cos x\, ci(x) - \sin x\, si(x) = \frac{1}{2}\lim_{\nu\to\infty}\left\{\nu^{\nu+1}\int_0^{\infty}\frac{\ln\left[1+\left(\frac{\xi}{x}\right)^2\right]}{\sqrt{\nu^2+\xi^2}\left[\xi+\sqrt{\nu^2+\xi^2}\right]^{\nu}}\,d\xi\right\} \tag{B.2.7}$$

$$\cos x\, ci(x) - \sin x\, si(x) = \lim_{\nu\to\infty}\left\{\nu^{\nu+1}\int_0^{\infty}\frac{1}{(\xi^2+x^2)\sqrt{\nu^2+\xi^2}\left[\xi+\sqrt{\nu^2+\xi^2}\right]^{\nu}}\,d\xi\right\} \tag{B.2.8}$$

$$\sin x\, ci(x) + \cos x\, si(x) = -\lim_{\nu\to\infty}\left\{\nu^{\nu+1}\int_0^\infty \frac{\tan^{-1}\left(\frac{\xi}{x}\right)}{\sqrt{\nu^2+\xi^2}\left[\xi+\sqrt{\nu^2+\xi^2}\right]^\nu}d\xi\right\} \tag{B.2.9}$$

$$\sin x\, ci(x) + \cos x\, si(x) =$$

$$-x\lim_{\nu\to\infty}\left\{\nu^{\nu+1}\int_0^\infty \frac{1}{(\xi^2+x^2)\sqrt{\nu^2+\xi^2}\left[\xi+\sqrt{\nu^2+\xi^2}\right]^\nu}d\xi\right\} \tag{B.2.10}$$

$$\sin x\, ci(x) + \cos x\, si(x) = -x\lim_{\nu\to\infty}\left\{\nu\int_0^\infty \frac{e^{-\nu\xi}}{x^2+(\nu\sinh\xi)^2}d\xi\right\} \tag{B.2.11}$$

$$\cos x\, ci(x) - \sin x\, si(x) = \lim_{\nu\to\infty}\left\{\nu\int_0^\infty \frac{e^{-\nu\xi}\nu\sinh\xi}{x^2+(\nu\sinh\xi)^2}d\xi\right\} \tag{B.2.12}$$

$$\sin x\, ci(x) + \cos x\, si(x) = -\frac{\sqrt{\pi}}{2}\lim_{\nu\to\infty}\left\{\nu\int_0^\infty \frac{e^{-x^2/4q}}{\sqrt{q}}\, erfc\left(\frac{x}{2q}\right)d\xi\right\} \tag{B.2.13}$$

$$q = \nu\sinh\xi$$

$$[si(x)]^2 + [ci(x)]^2 = \lim_{\nu\to\infty}\left\{\nu^{\nu+1}\int_0^\infty \frac{\ln\left[1+\left(\frac{\xi}{x}\right)^2\right]}{\xi\sqrt{\nu^2+\xi^2}\left[\xi+\sqrt{\nu^2+\xi^2}\right]^\nu}d\xi\right\} \tag{B.2.14}$$

B.3 Gamma Function and Related Functions

$$\Gamma(x+1) = \lim_{\nu\to\infty}\left\{\nu^{\nu+1}\int_0^\infty \frac{\xi^x}{\sqrt{\nu^2+\xi^2}\left[\xi+\sqrt{\nu^2+\xi^2}\right]^\nu}d\xi\right\} \tag{B.3.1}$$

$$\Gamma(x,\alpha) = \lim_{\nu\to\infty}\left\{\nu^{\nu+1}\int_\alpha^\infty \frac{\xi^{x-1}}{\sqrt{\nu^2+\xi^2}\left[\xi+\sqrt{\nu^2+\xi^2}\right]^\nu}d\xi\right\} \tag{B.3.2}$$

$$\Gamma(-x,\alpha) = \alpha^x x\lim_{\nu\to\infty}\left\{\nu^{\nu+1}\int_0^x \frac{e^{-\alpha e^{\xi/x}}}{\sqrt{\nu^2+\xi^2}\left[\xi+\sqrt{\nu^2+\xi^2}\right]^\nu}d\xi\right\};\quad \mathrm{Re}\,\alpha>0 \tag{B.3.3}$$

$$\Gamma(\alpha, x) = e^{-x} \lim_{\nu \to \infty} \left\{ \nu^{\nu+1} \int_0^\infty \frac{(x+\xi)^{\alpha-1}}{\sqrt{\nu^2+\xi^2} \left[\xi + \sqrt{\nu^2+\xi^2}\right]^\nu} d\xi \right\} \tag{B.3.4}$$

$$\Gamma(\alpha, x) = \lim_{\nu \to \infty} \left\{ \nu^{\nu+1} \int_x^\infty \frac{\xi^{\alpha-1}}{\sqrt{\nu^2+\xi^2} \left[\xi + \sqrt{\nu^2+\xi^2}\right]^\nu} d\xi \right\} \tag{B.3.5}$$

$$\Gamma(\alpha, x) = \frac{x^\alpha}{\Gamma(1-\alpha)} \lim_{\nu \to \infty} \left\{ \nu^{\nu+1} \int_x^\infty \frac{1}{\xi\,(\xi - x)^\alpha \sqrt{\nu^2+\xi^2} \left[\xi + \sqrt{\nu^2+\xi^2}\right]^\nu} d\xi \right\} \tag{B.3.6}$$

$\operatorname{Re}\alpha < 1; \quad x > 0$

$$\Gamma(\alpha, x) = e^{-x} \lim_{\nu \to \infty} \left\{ \nu \int_0^\infty \frac{e^{-\nu\xi}}{(x + \nu \sinh\xi)^{1-\alpha}} d\xi \right\} \tag{B.3.7}$$

$-1 < \operatorname{Re}\alpha < 1; \quad x > 0$

$$\Gamma(\alpha+1, x) = e^{-x} \lim_{\nu \to \infty} \left\{ \nu^{\nu+1} \int_x^\infty \frac{\xi^\alpha}{\sqrt{\nu^2+\xi^2} \left[\xi + \sqrt{\nu^2+\xi^2}\right]^\nu} d\xi \right\} \tag{B.3.8}$$

$$\Gamma\left(\alpha, \frac{1}{x}\right) = \frac{2e^{-x} x^{\alpha/2}}{\Gamma(1-\alpha)} \lim_{\nu \to \infty} \left\{ \nu \int_0^\infty \frac{e^{-\nu\xi} K_\alpha\left(\sqrt{\nu x \sinh\xi}\right)}{(\nu \sinh\xi)^{\alpha/2}} d\xi \right\} \tag{B.3.9}$$

$-1 < \operatorname{Re}\alpha < 1; \quad x > 0$

$$\Gamma(\alpha, -x) = \frac{e^{-x}}{\Gamma(\alpha+1)\, x^\alpha} \lim_{\nu \to \infty} \left\{ \nu^{\nu+1} \int_0^\infty \frac{\xi^\alpha}{(\xi + x) \sqrt{\nu^2+\xi^2} \left[\xi + \sqrt{\nu^2+\xi^2}\right]^\nu} d\xi \right\} \tag{B.3.10}$$

$\operatorname{Re}\alpha > -1$

$$\Gamma(-\alpha, x) = \frac{1}{\Gamma(\alpha+1)\, x^\alpha} \lim_{\nu \to \infty} \left\{ \nu^{\nu+1} \int_x^\infty \frac{1}{\xi\,(\xi - x)^\alpha \sqrt{\nu^2+\xi^2} \left[\xi + \sqrt{\nu^2+\xi^2}\right]^\nu} d\xi \right\}$$

$\operatorname{Re}\alpha > -1$

(B.3.11)

$$\Gamma(-\alpha, x) = \frac{1}{\Gamma(\alpha+1)\, x^{\alpha}} \lim_{\nu\to\infty} \left\{ \nu^{\nu+1} \int_{x}^{\infty} \frac{(\xi - x)^{\alpha}}{\xi\, \sqrt{\nu^2+\xi^2}\left[\xi + \sqrt{\nu^2+\xi^2}\right]^{\nu}}\, d\xi \right\} \tag{B.3.12}$$

$$\Gamma(-\alpha, x) = \frac{2e^{x}}{\Gamma(\alpha+1)\, x^{\alpha/2}} \lim_{\nu\to\infty} \left\{ \nu^{\nu+1} \int_{0}^{\infty} \frac{\xi^{\alpha/2} K_{\alpha}(2\sqrt{x\xi})}{\sqrt{\nu^2+\xi^2}\left[\xi + \sqrt{\nu^2+\xi^2}\right]^{\nu}}\, d\xi \right\} \tag{B.3.13}$$

$$\gamma(\alpha, x) = \frac{}{\Gamma(\alpha)\, x^{\alpha}} \lim_{\nu\to\infty} \left\{ \nu^{\nu+1} \int_{0}^{\infty} \frac{\xi^{\alpha-1}}{(1 - e^{\xi/x})\sqrt{\nu^2+\xi^2}\left[\xi + \sqrt{\nu^2+\xi^2}\right]^{\nu}}\, d\xi \right\} \tag{B.3.14}$$

$\mathrm{Re}\,\alpha > 1$

$$\gamma(\alpha, x) = \lim_{\nu\to\infty} \left\{ \nu^{\nu+1} \int_{0}^{x} \frac{\xi^{\alpha-1}}{\sqrt{\nu^2+\xi^2}\left[\xi + \sqrt{\nu^2+\xi^2}\right]^{\nu}}\, d\xi \right\}; \quad \mathrm{Re}\,\alpha > 0 \tag{B.3.15}$$

$$\gamma(\alpha, x) = \lim_{\nu\to\infty} \left\{ \nu^{\nu+1} \int_{x}^{\infty} \frac{\xi^{\alpha-1}}{\sqrt{\nu^2+\xi^2}\left[\xi + \sqrt{\nu^2+\xi^2}\right]^{\nu}}\, d\xi \right\}; \quad \mathrm{Re}\,\alpha > 0 \tag{B.3.16}$$

$$\gamma(x, \alpha) = \frac{\alpha^{x}}{x} \lim_{\nu\to\infty} \left\{ \nu^{\nu+1} \int_{0}^{x} \frac{e^{-\alpha e^{-\xi/x}}}{\sqrt{\nu^2+\xi^2}\left[\xi + \sqrt{\nu^2+\xi^2}\right]^{\nu}}\, d\xi \right\}; \quad \mathrm{Re}\,\alpha > 0 \tag{B.3.17}$$

$$\gamma(\alpha, x) = x^{\alpha/2} \lim_{\nu\to\infty} \left\{ \nu^{\nu+1} \int_{0}^{\infty} \frac{\xi^{\alpha/2-1} J_{\alpha}(2\sqrt{x\xi})}{\sqrt{\nu^2+\xi^2}\left[\xi + \sqrt{\nu^2+\xi^2}\right]^{\nu}}\, d\xi \right\}; \quad \mathrm{Re}\,\alpha > 0 \tag{B.3.18}$$

$$\gamma(\alpha, x) = \Gamma(\alpha)\, x^{\alpha/2} e^{-x} \lim_{\nu\to\infty} \left\{ \nu^{\nu+1} \int_{0}^{\infty} \frac{I_{\alpha}(2\sqrt{x\xi})}{\xi^{\alpha/2}\sqrt{\nu^2+\xi^2}\left[\xi + \sqrt{\nu^2+\xi^2}\right]^{\nu}}\, d\xi \right\} \tag{B.3.19}$$

$$\gamma(\alpha, -x) = e^{x} \lim_{\nu\to\infty} \left\{ \nu^{\nu+1} \int_{0}^{x} \frac{(x-\xi)^{\alpha-1}}{\sqrt{\nu^2+\xi^2}\left[\xi + \sqrt{\nu^2+\xi^2}\right]^{\nu}}\, d\xi \right\}; \quad \mathrm{Re}\,\alpha > 0 \tag{B.3.20}$$

$$B(x,y) = \frac{1}{x} \lim_{\nu\to\infty} \left\{ \nu^{\nu+1} \int_0^\infty \frac{\left(1-e^{\xi/x}\right)^{y-1}}{\sqrt{\nu^2+\xi^2}\left[\xi+\sqrt{\nu^2+\xi^2}\right]^\nu} d\xi \right\} \tag{B.3.21}$$

$\operatorname{Re} y > 0$

$$\psi(x) = \ln x + \lim_{\nu\to\infty} \left\{ \nu^{\nu+1} \int_0^\infty \frac{e^{\xi/x}}{(e^{\xi/x}-1)\sqrt{\nu^2+\xi^2}\left[\xi+\sqrt{\nu^2+\xi^2}\right]^\nu} d\xi \right\} \tag{B.3.22}$$

$$\psi(x) = -\lim_{\nu\to\infty} \left\{ \nu^{\nu+1} \int_0^\infty \frac{\ln\left[y\,(e^{\xi/x}-1)\right]}{\sqrt{\nu^2+\xi^2}\left[\xi+\sqrt{\nu^2+\xi^2}\right]^\nu} d\xi \right\} \tag{B.3.23}$$

$$\psi(x) = -\frac{1}{2x} + \ln x + 2x \lim_{\nu\to\infty} \left\{ \nu^{\nu+1} \int_0^\infty \frac{\left[\left(\frac{2x}{\xi}\right) - \operatorname{csch}\left(\frac{\xi}{2x}\right)\right]}{\sqrt{\nu^2+\xi^2}\left[\xi+\sqrt{\nu^2+\xi^2}\right]^\nu} d\xi \right\} \tag{B.3.24}$$

$$\psi(x) = \frac{\Gamma(\mu)}{2x^\mu} + 2^\mu x \lim_{\nu\to\infty} \left\{ \nu^{\nu+1} \int_0^\infty \frac{\left[\left(\frac{\xi}{2x}\right)^{\mu-1} \operatorname{ctnh}\left(\frac{\xi}{2x}\right)\right]}{\sqrt{\nu^2+\xi^2}\left[\xi+\sqrt{\nu^2+\xi^2}\right]^\nu} d\xi \right\} \tag{B.3.25}$$

$$\psi(x) = -\frac{1}{2x} + \ln x + 2x \lim_{\nu\to\infty} \left\{ \nu^{\nu+1} \int_0^\infty \frac{\left[\left(\frac{2x}{\xi}\right) - \operatorname{ctnh}\left(\frac{\xi}{2x}\right)\right]}{\sqrt{\nu^2+\xi^2}\left[\xi+\sqrt{\nu^2+\xi^2}\right]^\nu} d\xi \right\} \tag{B.3.26}$$

$$\psi(x) = \frac{1}{4x} + \ln(2x) - \frac{1}{2x} \lim_{\nu\to\infty} \left\{ \nu^{\nu+1} \int_0^\infty \frac{\left[\ln\left[\sinh\left(\frac{\xi}{2x}\right)\right] - \ln\left(\frac{\xi}{2x}\right)\right]}{\sqrt{\nu^2+\xi^2}\left[\xi+\sqrt{\nu^2+\xi^2}\right]^\nu} d\xi \right\} \tag{B.3.27}$$

$$\psi\left(\frac{x}{2}\right) = \frac{1}{x} - \ln\left(\frac{x}{2}\right) + x \lim_{\nu\to\infty} \left\{ \nu^{\nu+1} \int_0^\infty \frac{\left\{\frac{x}{\xi} \ln\left[\sinh\left(\frac{\xi}{x}\right)\right]\right\}}{\sqrt{\nu^2+\xi^2}\left[\xi+\sqrt{\nu^2+\xi^2}\right]^\nu} d\xi \right\} \tag{B.3.28}$$

$$\psi\left(\frac{x}{2}\right) = \ln\left(\frac{x}{2}\right) - \frac{1}{x} + \frac{1}{x} \lim_{\nu\to\infty} \left\{ \nu^{\nu+1} \int_0^\infty \frac{\left[\frac{x}{\xi} - \cosh\left(\frac{\xi}{x}\right)\right]}{\sqrt{\nu^2+\xi^2}\left[\xi+\sqrt{\nu^2+\xi^2}\right]^\nu} d\xi \right\} \tag{B.3.29}$$

$$\psi\left(\frac{x}{2}\right) = \ln\left(\frac{x}{2}\right) - \frac{1}{x} + \frac{1}{x}\lim_{\nu\to\infty}\left\{\nu^{\nu+1}\int_0^\infty \frac{\left[\frac{x}{\xi} - \cosh\left(\frac{\xi}{x}\right)\right]}{\sqrt{\nu^2+\xi^2}\left[\xi+\sqrt{\nu^2+\xi^2}\right]^\nu} d\xi\right\} \tag{B.3.30}$$

$$\psi\left(\frac{x+1}{2}\right) - \psi\left(\frac{x}{2}\right) = \frac{2}{x}\lim_{\nu\to\infty}\left\{\nu^{\nu+1}\int_0^\infty \frac{1}{(1+e^{-\xi/x})\sqrt{\nu^2+\xi^2}\left[\xi+\sqrt{\nu^2+\xi^2}\right]^\nu} d\xi\right\} \tag{B.3.31}$$

$$\psi\left(\frac{x+1}{2}\right) - \psi\left(\frac{x}{2}\right) = 2x\lim_{\nu\to\infty}\left\{\nu^{\nu+1}\int_0^\infty \frac{\ln\left(\frac{1+e^{-\xi/x}}{2}\right)}{\sqrt{\nu^2+\xi^2}\left[\xi+\sqrt{\nu^2+\xi^2}\right]^\nu} d\xi\right\} \tag{B.3.32}$$

$$\psi\left(\frac{x+1}{2}\right) = \ln\left(\frac{x}{2}\right) + \frac{1}{x}\lim_{\nu\to\infty}\left\{\nu^{\nu+1}\int_0^\infty \frac{\left[\frac{x}{\xi} - \operatorname{csc h}\left(\frac{\xi}{x}\right)\right]}{\sqrt{\nu^2+\xi^2}\left[\xi+\sqrt{\nu^2+\xi^2}\right]^\nu} d\xi\right\} \tag{B.3.33}$$

$$\psi^{(n)}(x) = \left(-\frac{1}{x}\right)^{n+1}\lim_{\nu\to\infty}\left\{\nu^{\nu+1}\int_0^\infty \frac{\xi^n}{(e^{-\xi/x}-1)\sqrt{\nu^2+\xi^2}\left[\xi+\sqrt{\nu^2+\xi^2}\right]^\nu} d\xi\right\}$$

$$n = 1, 2, 3, \ldots \tag{B.3.34}$$

B.4 Error Functions and Fresnel Integrals

$$\operatorname{erf}(x) = \frac{1}{\pi}\lim_{\nu\to\infty}\left\{\nu^{\nu+1}\int_0^\infty \frac{\sin(2x\sqrt{\xi})}{\xi\sqrt{\nu^2+\xi^2}\left[\xi+\sqrt{\nu^2+\xi^2}\right]^\nu} d\xi\right\} \tag{B.4.1}$$

$$\operatorname{erf}(x) = \frac{2}{\pi}\lim_{\nu\to\infty}\left\{\nu\int_0^\infty \frac{e^{-\xi^2/4x^2}\sin\left[\nu\sin^{-1}\left(\frac{\xi}{\nu}\right)\right]}{\xi\sqrt{\nu^2-\xi^2}} d\xi\right\} \tag{B.4.2}$$

$$\operatorname{erf}(\sqrt{x}) = \frac{e^{-x}}{\pi\sqrt{x}}\lim_{\nu\to\infty}\left\{\nu^{\nu+1}\int_0^\infty \frac{\left[\cosh(2\sqrt{x\xi})-1\right]}{\sqrt{\nu^2+\xi^2}\left[\xi+\sqrt{\nu^2+\xi^2}\right]^\nu} d\xi\right\} \tag{B.4.3}$$

$$\operatorname{erfc}(x) = \frac{e^{x^2}}{\sqrt{\pi}} \lim_{\nu\to\infty} \left\{ \nu^{\nu+1} \int_0^\infty \frac{1}{\sqrt{\xi + x^2}\,\sqrt{\nu^2+\xi^2}\left[\xi + \sqrt{\nu^2+\xi^2}\right]^\nu}\, d\xi \right\} \tag{B.4.4}$$

$$\operatorname{erfc}(x) = \frac{2e^{-x^2}}{\sqrt{\pi}} \lim_{\nu\to\infty} \left\{ \nu^{\nu+1} \int_0^\infty \frac{\left(\sqrt{x^2+\xi^2} - x\right)}{\sqrt{\nu^2+\xi^2}\left[\xi + \sqrt{\nu^2+\xi^2}\right]^\nu}\, d\xi \right\} \tag{B.4.5}$$

$$\operatorname{erfc}(x) = \frac{e^{-x^2}}{\sqrt{\pi}} \lim_{\nu\to\infty} \left\{ \nu^{\nu+1} \int_0^\infty \frac{e^{-2x^2\sqrt{\xi}}}{\sqrt{\xi}\,\sqrt{\nu^2+\xi^2}\left[\xi + \sqrt{\nu^2+\xi^2}\right]^\nu}\, d\xi \right\} \tag{B.4.6}$$

$$\operatorname{erfc}(x) = \frac{e^{-x^2}}{\sqrt{\pi}} \lim_{\nu\to\infty} \left\{ \nu^{\nu+1} \int_0^\infty \frac{\sinh(2x\sqrt{\xi})}{\sqrt{\nu^2+\xi^2}\left[\xi + \sqrt{\nu^2+\xi^2}\right]^\nu}\, d\xi \right\} \tag{B.4.7}$$

$$\operatorname{erfc}(x) = e^{-x^2} \left[1 - \lim_{\nu\to\infty} \left\{ \nu^{\nu+1} \int_0^\infty \frac{\operatorname{erfc}\left(\frac{\xi}{2x}\right)}{\sqrt{\nu^2+\xi^2}\left[\xi + \sqrt{\nu^2+\xi^2}\right]^\nu}\, d\xi \right\} \right] \tag{B.4.8}$$

$$\operatorname{erfc}(\sqrt{x}) = \frac{\sqrt{x}\,e^{-x}}{\pi} \lim_{\nu\to\infty} \left\{ \nu^{\nu+1} \int_0^\infty \frac{1}{\sqrt{\xi}\,(\xi + x)\,\sqrt{\nu^2+\xi^2}\left[\xi + \sqrt{\nu^2+\xi^2}\right]^\nu}\, d\xi \right\} \tag{B.4.9}$$

$$\operatorname{erfc}(\sqrt{x}) = \frac{e^{-x}}{\sqrt{\pi}} \lim_{\nu\to\infty} \left\{ \nu^{\nu+1} \int_0^\infty \frac{e^{-2\sqrt{x\xi}}}{\sqrt{\xi}\,\sqrt{\nu^2+\xi^2}\left[\xi + \sqrt{\nu^2+\xi^2}\right]^\nu}\, d\xi \right\} \tag{B.4.10}$$

$$\operatorname{erfc}(\sqrt{x}) = \frac{e^{-x}}{\sqrt{\pi}} \lim_{\nu\to\infty} \left\{ \nu \int_0^\infty \frac{e^{-\nu\xi}}{\sqrt{x + \nu \sinh\xi}}\, d\xi \right\} \tag{B.4.11}$$

$$\operatorname{erfc}(\sqrt{x}) = e^{-x} \left[1 - 2 \lim_{\nu\to\infty} \left\{ \nu \int_0^\infty \frac{e^{-\nu\xi}}{(x + \nu \sinh\xi)^{3/2}}\, d\xi \right\} \right] \tag{B.4.12}$$

$$C(x) = \frac{1}{2} - \frac{1}{\sqrt{2\pi x}} \lim_{\nu\to\infty} \left\{ \nu^{\nu+1} \int_0^\infty \frac{\sin\left(\frac{\xi^2}{4x} - x\right)}{\sqrt{\xi^2+\nu^2}\left[\xi + \sqrt{\xi^2+\nu^2}\right]^\nu}\, d\xi \right\} \tag{B.4.13}$$

$$S(x) = \frac{1}{2} - \frac{1}{\sqrt{2\pi x}} \lim_{\nu\to\infty} \left\{ \nu^{\nu+1} \int_0^\infty \frac{\cos(\frac{\xi^2}{4x} - x)}{\sqrt{\xi^2+\nu^2}\left[\xi + \sqrt{\xi^2+\nu^2}\right]^\nu} d\xi \right\} \tag{B.4.14}$$

$$\left[\frac{1}{2} - S(x)\right]^2 + \left[\frac{1}{2} - C(x)\right]^2 =$$

$$\frac{2}{\pi} \lim_{\nu\to\infty} \left\{ \nu^{\nu+1} \int_0^\infty \frac{\sin\left[\left(\frac{\xi}{2x}\right)^2\right]}{\xi\sqrt{\xi^2+\nu^2}\left[\xi + \sqrt{\xi^2+\nu^2}\right]^\nu} d\xi \right\} \tag{B.4.15}$$

$$\left[\frac{1}{2} - C\left(\frac{x^2}{4}\right)\right]^2 - \left[\frac{1}{2} - S\left(\frac{x^2}{4}\right)\right]^2 =$$

$$\frac{1}{\pi} \lim_{\nu\to\infty} \left\{ \nu^{\nu+1} \int_0^\infty \frac{Si(x\xi^2)}{\sqrt{\xi^2+\nu^2}\left[\xi + \sqrt{\xi^2+\nu^2}\right]^\nu} d\xi \right\} \tag{B.4.16}$$

$$\sin\left(x + \frac{\pi}{4}\right) C(x) - \cos\left(x + \frac{\pi}{4}\right) S(x) = \frac{\sin x}{\sqrt{2\pi}}$$

$$+ \frac{1}{2\sqrt{\pi}} \lim_{\nu\to\infty} \left\{ \nu \int_0^\infty e^{-\nu\xi} \sqrt{1 - \frac{\nu \sinh\xi}{x^2 + (\nu\sinh\xi)^2}} d\xi \right\} \tag{B.4.17}$$

$$\sin\left(x + \frac{\pi}{4}\right) C(x) - \cos\left(x + \frac{\pi}{4}\right) S(x) = \frac{\sin x}{\sqrt{2}} - \frac{1}{2\sqrt{\pi x}} -$$

$$\frac{1}{2^{3/2}\pi} \lim_{\nu\to\infty} \left\{ \nu \int_0^\infty \frac{e^{-\nu\xi}}{\left[x^2 + (\nu\sinh\xi)^2\right]^{3/4}} \sin\left[\frac{3}{2}\tan^{-1}\left(\frac{x}{\nu\sinh\xi}\right)\right] d\xi \right\} \tag{B.4.18}$$

B.5 Legendre Functions

$$P_\lambda^\mu\left(\frac{1}{\sqrt{1-x^2}}\right) = \frac{1}{\Gamma(\lambda-\mu+1)} \lim_{\nu\to\infty} \left\{ \nu^{\nu+1} \int_0^\infty \frac{\xi^\lambda I_{-\mu}(x\xi)}{\sqrt{\nu^2+\xi^2}\left[\xi + \sqrt{\nu^2+\xi^2}\right]^\nu} d\xi \right\} \tag{B.5.1}$$

$$P_{\mu}^{-\lambda}(x)=\frac{1}{\Gamma(\lambda+\mu+1)}\lim_{\nu\to\infty}\left\{\nu^{\nu+1}\int_0^{\infty}\frac{\xi^{\mu}I_{\lambda}\left(\frac{\xi}{x}\right)}{\sqrt{\nu^2+\xi^2}\left[\xi+\sqrt{\nu^2+\xi^2}\right]^{\nu}}d\xi\right\} \tag{B.5.2}$$

$\mathrm{Re}(\lambda+\mu)-1$

$$P_{\lambda}^{-\mu}(x)=\frac{\sqrt{\frac{2}{\pi x}}}{\Gamma(\lambda+\mu+1)\,\Gamma(\mu-\lambda)}\lim_{\nu\to\infty}\left\{\nu^{\nu+1}\int_0^{\infty}\frac{\xi^{\mu-1/2}K_{\lambda+1/2}\left(\frac{\xi}{x}\right)}{\sqrt{\nu^2+\xi^2}\left[\xi+\sqrt{\nu^2+\xi^2}\right]^{\nu}}d\xi\right\} \tag{B.5.3}$$

$\mathrm{Re}(\lambda+\mu)-1;\quad \mathrm{Re}(\mu-\lambda)>0$

$$Q_{\lambda}^{\mu}(x)=\frac{\sqrt{\frac{\pi}{2x}}\,\sin[\pi(\lambda+\mu)]}{\sin(\pi\lambda)}\lim_{\nu\to\infty}\left\{\nu^{\nu+1}\int_0^{\infty}\frac{\xi^{\mu-1/2}I_{\lambda+1/2}\left(\frac{\xi}{x}\right)}{\sqrt{\nu^2+\xi^2}\left[\xi+\sqrt{\nu^2+\xi^2}\right]^{\nu}}d\xi\right\} \tag{B.5.4}$$

$\mathrm{Re}(\lambda+\mu)-1$

$$Q_{\lambda}^{\mu}(x)=\frac{\sin[\pi(\lambda+\mu)]}{\Gamma(\lambda+\mu+1)\,\sin(\pi\lambda)}\lim_{\nu\to\infty}\left\{\nu^{\nu+1}\int_0^{\infty}\frac{\xi^{\mu}K_{\lambda}\left(\frac{\xi}{x}\right)}{\sqrt{\nu^2+\xi^2}\left[\xi+\sqrt{\nu^2+\xi^2}\right]^{\nu}}d\xi\right\} \tag{B.5.5}$$

$\mathrm{Re}(\mu\pm\lambda)-1$

B.6 Hypergeometric and Confluent Hypergeometric Functions

$${}_2F_1(\mu,\alpha;\alpha+\beta+1;x)=$$

$$\frac{1}{\alpha B(\alpha,\beta+1)}\lim_{\nu\to\infty}\left\{\nu^{\nu+1}\int_0^{\infty}\frac{(1-e^{-\xi/\alpha})^{\beta}}{(1-xe^{-\xi/\alpha})^{\mu}\sqrt{\nu^2+\xi^2}\left[\xi+\sqrt{\nu^2+\xi^2}\right]^{\nu}}d\xi\right\} \tag{B.6.1}$$

$\beta>-1,\quad |x|<1$

$${}_2F_1(\alpha,\alpha+\frac{1}{2};\beta;x^2)=$$

$$\frac{2^{2\alpha-\beta}\Gamma(\alpha)\,x^{1-\alpha}}{\Gamma(2\alpha)}\lim_{\nu\to\infty}\left\{\nu^{\nu+1}\int_0^{\infty}\frac{\xi^{2\alpha-\beta}I_{\beta-1}(x\xi)}{\sqrt{\nu^2+\xi^2}\left[\xi+\sqrt{\nu^2+\xi^2}\right]^{\nu}}d\xi\right\} \tag{B.6.2}$$

$\mathrm{Re}\,\alpha>0;\quad \mathrm{Re}\,\beta>0$

$$
{}_2F_1(\alpha, \alpha + \frac{1}{2}; \beta; -x^2) =
$$

$$
\frac{2^{2\alpha-\beta}\Gamma(\alpha)\,x^{1-\alpha}}{\Gamma(2\alpha)} \lim_{\nu\to\infty} \left\{ \nu^{\nu+1} \int_0^\infty \frac{\xi^{2\alpha-\beta} J_{\beta-1}(x\xi)}{\sqrt{\nu^2+\xi^2}\left[\xi + \sqrt{\nu^2+\xi^2}\right]^\nu} d\xi \right\} \tag{B.6.3}
$$

$\mathrm{Re}\,\alpha > 0; \quad \mathrm{Re}\,\beta > 0$

$$
{}_2F_1(\alpha, \alpha + 1; 2\alpha; -x^2) =
$$

$$
\frac{\pi x^{1-2\alpha}}{2\alpha B(\alpha,\alpha)} \lim_{\nu\to\infty} \left\{ \nu^{\nu+1} \int_0^\infty \frac{\xi \left[J_{\alpha-1/2}\left(\frac{x\xi}{2}\right)\right]^2}{\sqrt{\nu^2+\xi^2}\left[\xi + \sqrt{\nu^2+\xi^2}\right]^\nu} d\xi \right\} \tag{B.6.4}
$$

$\mathrm{Re}\,\alpha > -\frac{1}{2}$

$$
{}_2F_1\left(\frac{3}{2} - \mu, \frac{3}{2} - 2\mu; 2 - \mu; -\frac{1}{4x^2}\right) =
$$

$$
\frac{8\,\Gamma(2-\mu)}{\pi\,\Gamma(\frac{3}{2}-\mu)\,\Gamma(\frac{3}{2}-2\mu)} \lim_{\nu\to\infty} \left\{ \nu \int_0^\infty \frac{\xi^{1-2\mu}\left[K_\mu(x\xi)\right]^2 \sin\left[\nu \sin^{-1}\left(\frac{\xi}{\nu}\right)\right]}{\sqrt{\nu^2-\xi^2}} d\xi \right\} \tag{B.6.5}
$$

$\mathrm{Re}\,\mu < \frac{3}{4}$

$$
{}_2F_1\left(\alpha, 2\alpha - \frac{1}{2}; \alpha + \frac{1}{2}; -x^2\right) =
$$

$$
\frac{\pi\,\Gamma(\alpha+\frac{1}{2})\,(2x)^{1-2\alpha}}{\Gamma(\alpha)\,\Gamma(2\alpha-\frac{1}{2})} \lim_{\nu\to\infty} \left\{ \nu^{\nu+1} \int_0^\infty \frac{\xi^{2\alpha-1}\left[J_{\alpha-1/2}\left(\frac{x\xi}{2}\right)\right]^2}{\sqrt{\nu^2+\xi^2}\left[\xi + \sqrt{\nu^2+\xi^2}\right]^\nu} d\xi \right\} \tag{B.6.6}
$$

$\mathrm{Re}\,\alpha > \frac{1}{2}$

$$
{}_1F_1(\lambda+\mu; 2\lambda+1; -x) =
$$

$$
\frac{\Gamma(2\lambda+1)\,e^{x/2}\sqrt{x}}{\Gamma(\lambda+\mu)\,x^\lambda} \lim_{\nu\to\infty} \left\{ \nu^{\nu+1} \int_0^\infty \frac{\xi^{\mu-1} J_{2\lambda}(2\sqrt{x\xi})}{(x-\xi)^{\kappa-\mu+\frac{1}{2}}\sqrt{\nu^2+\xi^2}\left[\xi + \sqrt{\nu^2+\xi^2}\right]^\nu} d\xi \right\}
$$

$\mathrm{Re}(\lambda+\mu) > 0$

(B.6.7)

$$
{}_1F_1\left(\frac{1}{2}-\mu, \frac{3}{2}-2\mu; 2-\mu; -\frac{1}{4x^2}\right) =
$$

$$
\frac{2^{2\mu-3/2}\,\Gamma(1-\mu)\,e^{1/8x}}{\Gamma(\mu+\frac{1}{2})\,\sqrt{x}} \lim_{\nu\to\infty}\left\{\nu \int_0^{\infty} \frac{\xi^{2\mu}\,e^{-x\xi^2} I_\mu(x\xi^2)\sin\left[\nu\sin^{-1}\left(\frac{\xi}{\nu}\right)\right]}{\sqrt{\nu^2-\xi^2}}\,d\xi\right\} \tag{B.6.8}
$$

$$
\mathrm{Re}|\mu| < \frac{1}{2}
$$

B.7 Complete Elliptic Integrals

$$
E(k) = \frac{\pi}{2} - \frac{\pi k}{2} \lim_{\nu\to\infty}\left\{\nu^{\nu+1} \int_0^{\infty} \frac{I_0(k\xi) I_1(k\xi)}{\xi\sqrt{\nu^2+\xi^2}\left[\xi+\sqrt{\nu^2+\xi^2}\right]^{\nu}}\,d\xi\right\} \tag{B.7.1}
$$

$$
E(k) = \frac{\pi}{2} \lim_{\nu\to\infty}\left\{\nu^{\nu+1} \int_0^{\infty} \frac{\left[I_0\left(\frac{k\xi}{2}\right)\right]^2}{\sqrt{\nu^2+\xi^2}\left[\xi+\sqrt{\nu^2+\xi^2}\right]^{\nu}}\,d\xi\right\} \tag{B.7.2}
$$

$$
E(2k\sqrt{1+k}) = \sqrt{\pi(1-k)} \lim_{\nu\to\infty}\left\{\nu^{\nu+1} \int_0^{\infty} \frac{\sqrt{\xi}\,J_0(k\xi)}{\sqrt{\nu^2+\xi^2}\left[\xi+\sqrt{\nu^2+\xi^2}\right]^{\nu}}\,d\xi\right\} \tag{B.7.3}
$$

$$
E(k) = \frac{\pi(1-k^2)}{2} \lim_{\nu\to\infty}\left\{\nu^{\nu+1} \int_0^{\infty} \frac{I_0\left(\frac{k\xi}{2}\right)\left[I_0\left(\frac{k\xi}{2}\right)+k\xi I_1\left(\frac{k\xi}{2}\right)\right]}{\sqrt{\nu^2+\xi^2}\left[\xi+\sqrt{\nu^2+\xi^2}\right]^{\nu}}\,d\xi\right\} \tag{B.7.4}
$$

$$
K(\sqrt{1-k^2}) = \lim_{\nu\to\infty}\left\{\nu^{\nu+1} \int_0^{\infty} \frac{I_0\left(\frac{k\xi}{2}\right) K_0\left(\frac{k\xi}{2}\right)}{\sqrt{\nu^2+\xi^2}\left[\xi+\sqrt{\nu^2+\xi^2}\right]^{\nu}}\,d\xi\right\} \tag{B.7.5}
$$

$$
K\left(\frac{k}{\sqrt{1+k^2}}\right) = \frac{\pi\sqrt{1+k^2}}{2} \lim_{\nu\to\infty}\left\{\nu^{\nu+1} \int_0^{\infty} \frac{\left[I_0\left(\frac{k\xi}{2}\right)\right]^2}{\sqrt{\nu^2+\xi^2}\left[\xi+\sqrt{\nu^2+\xi^2}\right]^{\nu}}\,d\xi\right\} \tag{B.7.6}
$$

$$
K\left(\frac{1}{\sqrt{1+k^2}}\right) = -\frac{\pi\sqrt{1+k^2}}{2} \lim_{\nu\to\infty}\left\{\nu^{\nu+1} \int_0^{\infty} \frac{J_0\left(\frac{k\xi}{2}\right) Y_0\left(\frac{k\xi}{2}\right)}{\sqrt{\nu^2+\xi^2}\left[\xi+\sqrt{\nu^2+\xi^2}\right]^{\nu}}\,d\xi\right\} \tag{B.7.7}
$$

$$E\left(\frac{k}{\sqrt{1+k^2}}\right) = \frac{\pi\sqrt{1+k^2}}{2} \lim_{\nu\to\infty}\left\{\nu^{\nu+1}\int_0^\infty \frac{\left[J_0\left(\frac{k\xi}{2}\right)\right]^2}{\sqrt{\nu^2+\xi^2}\left[\xi+\sqrt{\nu^2+\xi^2}\right]^\nu}\,d\xi\right\} \tag{B.7.8}$$

$$K(2k\sqrt{1+k}) = \frac{\sqrt{\pi(1+k)}}{2} \lim_{\nu\to\infty}\left\{\nu^{\nu+1}\int_0^\infty \frac{I_0(k\xi)}{\xi\sqrt{\nu^2+\xi^2}\left[\xi+\sqrt{\nu^2+\xi^2}\right]^\nu}\,d\xi\right\}$$

$$E(k)-K(k) = \frac{\pi k^2}{4} \lim_{\nu\to\infty}\left\{\nu^{\nu+1}\int_0^\infty \frac{\left[I_0\left(\frac{k\xi}{2}\right)\right]^2+\left[I_1\left(\frac{k\xi}{2}\right)\right]^2}{\sqrt{\nu^2+\xi^2}\left[\xi+\sqrt{\nu^2+\xi^2}\right]^\nu}\,d\xi\right\} \tag{B.7.9}$$

$$E(k)-(1-k^2)\,K(k) =$$

$$\frac{\pi k(1-k^2)}{2} \lim_{\nu\to\infty}\left\{\nu^{\nu+1}\int_0^\infty \frac{\xi\, I_0\left(\frac{k\xi}{2}\right) I_1\left(\frac{k\xi}{2}\right)}{\sqrt{\nu^2+\xi^2}\left[\xi+\sqrt{\nu^2+\xi^2}\right]^\nu}\,d\xi\right\} \tag{B.7.10}$$

$$(2-k^2)K(k)-2E(k) = \frac{\pi k^2}{2} \lim_{\nu\to\infty}\left\{\nu^{\nu+1}\int_0^\infty \frac{\left[I_1\left(\frac{k\xi}{2}\right)\right]^2}{\sqrt{\nu^2+\xi^2}\left[\xi+\sqrt{\nu^2+\xi^2}\right]^\nu}\,d\xi\right\} \tag{B.7.11}$$

$$\frac{1}{\sqrt{1+k^2}}K\left(\frac{k}{\sqrt{1+k^2}}\right) - E\left(\frac{k}{\sqrt{1+k^2}}\right) =$$

$$\frac{\pi k}{2} \lim_{\nu\to\infty}\left\{\nu^{\nu+1}\int_0^\infty \frac{\xi\, J_0\left(\frac{k\xi}{2}\right) J_1\left(\frac{k\xi}{2}\right)}{\sqrt{\nu^2+\xi^2}\left[\xi+\sqrt{\nu^2+\xi^2}\right]^\nu}\,d\xi\right\} \tag{B.7.12}$$

B.8 Parabolic Cylinder and Whittaker Functions

$$D_\mu(x) = \frac{e^{-x^2/4}}{\Gamma\left(\frac{1-\mu}{2}\right)} \lim_{\nu\to\infty}\left\{\nu^{\nu+1}\int_0^\infty \frac{\xi^{-(\mu+1)/2}(x^2+2\xi)^{\mu/2}}{\sqrt{\nu^2+\xi^2}\left[\xi+\sqrt{\nu^2+\xi^2}\right]^\nu}\,d\xi\right\} \tag{B.8.1}$$

$\mathrm{Re}\,\mu < 1$

$$D_\mu(x) = \frac{2^{\mu/2} e^{x^2/4}}{\sqrt{\pi}} \lim_{\nu\to\infty} \left\{ \nu^{\nu+1} \int_0^\infty \frac{\xi^{(\mu-1)/2} \cos\left(\frac{\pi\mu}{2} - x\sqrt{2\xi}\right)}{\sqrt{\nu^2+\xi^2}\left[\xi + \sqrt{\nu^2+\xi^2}\right]^\nu} d\xi \right\} \tag{B.8.2}$$

$\operatorname{Re}\mu > -1$

$$D_{-\mu}(x) = \frac{e^{-x^2/4}}{\Gamma(\mu)\, x^\mu} \lim_{\nu\to\infty} \left\{ \nu^{\nu+1} \int_0^\infty \frac{\xi^{\mu-1} e^{-\xi^2/2x^2}}{\sqrt{\nu^2+\xi^2}\left[\xi + \sqrt{\nu^2+\xi^2}\right]^\nu} d\xi \right\}; \quad \operatorname{Re}\mu > 0 \tag{B.8.3}$$

$$D_{2\mu}(x) = \frac{2^{\mu-1} e^{-x^2}}{\Gamma(2\mu)} \lim_{\nu\to\infty} \left\{ \nu^{\nu+1} \int_0^\infty \frac{\xi^{\mu-1} e^{-x^2\sqrt{\xi}}}{\sqrt{\nu^2+\xi^2}\left[\xi + \sqrt{\nu^2+\xi^2}\right]^\nu} d\xi \right\} \tag{B.8.4}$$

$\operatorname{Re}\mu > 0$

$$D_{-2\mu}(\sqrt{x}) =$$

$$\frac{x\, e^{-x^2/2}}{2^\mu\, \Gamma(\mu)} \lim_{\nu\to\infty} \left\{ \nu^{\nu+1} \int_x^\infty \frac{\xi^{\mu-1}}{(\xi+x)^{\mu+1/2}\sqrt{\nu^2+\xi^2}\left[\xi + \sqrt{\nu^2+\xi^2}\right]^\nu} d\xi \right\} \tag{B.8.5}$$

$\operatorname{Re}\mu > 0$

$$D_{-2\mu}(2\sqrt{x}) =$$

$$\frac{e^{-x/}\sqrt{x}}{2^{\mu-1}\,\Gamma(\mu)} \lim_{\nu\to\infty} \left\{ \nu^{\nu+1} \int_0^\infty \frac{\xi^{\mu-1}}{(\xi+x)^{\mu+1/2}\sqrt{\nu^2+\xi^2}\left[\xi + \sqrt{\nu^2+\xi^2}\right]^\nu} d\xi \right\} \tag{B.8.6}$$

$\operatorname{Re}\mu > 0$

$$D_{-2\mu}(2\sqrt{x}) =$$

$$\frac{\sqrt{x}}{2^{\mu-1/2}\,\Gamma(\mu)} \lim_{\nu\to\infty} \left\{ \nu^{\nu+1} \int_x^\infty \frac{(\xi-x)^{\mu-1}}{(\xi+x)^{\mu+1/2}\sqrt{\nu^2+\xi^2}\left[\xi + \sqrt{\nu^2+\xi^2}\right]^\nu} d\xi \right\} \tag{B.8.7}$$

$\operatorname{Re}\mu > 0$

$$D_{-2\mu}(\sqrt{x}) =$$

$$\frac{2^{\mu-1}e^{x/4}}{\Gamma(2\mu)} \lim_{\nu\to\infty}\left\{\nu^{\nu+1}\int_0^\infty \frac{\xi^{\mu-1}e^{-\sqrt{2x\xi}}}{\sqrt{\nu^2+\xi^2}\left[\xi+\sqrt{\nu^2+\xi^2}\right]^\nu}\,d\xi\right\} \tag{B.8.8}$$

$\operatorname{Re}\mu > 0$

$$D_{-2\mu}(x) = \frac{2^{\mu+1}e^{-x^2}}{\Gamma(2\mu)} \lim_{\nu\to\infty}\left\{\nu^{\nu+1}\int_0^\infty \frac{\gamma\left(\mu,\frac{\xi^2}{8x^2}\right)}{\sqrt{\nu^2+\xi^2}\left[\xi+\sqrt{\nu^2+\xi^2}\right]^\nu}\,d\xi\right\} \tag{B.8.9}$$

$x > -\frac{1}{2}$

$$D_{2\mu-1}(\sqrt{x}) = \frac{2^{\mu+1/2}e^{x/4}}{\sqrt{\pi}} \lim_{\nu\to\infty}\left\{\nu^{\nu+1}\int_0^\infty \frac{\xi^{\mu-1}\sin(\pi\mu-\sqrt{2x\xi})}{\sqrt{\nu^2+\xi^2}\left[\xi+\sqrt{\nu^2+\xi^2}\right]^\nu}\,d\xi\right\} \tag{B.8.10}$$

$\operatorname{Re}\mu > 0$

$$D_{1-2\mu}(2\sqrt{x}) =$$

$$\frac{1}{2^{\mu-1/2}\Gamma(\mu)} \lim_{\nu\to\infty}\left\{\nu^{\nu+1}\int_x^\infty \frac{(\xi-x)^{\mu-1}}{(\xi+x)^{\mu-1/2}\sqrt{\nu^2+\xi^2}\left[\xi+\sqrt{\nu^2+\xi^2}\right]^\nu}\,d\xi\right\} \tag{B.8.11}$$

$\operatorname{Re}\mu > 0$

$$D_{1-2\mu}(2\sqrt{x}) =$$

$$\frac{e^{-x/2}}{2^{\mu-1}\Gamma(\mu)} \lim_{\nu\to\infty}\left\{\nu^{\nu+1}\int_0^\infty \frac{\xi^{\mu-1}}{(\xi+x)^{\mu-1/2}\sqrt{\nu^2+\xi^2}\left[\xi+\sqrt{\nu^2+\xi^2}\right]^\nu}\,d\xi\right\} \tag{B.8.12}$$

$\operatorname{Re}\mu > 0$

$$M_{\kappa,\mu}(x) = \frac{x^{1-\mu} e^{x/2}}{B\left(\kappa+\mu+\frac{1}{2}, \kappa-\mu+\frac{1}{2}\right)} I$$

$$I = \lim_{\nu\to\infty} \left\{ \nu^{\nu+1} \int_0^\infty \frac{\xi^{\kappa+\mu+\frac{1}{2}}}{(x-\xi)^{\kappa-\mu+\frac{1}{2}} \sqrt{\nu^2+\xi^2}\left[\xi+\sqrt{\nu^2+\xi^2}\right]^\nu} d\xi \right\} \tag{B.8.13}$$

$$\mu \pm \kappa > -\frac{1}{2}; \quad x > 0$$

$$M_{\kappa,\mu}(x) =$$

$$\frac{\Gamma(2\mu+1)\sqrt{x}\,e^{-x/2}}{\Gamma(\mu-\kappa+\frac{1}{2})} \lim_{\nu\to\infty} \left\{ \nu^{\nu+1} \int_0^\infty \frac{I_{2\mu}(2\sqrt{x\xi})}{\xi^{\kappa+\frac{1}{2}} \sqrt{\nu^2+\xi^2}\left[\xi+\sqrt{\nu^2+\xi^2}\right]^\nu} d\xi \right\} \tag{B.8.14}$$

$$\mathrm{Re}(\kappa-\mu) < \frac{1}{2}$$

$$M_{\kappa,\mu}(x) =$$

$$\frac{\Gamma(2\mu+1)\,e^{x/2}\sqrt{x}}{\Gamma\left(\mu+\kappa+\frac{1}{2}\right)} \lim_{\nu\to\infty} \left\{ \nu^{\nu+1} \int_0^\infty \frac{\xi^{\kappa-\frac{1}{2}} J_\mu(2\sqrt{x\xi})}{(x-\xi)^{\kappa-\mu+\frac{1}{2}} \sqrt{\nu^2+\xi^2}\left[\xi+\sqrt{\nu^2+\xi^2}\right]^\nu} d\xi \right\}$$

$$\mathrm{Re}(\mu+\kappa) > -\frac{1}{2}$$

(B.8.15)

$$M_{(\mu,-1)/2,\mu/2}(x) = \frac{e^{-(\mu-1)/2}}{\mu} \lim_{\nu\to\infty} \left\{ \nu^{\nu+1} \int_0^\infty \frac{e^{xe^{-\xi}}(1-e^{-\xi})^{\mu-1}}{\sqrt{\nu^2+\xi^2}\left[\xi+\sqrt{\nu^2+\xi^2}\right]^\nu} d\xi \right\} \tag{B.8.16}$$

$$x > 0; \quad \mathrm{Re}\,\mu > 0$$

$$W_{\kappa,\mu}(x) = \frac{e^{-\alpha/2} x^{1/2-\mu}}{\Gamma\left(\mu-\kappa+\frac{1}{2}\right)} \lim_{\nu\to\infty} \left\{ \nu^{\nu+1} \int_0^\infty \frac{\xi^{\mu-\kappa-\frac{1}{2}}(x+\xi)^{\mu+\kappa-\frac{1}{2}}}{\sqrt{\nu^2+\xi^2}\left[\xi+\sqrt{\nu^2+\xi^2}\right]^\nu} d\xi \right\} \tag{B.8.17}$$

$$\mathrm{Re}\left(\mu-\kappa+\frac{1}{2}\right) > 0$$

$$W_{-\mu,\,\lambda}(x) =$$

$$\frac{2\sqrt{x}}{\Gamma(\lambda+\mu+\frac{1}{2})\,\Gamma(\mu-\lambda+\frac{1}{2})}\lim_{\nu\to\infty}\left\{\nu^{\nu+1}\int_0^\infty \frac{\xi^{\mu-1/2}K_{2\lambda}(2\sqrt{x\xi})}{\sqrt{\nu^2+\xi^2}\left[\xi+\sqrt{\nu^2+\xi^2}\right]^\nu}\,d\xi\right\} \tag{B.8.18}$$

$\mathrm{Re}\,\lambda > -1$

$$W_{\mu,\,\kappa+1/2}(x) = \frac{e^{-x/2}}{x}\lim_{\nu\to\infty}\left\{\nu^{\nu+1}\int_0^\infty \frac{\left(1+\frac{\xi}{x}\right)^{\mu/2}P_\kappa^\mu\left(1+\frac{2\xi}{x}\right)}{\sqrt{\nu^2+\xi^2}\left[\xi+\sqrt{\nu^2+\xi^2}\right]^\nu}\,d\xi\right\} \tag{B.8.19}$$

$\mathrm{Re}\,\mu < 1$

$$W_{\mu,\,\kappa+1/2}(2x) = \lim_{\nu\to\infty}\left\{\nu^{\nu+1}\int_x^\infty \frac{\left(\frac{\xi+x}{\xi-x}\right)^{\mu/2}P_\kappa^\mu\left(\frac{\xi}{x}\right)}{\sqrt{\nu^2+\xi^2}\left[\xi+\sqrt{\nu^2+\xi^2}\right]^\nu}\,d\xi\right\} \tag{B.8.20}$$

$\mathrm{Re}\,\mu > 1; \quad \kappa \neq 0, \pm 1, \pm 2, \pm 3, \ldots$

$$W_{-\mu,\,-1/2}(x) = \frac{e^{x/2}}{\Gamma(\mu)}\lim_{\nu\to\infty}\left\{\nu^{\nu+1}\int_0^\infty \frac{e^{-xe^{-\xi}}(1-e^{-\xi})^{\mu-1}}{\sqrt{\nu^2+\xi^2}\left[\xi+\sqrt{\nu^2+\xi^2}\right]^\nu}\,d\xi\right\} \tag{B.8.21}$$

$x > 0; \quad \mathrm{Re}\,\mu > 0$

B.9 Legendre, Laguerre and Hermite Orthogonal Polynomials

$$P_n(1-x) = \lim_{\nu\to\infty}\left\{\nu^{\nu+1}\int_0^\infty \frac{{}_2F_2\left(-n, n+1; 1, 1; \frac{x\xi}{2}\right)}{\sqrt{\nu^2+\xi^2}\left[\xi+\sqrt{\nu^2+\xi^2}\right]^\nu}\,d\xi\right\} \tag{B.9.1}$$

$$P_n\left(\frac{1}{\sqrt{1+x^2}}\right) = \frac{(1+x^2)^{(n-1)/2}}{n!}\lim_{\nu\to\infty}\left\{\nu^{\nu+1}\int_0^\infty \frac{\xi^n J_0(x\xi)}{\sqrt{\nu^2+\xi^2}\left[\xi+\sqrt{\nu^2+\xi^2}\right]^\nu}\,d\xi\right\} \tag{B.9.2}$$

$$P_n\left(\frac{1}{\sqrt{1-x^2}}\right) = \frac{(1-x^2)^{(n+1)/2}}{n!}\lim_{\nu\to\infty}\left\{\nu^{\nu+1}\int_0^\infty \frac{\xi^n I_0(x\xi)}{\sqrt{\nu^2+\xi^2}\left[\xi+\sqrt{\nu^2+\xi^2}\right]^\nu}\,d\xi\right\} \tag{B.9.3}$$

$$L_n^{\alpha}(x) = \frac{1}{n!x^{\alpha/2}} \lim_{\nu\to\infty} \left\{ \nu^{\nu+1} \int_0^{\infty} \frac{\xi^{n+\alpha/2} J_{\alpha}(2\sqrt{x\xi})}{\sqrt{\nu^2+\xi^2}\left[\xi+\sqrt{\nu^2+\xi^2}\right]^{\nu}} d\xi \right\} \tag{B.9.4}$$

$\mathrm{Re}(\alpha+n) > -1$

$$H_n(x) = \frac{2^n e^{x^2}}{\sqrt{\pi}} \lim_{\nu\to\infty} \left\{ \nu^{\nu+1} \int_0^{\infty} \frac{\xi^{(n-1)/2} \cos\left(\frac{\pi n}{2} - 2x\sqrt{\xi}\right)}{\sqrt{\nu^2+\xi^2}\left[\xi+\sqrt{\nu^2+\xi^2}\right]^{\nu}} d\xi \right\} \tag{B.9.5}$$

$n = 0, 1, 2, 3, \ldots$

$$He_{2n}(\sqrt{2x}) = \frac{(-1)^n 2^n e^{x}}{\sqrt{\pi}} \lim_{\nu\to\infty} \left\{ \nu^{\nu+1} \int_0^{\infty} \frac{\xi^{n-1/2} \cos(2\sqrt{x\xi})}{\sqrt{\nu^2+\xi^2}\left[\xi+\sqrt{\nu^2+\xi^2}\right]^{\nu}} d\xi \right\} \tag{B.9.6}$$

$n = 0, 1, 2, 3, \ldots$

$$He_{2n+1}\left(\sqrt{2x}\right) = \frac{(-1)^n 2^{n+1/2} e^{x}}{\sqrt{\pi}} \lim_{\nu\to\infty} \left\{ \nu^{\nu+1} \int_0^{\infty} \frac{\xi^{n-1/2} \sin(2\sqrt{x\xi})}{\sqrt{\nu^2+\xi^2}\left[\xi+\sqrt{\nu^2+\xi^2}\right]^{\nu}} d\xi \right\}$$

$n = 0, 1, 2, 3, \ldots$

(B.9.7)

$$He_{2n}(x) = \frac{(-1)^n 2^{3/2} x^{1-2n} e^{x^2}}{\sqrt{\pi}} \lim_{\nu\to\infty} \left\{ \nu \int_0^{\infty} \frac{\xi^{2n} e^{-\xi^2/4x^2} \cos\left[\nu \sin^{-1}\left(\frac{\xi}{\nu}\right)\right]}{\sqrt{\nu^2-\xi^2}} d\xi \right\}$$

(B.9.8)

B.10 Riemann Zeta Functions

$$\zeta(z-1) = \frac{2^z}{z} \lim_{\nu\to\infty} \left\{ \nu^{\nu+1} \int_0^{\infty} \frac{\tanh\left[\frac{\pi}{2}\sqrt{e^{2\xi/z}-1}\right]}{\sqrt{\nu^2+\xi^2}\left[\xi+\sqrt{\nu^2+\xi^2}\right]^{\nu}} d\xi \right\} \tag{B.10.1}$$

$$\zeta(\mu,x) =$$

$$\frac{1}{\Gamma(\mu)\,(x-1)^{\mu-1}} \lim_{\nu\to\infty} \left\{ \nu^{\nu+1} \int_0^\infty \frac{\xi^{\mu-1}}{\sqrt{\nu^2+\xi^2}\,\left[\xi+\sqrt{\nu^2+\xi^2}\right]^\nu \left(e^{\xi/(x-1)}-1\right)} \, d\xi \right\} \tag{B.10.2}$$

$\operatorname{Re}\mu > 0, \quad x > 1$

$$\zeta(\mu,x) =$$

$$\frac{(2x-1)}{2^{\mu-1}\Gamma(\mu)} \lim_{\nu\to\infty} \left\{ \nu^{\nu+1} \int_0^\infty \frac{\xi^{\mu-1}\operatorname{csch}\left(\frac{\xi}{2x-1}\right)}{(2x-1)^{\mu-1}\sqrt{\nu^2+\xi^2}\,\left[\xi+\sqrt{\nu^2+\xi^2}\right]^\nu} \, d\xi \right\} \tag{B.10.3}$$

$\operatorname{Re}\mu > 1$

$$\zeta(\mu,x) =$$

$$\frac{1}{2^{2\mu-1}\Gamma(\mu)\,(x-1)} \lim_{\nu\to\infty} \left\{ \nu^{\nu+1} \int_0^\infty \frac{\xi^{\mu-1}\left[\operatorname{ctnh}\left(\frac{\xi}{2x-1}\right)-1\right]}{(2x-1)^{\mu-1}\sqrt{\nu^2+\xi^2}\,\left[\xi+\sqrt{\nu^2+\xi^2}\right]^\nu} \, d\xi \right\} \tag{B.10.4}$$

$\operatorname{Re}\mu > 1$

$$\zeta\left(\mu, \frac{1+x}{2}\right) = \frac{(2x)^\mu}{2} \lim_{\nu\to\infty} \left\{ \nu^{\nu+1} \int_0^\infty \frac{\xi^{\mu-1}\operatorname{csch}\left(\frac{\xi}{x}\right)}{\sqrt{\nu^2+\xi^2}\,\left[\xi+\sqrt{\nu^2+\xi^2}\right]^\nu} \, d\xi \right\} \tag{B.10.5}$$

$\operatorname{Re}\mu > 1$

B.11 Volterra Functions

$$\nu(x) = e^x - \lim_{\lambda\to\infty} \left\{ \lambda \int_0^\infty \frac{1}{\xi} J_\lambda(\lambda x\xi) \left[e^{x/\xi} - \nu\left(\frac{1}{\xi}\right)\right] d\xi \right\} \tag{B11.1}$$

$$\nu(x) = \frac{1}{2\sqrt{\pi}} \lim_{\nu\to\infty} \left\{ \nu^{\nu+1} \int_0^\infty \frac{\nu(2\sqrt{x\xi})}{\sqrt{\xi}\,\sqrt{\nu^2+\xi^2}\,\left[\xi+\sqrt{\nu^2+\xi^2}\right]^\nu} \, d\xi \right\} \tag{B11.2}$$

$$\nu(x,\alpha) = \frac{1}{2\sqrt{\pi}} \lim_{\nu\to\infty} \left\{ \nu^{\nu+1} \int_0^\infty \frac{\nu(2\sqrt{x\xi}, 2\alpha)}{\sqrt{\xi}\,\sqrt{\nu^2+\xi^2}\,\left[\xi+\sqrt{\nu^2+\xi^2}\right]^\nu} \, d\xi \right\} \tag{B11.3}$$

$$v(x,\alpha) = \frac{1}{2\sqrt{\pi x}} \lim_{\nu\to\infty} \left\{ \nu^{\nu+1} \int_0^\infty \frac{v(2\sqrt{x\xi}, 1+2\alpha)}{\sqrt{\nu^2+\xi^2}\left[\xi+\sqrt{\nu^2+\xi^2}\right]^\nu} d\xi \right\} \tag{B11.4}$$

$$v(x,\alpha) = x^\alpha \lim_{\lambda\to\infty} \left\{ \lambda \int_0^\infty \frac{1}{\xi x^{\alpha/\xi}} J_\lambda(\lambda\xi)\, v(x^{1/\xi},\alpha)\, d\xi \right\}; \quad \operatorname{Re}\alpha > -1 \tag{B11.5}$$

$$v(e^{-x},\alpha) = e^{-\alpha x} \lim_{\lambda\to\infty} \left\{ \lambda \int_0^\infty \frac{1}{\xi} J_\lambda(\lambda\,\xi)\, e^{\alpha x/\xi}\, v(e^{-x/\xi},\alpha)\, d\xi \right\} \tag{B11.6}$$

$$v(\alpha,x) = \lim_{\lambda\to\infty} \left\{ \lambda \int_0^\infty \alpha^{x(1-\xi)} J_\lambda(\lambda\xi)\, v(\alpha, x\xi)\, d\xi \right\}; \quad x > -1 \tag{B11.7}$$

$$\mu(x,\alpha) = \frac{1}{2^{\alpha+1}\sqrt{\pi}} \lim_{\nu\to\infty} \left\{ \nu^{\nu+1} \int_0^\infty \frac{\mu(2\sqrt{x\xi},\alpha)}{\sqrt{\xi}\sqrt{\nu^2+\xi^2}\left[\xi+\sqrt{\nu^2+\xi^2}\right]^\nu} d\xi \right\} \tag{B11.8}$$

B.12 Bessel and Related Functions

$$J_0(x) = \frac{4}{\pi x} \lim_{\nu\to\infty} \left\{ \nu \int_0^\infty \frac{\sin\left[\nu \sin^{-1}\left(\frac{\xi}{\nu}\right)\right] \sin\left(\frac{x^2}{4\xi}\right)}{\sqrt{\nu^2-\xi^2}} d\xi \right\} \tag{B.12.1}$$

$$Y_0(x) = -\frac{2}{\pi} K_0(x) + \frac{2}{\pi} \lim_{\nu\to\infty} \left\{ \nu \int_0^\infty \frac{\sin\left[\nu \sin^{-1}\left(\frac{\xi}{\nu}\right)\right] \sin\left(\frac{x^2}{4\xi}\right)}{\xi\sqrt{\nu^2-\xi^2}} d\xi \right\} \tag{B.12.2}$$

$$J_1(x) = \frac{x}{\pi} \lim_{\nu\to\infty} \left\{ \nu \int_0^\infty \frac{\sin\left[\nu \sin^{-1}\left(\frac{\xi}{\nu}\right)\right] \sin\left(\frac{x^2}{4\xi}\right)}{\xi^2\sqrt{\nu^2-\xi^2}} d\xi \right\} \tag{B.12.3}$$

$$\sin x\, J_0(x) - \cos x\, Y_0(x) =$$

$$\frac{\sqrt{2}}{\pi} \lim_{\nu\to\infty} \left\{ \nu^{\nu+1} \int_0^\infty \frac{\sqrt{\xi+\sqrt{\xi^2+4x^2}}}{\sqrt{\xi(\xi^2+4x^2)}\sqrt{\nu^2+\xi^2}\left[\xi+\sqrt{\nu^2+\xi^2}\right]^\nu} d\xi \right\} \tag{B.12.4}$$

$$\cos x J_0(x) + \sin x Y_0(x) = \frac{\sqrt{2}}{\pi} \lim_{\nu\to\infty} \left\{ \nu^{\nu+1} \int_0^\infty \frac{\sqrt{\sqrt{\xi^2+4x^2}-\xi}}{\sqrt{\xi\,(\xi^2+4x^2)}\,\sqrt{\nu^2+\xi^2}\left[\xi+\sqrt{\nu^2+\xi^2}\right]^\nu}\,d\xi \right\} \tag{B.12.5}$$

$$\sin x J_1(x) - \cos x Y_1(x) = \frac{1}{\pi\sqrt{2}x} \lim_{\nu\to\infty} \left\{ \nu^{\nu+1} \int_0^\infty \frac{\left(\xi+\sqrt{\xi^2+4x^2}\right)^{3/2}}{\sqrt{\xi\,(\xi^2+4x^2)}\,\sqrt{\nu^2+\xi^2}\left[\xi+\sqrt{\nu^2+\xi^2}\right]^\nu}\,d\xi \right\} \tag{B.12.6}$$

$$\cos x J_1(x) + \sin x Y_1(x) = -\frac{1}{\pi\sqrt{2}x} \lim_{\nu\to\infty} \left\{ \nu^{\nu+1} \int_0^\infty \frac{\left(\sqrt{\xi^2+4x^2}-\xi\right)^{3/2}}{\sqrt{\xi\,(\xi^2+4x^2)}\,\sqrt{\nu^2+\xi^2}\left[\xi+\sqrt{\nu^2+\xi^2}\right]^\nu}\,d\xi \right\} \tag{B.12.7}$$

$$J_0^2(x) + Y_0^2(x) = \frac{8}{\pi^2} \lim_{\nu\to\infty} \left\{ \nu^{\nu+1} \int_0^\infty \frac{K\left[\xi\sqrt{\xi^2+4x^2}\right]}{\sqrt{\xi^2+4x^2}\,\sqrt{\nu^2+\xi^2}\left[\xi+\sqrt{\nu^2+\xi^2}\right]^\nu}\,d\xi \right\} \tag{B.12.8}$$

$$J_\mu(x) = \frac{1}{2^{\mu-1}\sqrt{\pi}\,\Gamma\left(\mu+\frac{1}{2}\right)x^\mu} \lim_{\nu\to\infty} \left\{ \nu \int_0^x \frac{\cos\left[\nu\sin^{-1}\left(\frac{\xi}{\nu}\right)\right](x^2-\xi^2)^{\mu-1/2}}{\sqrt{\nu^2-\xi^2}}\,d\xi \right\}$$

$$\operatorname{Re}\mu > -\frac{1}{2} \tag{B.12.9}$$

$$J_\mu(x) = \frac{1}{\sqrt{\pi}\,\Gamma\left(\mu+\frac{1}{2}\right)(2x)^\mu\cos x}\, I$$

$$I = \lim_{\nu\to\infty} \left\{ \nu \int_0^{2x} \frac{\cos\left[\nu\sin^{-1}\left(\frac{\xi}{\nu}\right)\right](2x\xi-\xi^2)^{\mu-1/2}}{\sqrt{\nu^2-\xi^2}}\,d\xi \right\} \tag{B.12.10}$$

$$\operatorname{Re}\mu > -\frac{1}{2}$$

$$J_\mu(x) = \frac{1}{2^{\mu-1}\sqrt{\pi}\,\Gamma\left(\mu+\frac{1}{2}\right)} I$$

$$I = \lim_{\nu\to\infty}\left\{\nu\int_0^x \frac{\sin\left[\nu\sin^{-1}\left(\frac{\xi}{\nu}\right)\right](x^2-\xi^2)^{\mu-1/2}}{\sqrt{\nu^2-\xi^2}}\,d\xi\right\} \tag{B.12.11}$$

$$\operatorname{Re}\mu > -\frac{1}{2}$$

$$J_\mu(x) = \frac{1}{\sqrt{\pi}\,\Gamma\left(\mu+\frac{1}{2}\right)(2x)^\mu \sin x} I$$

$$I = \lim_{\nu\to\infty}\left\{\nu\int_0^x \frac{\sin\left[\nu\sin^{-1}\left(\frac{\xi}{\nu}\right)\right](2x\xi-\xi^2)^{\mu-1/2}}{\sqrt{\nu^2-\xi^2}}\,d\xi\right\} \tag{B.12.12}$$

$$\operatorname{Re}\mu > -\frac{1}{2}$$

$$J_{\mu+1}(x) = \frac{1}{2^{\mu-1}\sqrt{\pi}\,\Gamma\left(\mu+\frac{1}{2}\right)x^{\mu+1}} I$$

$$I = \lim_{\nu\to\infty}\left\{\nu\int_0^x \frac{\sin\left[\nu\sin^{-1}\left(\frac{\xi}{\nu}\right)\right]\xi\,(x^2-\xi^2)^{\mu-1/2}}{\sqrt{\nu^2-\xi^2}}\,d\xi\right\} \tag{B.12.13}$$

$$\operatorname{Re}\mu > -\frac{1}{2}$$

$$J_{-\mu}(x) = \frac{1}{2^{\mu-1}\sqrt{\pi}\,\Gamma\left(\mu+\frac{1}{2}\right)x^\mu}\lim_{\nu\to\infty}\left\{\nu\int_x^\infty \frac{\sin\left[\nu\sin^{-1}\left(\frac{\xi}{\nu}\right)\right](\xi^2-x^2)^{\mu-1/2}}{\sqrt{\nu^2-\xi^2}}\,d\xi\right\}$$

$$|\operatorname{Re}\mu| < \frac{1}{2} \tag{B.12.14}$$

$$Y_\mu(x) = -\frac{2^{\mu+1}x^\mu}{\Gamma\left(\frac{1}{2}-\mu\right)\sqrt{\pi}}\lim_{\nu\to\infty}\left\{\nu\int_x^\infty \frac{\cos\left[\nu\sin^{-1}\left(\frac{\xi}{\nu}\right)\right]}{(\xi^2-x^2)^{\mu+1/2}\sqrt{\nu^2-\xi^2}}\,d\xi\right\} \tag{B.12.15}$$

$$|\operatorname{Re}\mu| < \frac{1}{2}$$

$$Y_{-(\mu+1)}(x) =$$

$$\frac{1}{2^{\mu-1}\sqrt{\pi}\,\Gamma(\mu+\frac{1}{2})\,x^{\mu+1}} \lim_{\nu\to\infty}\left\{\nu\int_x^\infty \frac{\sin\left[\nu\sin^{-1}\left(\frac{\xi}{\nu}\right)\right]\,\xi\,(\xi^2-x^2)^{\mu-1/2}}{\sqrt{\nu^2-\xi^2}}\,d\xi\right\}$$

$$-\frac{1}{2} < \operatorname{Re}\mu < 0$$

(B.12.16)

$$J_\mu^2(x) + Y_\mu^2(x) = \frac{2}{\pi}\lim_{\nu\to\infty}\left\{\nu^{\nu+1}\int_0^\infty \frac{e^{-x^2/2\xi}K_\mu\left(\frac{x^2}{2\xi}\right)}{\xi\sqrt{\nu^2+\xi^2}\left[\xi+\sqrt{\nu^2+\xi^2}\right]^\nu}\,d\xi\right\} \tag{B.12.17}$$

$$I_{-5/4}(x) = \frac{1}{2^{1/4}\pi}\lim_{\nu\to\infty}\left\{\nu^{\nu+1}\int_0^\infty \frac{\left[\left(\frac{3}{8\xi}-1\right)\frac{\cosh\left(\sqrt{8x\xi}\right)}{\xi^{3/4}} - \frac{3\sinh\left(\sqrt{8x\xi}\right)}{\xi^{5/4}}\right]}{\sqrt{\nu^2+\xi^2}\left[\xi+\sqrt{\nu^2+\xi^2}\right]^\nu}\,d\xi\right\}$$

(B.12.18)

$$I_{-3/4}(x) = \frac{e^{-x}}{2^{7/4}\pi\sqrt{x}}\lim_{\nu\to\infty}\left\{\nu^{\nu+1}\int_0^\infty \frac{\left[2^{3/2}\sqrt{x}\sinh\left(\sqrt{8x\xi}\right)-\cosh\left(\sqrt{8x\xi}\right)\right]}{\xi^{3/4}\sqrt{\nu^2+\xi^2}\left[\xi+\sqrt{\nu^2+\xi^2}\right]^\nu}\,d\xi\right\}$$

(B.12.19)

$$I_{-1/4}(x) = \frac{e^{-x}}{\pi\,(2x)^{1/4}}\lim_{\nu\to\infty}\left\{\nu^{\nu+1}\int_0^\infty \frac{\cosh\left(\sqrt{8x\xi}\right)}{\xi^{3/4}\sqrt{\nu^2+\xi^2}\left[\xi+\sqrt{\nu^2+\xi^2}\right]^\nu}\,d\xi\right\} \tag{B.12.20}$$

$$I_{-1/4}(x) = \frac{e^x}{\pi\sqrt{x}}\lim_{\nu\to\infty}\left\{\nu\int_0^\infty \frac{e^{-1/8x\xi}\cos\left[\nu\sin^{-1}\left(\frac{\xi}{\nu}\right)\right]}{\sqrt{\xi}\sqrt{\nu^2-\xi^2}}\,d\xi\right\} \tag{B.12.21}$$

$$I_0(x) = \frac{\sqrt{2}\,e^{2x}}{\pi}\lim_{\nu\to\infty}\left\{\nu\int_0^\infty \frac{\sin\left[\nu\sin^{-1}\left(\frac{\xi}{\nu}\right)\right]\sqrt{\xi+\sqrt{\xi^2+4x^2}}}{\sqrt{\xi\,(\xi^2+4x^2)}\sqrt{\nu^2-\xi^2}}\,d\xi\right\} \tag{B.12.22}$$

$$I_{1/4}(x) = \frac{e^{-x}}{\pi\,(2x)^{1/4}}\lim_{\nu\to\infty}\left\{\nu^{\nu+1}\int_0^\infty \frac{\sinh\left(\sqrt{8x\xi}\right)}{\xi^{3/4}\sqrt{\nu^2+\xi^2}\left[\xi+\sqrt{\nu^2+\xi^2}\right]^\nu}\,d\xi\right\} \tag{B.12.23}$$

$$I_{3/4}(x) = \frac{e^{-x}}{2^{7/4}\pi\sqrt{x}} \lim_{\nu\to\infty} \left\{ \nu^{\nu+1} \int_0^\infty \frac{\left[2^{3/2}\sqrt{x}\cosh(\sqrt{8x\xi}) - \sinh(\sqrt{8x\xi})\right]}{\xi^{3/4}\sqrt{\nu^2+\xi^2}\left[\xi+\sqrt{\nu^2+\xi^2}\right]^\nu} d\xi \right\}$$

(B.12.24)

$$I_{5/4}(x) = \frac{1}{2^{1/4}\pi} \lim_{\nu\to\infty} \left\{ \nu^{\nu+1} \int_0^\infty \frac{\left[\left(\frac{3}{8\xi}+1\right)\frac{\sinh\left(\sqrt{8x\xi}\right)}{\xi^{3/4}} - \frac{3\cosh\left(\sqrt{8x\xi}\right)}{\xi^{5/4}}\right]}{\sqrt{\nu^2+\xi^2}\left[\xi+\sqrt{\nu^2+\xi^2}\right]^\nu} d\xi \right\} \quad \text{(B.12.25)}$$

$$I_0(x) = \frac{e^x}{\pi} \lim_{\nu\to\infty} \left\{ \nu^{\nu+1} \int_0^{2x} \frac{1}{\sqrt{2x\xi-\xi^2}\sqrt{\nu^2+\xi^2}\left[\xi+\sqrt{\nu^2+\xi^2}\right]^\nu} d\xi \right\} \quad \text{(B.12.26)}$$

$$I_1(x) = \frac{e^x}{\pi x} \lim_{\nu\to\infty} \left\{ \nu^{\nu+1} \int_0^{2x} \frac{(x-\xi)}{\sqrt{2x\xi-\xi^2}\sqrt{\nu^2+\xi^2}\left[\xi+\sqrt{\nu^2+\xi^2}\right]^\nu} d\xi \right\} \quad \text{(B.12.27)}$$

$$I_1(x) = \frac{x}{\pi} \lim_{\nu\to\infty} \left\{ \nu^{\nu+1} \int_0^{x} \frac{\cosh\left(x\sqrt{1-e^{-\xi}}\right)}{\sqrt{e^\xi-1}\sqrt{\nu^2+\xi^2}\left[\xi+\sqrt{\nu^2+\xi^2}\right]^\nu} d\xi \right\} \quad \text{(B.12.28)}$$

$$I_\mu(x) = \frac{e^x}{\sqrt{\pi}\,\Gamma\left(\mu+\frac{1}{2}\right)(2x)^\mu} \lim_{\nu\to\infty} \left\{ \nu^{\nu+1} \int_0^{2x} \frac{(2x\xi-\xi^2)^{\mu-1/2}}{\sqrt{\nu^2+\xi^2}\left[\xi+\sqrt{\nu^2+\xi^2}\right]^\nu} d\xi \right\}$$

(B.12.29)

$$\operatorname{Re}\mu > -\frac{1}{2}$$

$$K_{1/4}(x) = \frac{e^{-x}}{\sqrt{2x}} \lim_{\nu\to\infty} \left\{ \nu^{\nu+1} \int_0^\infty \frac{e^{-\xi^2/8x}}{\sqrt{\xi}\sqrt{\nu^2+\xi^2}\left[\xi+\sqrt{\nu^2+\xi^2}\right]^\nu} d\xi \right\} \quad \text{(B.12.30)}$$

$$K_{1/4}(x) = \frac{e^{-x}}{\sqrt{2x}} \lim_{\nu\to\infty} \left\{ \nu^{\nu+1} \int_x^\infty \frac{e^{-\xi^2/8x}}{\sqrt{\xi}\sqrt{\xi^2-x^2}\sqrt{\nu^2+\xi^2}\left[\xi+\sqrt{\nu^2+\xi^2}\right]^\nu} d\xi \right\} \quad \text{(B.12.31)}$$

$$K_0(x) = e^{-x} \lim_{\nu\to\infty} \left\{ \nu^{\nu+1} \int_0^\infty \frac{1}{\sqrt{\xi^2+2x\xi}\,\sqrt{\nu^2+\xi^2}\,\left[\xi+\sqrt{\nu^2+\xi^2}\right]^\nu}\, d\xi \right\} \tag{B.12.32}$$

$$K_0(x) = \frac{1}{2} \lim_{\nu\to\infty} \left\{ \nu^{\nu+1} \int_0^\infty \frac{e^{-x^2/4\xi}}{\xi\,\sqrt{\nu^2+\xi^2}\,\left[\xi+\sqrt{\nu^2+\xi^2}\right]^\nu}\, d\xi \right\}; \quad x\neq 0 \tag{B.12.33}$$

$$K_0(x) = \frac{\sqrt{2}x}{\sinh x} \lim_{\nu\to\infty} \left\{ \nu \int_0^\infty \frac{\sin\left[\nu \sin^{-1}\left(\frac{\xi}{\nu}\right)\right]}{\sqrt{\xi\,(\xi^2+4x^2)}\,\sqrt{\xi+\sqrt{\xi^2+4x^2}}\,\sqrt{\nu^2-\xi^2}}\, d\xi \right\} \tag{B.12.34}$$

$$K_0(x) = e^{-x} \lim_{\nu\to\infty} \left\{ \nu^{\nu+1} \int_0^\infty \frac{\cosh\left[\mu \cosh^{-1}\left(1+\frac{\xi}{x}\right)\right]}{\sqrt{\nu^2+\xi^2}\,\left[\xi+\sqrt{\nu^2+\xi^2}\right]^\nu}\, d\xi \right\} \tag{B.12.35}$$

$$K_0(x) = x\,e^{-x} \lim_{\nu\to\infty} \left\{ \nu^{\nu+1} \int_0^\infty \frac{\sinh^{-1}\left(\sqrt{\frac{\xi}{x}}\right)}{\sqrt{\nu^2+\xi^2}\,\left[\xi+\sqrt{\nu^2+\xi^2}\right]^\nu}\, d\xi \right\} \tag{B.12.36}$$

$$K_0(x) = e^{-x} \lim_{\nu\to\infty} \left\{ \nu^{\nu+1} \int_0^\infty \frac{\cosh^{-1}\left[1+\left(\frac{\xi}{x}\right)\right]}{\sqrt{\nu^2+\xi^2}\,\left[\xi+\sqrt{\nu^2+\xi^2}\right]^\nu}\, d\xi \right\} \tag{B.12.37}$$

$$K_0(x) = \lim_{\nu\to\infty} \left\{ \nu \int_0^\infty \frac{\cos\left[\nu \sin^{-1}\left(\frac{\xi}{\nu}\right)\right]}{\sqrt{t^2+\xi^2}\,\sqrt{\nu^2-\xi^2}}\, d\xi \right\}; \quad x>0 \tag{B.12.38}$$

$$K_0(x) = \lim_{\nu\to\infty} \left\{ \nu \int_0^\infty \frac{\xi\,\sin\left[\nu \sin^{-1}\left(\frac{\xi}{\nu}\right)\right]}{(\xi^2+x^2)^{3/2}\,\sqrt{\nu^2-\xi^2}}\, d\xi \right\} \tag{B.12.39}$$

$$K_0(x) = \frac{2}{\pi} \lim_{\nu\to\infty} \left\{ \nu \int_0^\infty \frac{\sin\left[\nu \sin^{-1}\left(\frac{\xi}{\nu}\right)\right]\,\sinh^{-1}\left(\frac{\xi}{x}\right)}{\sqrt{\xi^2+x^2}\,\sqrt{\nu^2-\xi^2}}\, d\xi \right\} \tag{B.12.40}$$

$$K_1(x) = e^{-x} \lim_{\nu\to\infty} \left\{ \nu^{\nu+1} \int_0^\infty \frac{1}{\sqrt{\xi^2+2x}\,\sqrt{\nu^2+\xi^2}\,\left[\xi+\sqrt{\nu^2+\xi^2}\right]^\nu}\, d\xi \right\} \tag{B.12.41}$$

$$K_1(x) = \frac{1}{x} \lim_{\nu\to\infty} \left\{ \nu^{\nu+1} \int_x^\infty \frac{\xi}{\sqrt{\xi^2 - x^2}\sqrt{\nu^2+\xi^2}\left[\xi + \sqrt{\nu^2+\xi^2}\right]^\nu} d\xi \right\} \tag{B.12.42}$$

$$K_1(x) = \frac{1}{x} \lim_{\nu\to\infty} \left\{ \nu^{\nu+1} \int_0^\infty \frac{(x+\xi)}{\sqrt{\xi^2 + 2x\xi}\sqrt{\nu^2+\xi^2}\left[\xi + \sqrt{\nu^2+\xi^2}\right]^\nu} d\xi \right\} \tag{B.12.43}$$

$$K_1(x) = \frac{1}{x} \lim_{\nu\to\infty} \left\{ \nu^{\nu+1} \int_0^\infty \frac{e^{-x^2/4\xi}}{\sqrt{\nu^2+\xi^2}\left[\xi + \sqrt{\nu^2+\xi^2}\right]^\nu} d\xi \right\} \tag{B.12.44}$$

$$K_1(x) = \frac{x}{4} \lim_{\nu\to\infty} \left\{ \nu^{\nu+1} \int_0^\infty \frac{e^{-x^2/4\xi}}{\xi^2\sqrt{\nu^2+\xi^2}\left[\xi + \sqrt{\nu^2+\xi^2}\right]^\nu} d\xi \right\} \tag{B.12.45}$$

$$K_1(x) = \frac{1}{x} \lim_{\nu\to\infty} \left\{ \nu^{\nu+1} \int_x^\infty \frac{\sin\left(\sqrt{\xi^2 - x^2}\right)}{\sqrt{\nu^2+\xi^2}\left[\xi + \sqrt{\nu^2+\xi^2}\right]^\nu} d\xi \right\} \tag{B.12.46}$$

$$K_1(x) = \frac{x^2}{2^{3/2}\sinh x} \lim_{\nu\to\infty} \left\{ \nu \int_0^\infty \frac{\cos\left[\nu \sin^{-1}\left(\frac{\xi}{\nu}\right)\right]}{\xi\sqrt{\xi^2+4x^2}\left[\xi + \sqrt{\xi^2+4x^2}\right]^{3/2}\sqrt{\nu^2-\xi^2}} d\xi \right\}$$

$x > 0$

(B.12.47)

$$K_\mu(x) = \frac{2\Gamma\left(\mu+\frac{1}{2}\right) x}{\sqrt{\pi}} \lim_{\nu\to\infty} \left\{ \nu \int_0^\infty \frac{\cos\left[\nu \sin^{-1}\left(\frac{\xi}{\nu}\right)\right]}{(\xi^2+x^2)^{\mu+1/2}\sqrt{\nu^2-\xi^2}} d\xi \right\} \tag{B.12.48}$$

$\operatorname{Re}\mu > -\frac{1}{2}$

$$K_\mu(x) = \frac{\sqrt{\pi}\, e^{-x}}{\Gamma\left(\mu+\frac{1}{2}\right)(2x)^\mu} \lim_{\nu\to\infty} \left\{ \nu^{\nu+1} \int_0^\infty \frac{(\xi^2+2x\xi)^{\mu-1/2}}{\sqrt{\nu^2+\xi^2}\left[\xi + \sqrt{\nu^2+\xi^2}\right]^\nu} d\xi \right\} \tag{B.12.49}$$

$\operatorname{Re}\mu > -\frac{1}{2}$

$$K_\mu(x) = \frac{\sqrt{\pi}\, e^{-x}}{\Gamma(\mu+\frac{1}{2})\,(2x)^\mu} \lim_{\nu\to\infty} \left\{ \nu^{\nu+1} \int_x^\infty \frac{(\xi^2 - x^2)^{\mu-1/2}}{\sqrt{\nu^2+\xi^2}\left[\xi + \sqrt{\nu^2+\xi^2}\right]^\nu} d\xi \right\} \tag{B.12.50}$$

$\mathrm{Re}\,\mu > -\frac{1}{2}$

$$K_\mu(x) = \frac{e^{-x}}{2^{\mu+1}\mu x^\mu} \lim_{\nu\to\infty} \left\{ \nu^{\nu+1} \int_0^\infty \frac{\left[(\sqrt{\xi+2x}+\sqrt{\xi})^{2\mu} - (\sqrt{\xi+2x}-\sqrt{\xi})^{2\mu}\right]}{\sqrt{\nu^2+\xi^2}\left[\xi + \sqrt{\nu^2+\xi^2}\right]^\nu} d\xi \right\} \tag{B.12.51}$$

$$K_\mu(x) = \frac{e^{-x}}{2^{\mu+1}\mu x^\mu} \lim_{\nu\to\infty} \left\{ \nu^{\nu+1} \int_x^\infty \frac{\left[(\sqrt{\xi+x}+\sqrt{\xi-x})^{2\mu} - (\sqrt{\xi+x}-\sqrt{\xi-x})^{2\mu}\right]}{\sqrt{\nu^2+\xi^2}\left[\xi + \sqrt{\nu^2+\xi^2}\right]^\nu} d\xi \right\} \tag{B.12.52}$$

$$K_\mu(x) = \frac{x^\mu}{2^{\mu+1}} \lim_{\nu\to\infty} \left\{ \nu^{\nu+1} \int_0^\infty \frac{e^{-x^2/4\xi}}{\xi^{\mu+1}\sqrt{\nu^2+\xi^2}\left[\xi + \sqrt{\nu^2+\xi^2}\right]^\nu} d\xi \right\} \tag{B.12.53}$$

$$K_\mu(x) = \lim_{\nu\to\infty} \left\{ \nu^{\nu+1} \int_x^\infty \frac{\cosh\left[\mu \cosh^{-1}\left(\frac{\xi}{x}\right)\right]}{\sqrt{\xi^2-x^2}\sqrt{\nu^2+\xi^2}\left[\xi + \sqrt{\nu^2+\xi^2}\right]^\nu} d\xi \right\} \tag{B.12.54}$$

$$K_\mu(x) = \frac{1}{\mu} \lim_{\nu\to\infty} \left\{ \nu^{\nu+1} \int_x^\infty \frac{\sinh\left[\mu \cosh^{-1}\left(\frac{\xi}{x}\right)\right]}{\sqrt{\xi^2-x^2}\sqrt{\nu^2+\xi^2}\left[\xi + \sqrt{\nu^2+\xi^2}\right]^\nu} d\xi \right\} \tag{B.12.55}$$

$$K_\mu(x) = \frac{e^{-x}}{\mu} \lim_{\nu\to\infty} \left\{ \nu^{\nu+1} \int_0^\infty \frac{\sinh\left[\mu \cosh^{-1}\left(1+\frac{\xi}{x}\right)\right]}{\sqrt{\nu^2+\xi^2}\left[\xi + \sqrt{\nu^2+\xi^2}\right]^\nu} d\xi \right\} \tag{B.12.56}$$

$$K_{\mu+1/2}(x) = \frac{\sqrt{\pi}}{\Gamma(\mu+\frac{1}{2})\,(2x)^{\mu+1/2}} \lim_{\nu\to\infty} \left\{ \nu^{\nu+1} \int_0^\infty \frac{(\xi^2-x^2)^\mu}{\sqrt{\nu^2+\xi^2}\left[\xi + \sqrt{\nu^2+\xi^2}\right]^\nu} d\xi \right\} \tag{B.12.57}$$

$$K_{\mu}(x) = \frac{1}{2x^{\mu}\cos\left(\frac{\pi\mu}{2}\right)}\, I$$

$$I = \lim_{\nu\to\infty}\left\{\nu\int_0^x \frac{\cos\left[\nu\sin^{-1}\left(\frac{\xi}{\nu}\right)\right]\left\{\left[\sqrt{x^2+\xi^2}+\xi\right]^{\mu}+\left[\sqrt{x^2+\xi^2}-\xi\right]^{\mu}\right\}}{\sqrt{(\nu^2-\xi^2)\,(x^2+\xi^2)}}\, d\xi\right\}$$

$|\mathrm{Re}\,\mu| < 1$

(B.12.58)

$$K_{\mu}(x) = \frac{\Gamma\left(\mu+\frac{1}{2}\right)(2x)^{\mu}}{\sqrt{\pi}} \lim_{\nu\to\infty}\left\{\nu\int_0^{\infty}\frac{\cos\left[\nu\sin^{-1}\left(\frac{\xi}{\nu}\right)\right]}{(\xi^2+x^2)^{\mu+1/2}\sqrt{\nu^2-\xi^2}}\, d\xi\right\} \tag{B.12.59}$$

$x > 0; \quad \mathrm{Re}\,\mu > -\frac{1}{2}$

$$K_{\mu}(x) = \frac{2\Gamma\left(\frac{3}{2}-\mu\right)}{\sqrt{\pi}\,(2x)^{\mu}} \lim_{\nu\to\infty}\left\{\nu\int_0^{\infty}\frac{\sin\left[\nu\sin^{-1}\left(\frac{\xi}{\nu}\right)\right]\,\xi(x^2+\xi^2)^{\mu-1/2}}{\sqrt{\nu^2-\xi^2}}\, d\xi\right\} \tag{B.12.60}$$

$\mathrm{Re}\,\mu > -1; \quad x > 0$

$$\ker(x) = -\frac{1}{2}\lim_{\nu\to\infty}\left\{\nu^{\nu+1}\int_0^{\infty}\frac{Ci\left(\frac{x^2}{4\xi}\right)}{\sqrt{\nu^2+\xi^2}\left[\xi+\sqrt{\nu^2+\xi^2}\right]^{\nu}}\, d\xi\right\} \tag{B.12.61}$$

$$\ker_{\mu}(x) = \frac{x^{\mu}}{2^{\mu+1}}\lim_{\nu\to\infty}\left\{\nu^{\nu+1}\int_0^{\infty}\frac{\xi^{\mu-1}\cos\left(\frac{\pi\mu}{2}+\frac{x^2}{4\xi}\right)}{\sqrt{\nu^2+\xi^2}\left[\xi+\sqrt{\nu^2+\xi^2}\right]^{\nu}}\, d\xi\right\} \tag{B.12.62}$$

$\mathrm{Re}\,\mu > -1$

$$\mathrm{kei}_{\mu}(x) = -\frac{x^{\mu}}{2^{\mu+1}}\lim_{\nu\to\infty}\left\{\nu^{\nu+1}\int_0^{\infty}\frac{\xi^{\mu-1}\sin\left(\frac{\pi\mu}{2}+\frac{x^2}{4\xi}\right)}{\sqrt{\nu^2+\xi^2}\left[\xi+\sqrt{\nu^2+\xi^2}\right]^{\nu}}\, d\xi\right\} \tag{B.12.63}$$

$\mathrm{Re}\,\mu > -1$

$$H_{-1}(x) = Y_{-1} - \frac{x}{4\pi^{3/2}}\lim_{\nu\to\infty}\left\{\nu^{\nu+1}\int_0^{\infty}\frac{e^{x^2/8\xi}\left[K_1\left(\frac{x^2}{8\xi}\right)-K_0\left(\frac{x^2}{8\xi}\right)\right]}{\sqrt{\nu^2+\xi^2}\left[\xi+\sqrt{\nu^2+\xi^2}\right]^{\nu}}\, d\xi\right\} \tag{B.12.64}$$

$$H_{-1/4}(x) = Y_{-1/4}(x) + \frac{1}{\sqrt{\pi x}} \lim_{\nu\to\infty} \left\{ \nu^{\nu+1} \int_0^\infty \frac{\sqrt{\xi}\, J_{1/4}\left(\frac{\xi^2}{4x}\right)}{\sqrt{\nu^2+\xi^2}\left[\xi+\sqrt{\nu^2+\xi^2}\right]^\nu}\, d\xi \right\} \tag{B.12.65}$$

$$H_{1/4}(x) = Y_{1/4}(x) + \frac{1}{\sqrt{\pi x}} \lim_{\nu\to\infty} \left\{ \nu^{\nu+1} \int_0^\infty \frac{\sqrt{\xi}\, J_{-1/4}\left(\frac{\xi^2}{4x}\right)}{\sqrt{\nu^2+\xi^2}\left[\xi+\sqrt{\nu^2+\xi^2}\right]^\nu}\, d\xi \right\} \tag{B.12.66}$$

$$H_0(x) = Y_0(x) + \frac{2}{\pi x} \lim_{\nu\to\infty} \left\{ \nu^{\nu+1} \int_0^\infty \frac{1}{\sqrt{1+\left(\frac{\xi}{x}\right)^2}\sqrt{\nu^2+\xi^2}\left[\xi+\sqrt{\nu^2+\xi^2}\right]^\nu}\, d\xi \right\} \tag{B.12.67}$$

$$H_0(x) = Y_0(x) + \frac{2}{\pi} \lim_{\nu\to\infty} \left\{ \nu^{\nu+1} \int_0^\infty \frac{\ln\left[\frac{\xi}{x}+\sqrt{1+\left(\frac{\xi}{x}\right)^2}\right]}{\sqrt{\nu^2+\xi^2}\left[\xi+\sqrt{\nu^2+\xi^2}\right]^\nu}\, d\xi \right\} \tag{B.12.68}$$

$$H_0(x) = Y_0(x) + \frac{2}{\pi^{3/2}} \lim_{\nu\to\infty} \left\{ \nu^{\nu+1} \int_0^\infty \frac{e^{-x^2/8\xi} K_0\left(\frac{x^2}{8\xi}\right)}{\sqrt{\xi}\sqrt{\nu^2+\xi^2}\left[\xi+\sqrt{\nu^2+\xi^2}\right]^\nu}\, d\xi \right\} \tag{B.12.69}$$

$$H_1(x) = \frac{2}{\pi x} + Y_1(x) + \frac{2x}{\pi} \lim_{\nu\to\infty} \left\{ \nu^{\nu+1} \int_0^\infty \frac{1}{\left[\xi+\sqrt{x^2+\xi^2}\right]\sqrt{\nu^2+\xi^2}\left[\xi+\sqrt{\nu^2+\xi^2}\right]^\nu}\, d\xi \right\} \tag{B.12.70}$$

$$H_1(x) = \frac{2}{\pi} + Y_1(x) + \frac{2}{\pi x} \lim_{\nu\to\infty} \left\{ \nu^{\nu+1} \int_0^\infty \frac{\xi}{\sqrt{\xi^2+x^2}\sqrt{\nu^2+\xi^2}\left[\xi+\sqrt{\nu^2+\xi^2}\right]^\nu}\, d\xi \right\} \tag{B.12.71}$$

$$H_\mu(x) = \frac{2}{\sqrt{\pi}\,\Gamma\left(\mu+\frac{1}{2}\right)(2x)^\mu} \lim_{\nu\to\infty} \left\{ \nu \int_0^x \frac{(x^2-\xi^2)^{\mu-1/2} \sin\left[\nu \sin^{-1}\left(\frac{\xi}{\nu}\right)\right]}{\sqrt{\nu^2-\xi^2}}\, d\xi \right\} \tag{B.12.72}$$

$$\mathrm{H}_{\mu}(x) = Y_{\mu}(x) +$$

$$\frac{1}{2^{\mu-1}\sqrt{\pi}\,\Gamma\left(\mu+\frac{1}{2}\right)x^{\mu}} \lim_{\nu\to\infty}\left\{\nu^{\nu+1}\int_0^{\infty}\frac{\left(x^2+\xi^2\right)^{\mu-1/2}}{\sqrt{\nu^2+\xi^2}\left[\xi+\sqrt{\nu^2+\xi^2}\right]^{\nu}}\,d\xi\right\} \tag{B.12.73}$$

$$\mu > -\frac{1}{2}, \quad x > 0$$

$$\mathrm{H}_{-\mu}(x) = Y_{-\mu}(x) +$$

$$\frac{2^{\mu}\cos(\pi\mu)}{\pi x^{\mu}} \lim_{\nu\to\infty}\left\{\nu^{\nu+1}\int_0^{\infty}\frac{\xi^{\mu-1}e^{x^2/4\xi}\,erfc\left(\frac{x}{2\sqrt{\xi}}\right)}{\sqrt{\nu^2+\xi^2}\left[\xi+\sqrt{\nu^2+\xi^2}\right]^{\nu}}\,d\xi\right\}; \quad \mathrm{Re}\,\mu > -\frac{1}{2} \tag{B.12.74}$$

$$\mathrm{H}_{-\mu}(x) = Y_{-\mu}(x) +$$

$$\frac{2^{\mu+1}\Gamma\left(\mu+\frac{1}{2}\right)x^{\mu}\cos(\pi\mu)}{\sqrt{\pi}} \lim_{\nu\to\infty}\left\{\nu\int_0^{\infty}\frac{e^{-\nu\xi}\nu\sinh\xi}{\left[x^2+(\nu\sinh\xi)^2\right]^{\mu+1/2}}\,d\xi\right\} \tag{B.12.75}$$

$$-\frac{1}{2} < \mathrm{Re}\,\mu < \frac{3}{2}$$

$$L_0(x) = I_0(x) - \frac{2}{\pi}\lim_{\nu\to\infty}\left\{\nu^{\nu+1}\int_0^{\infty}\frac{\sin^{-1}\left(\frac{\xi}{x}\right)}{\sqrt{\nu^2+\xi^2}\left[\xi+\sqrt{\nu^2+\xi^2}\right]^{\nu}}\,d\xi\right\}; \quad x > 0 \tag{B.12.76}$$

$$L_0(x) = I_0(x) - \frac{2}{\pi}\lim_{\nu\to\infty}\left\{\nu\int_0^{\infty}\frac{\sin\left[\nu\sin^{-1}\left(\frac{\xi}{\nu}\right)\right]}{\sqrt{\xi^2+x^2}\sqrt{\nu^2-\xi^2}}\,d\xi\right\} \tag{B.12.77}$$

$$L_0(x) = I_0(x) - \frac{1}{\sqrt{\pi}}\lim_{\nu\to\infty}\left\{\nu\int_0^{\infty}\frac{e^{-x^2/8\xi}\,I_0\left(\frac{x^2}{8\xi}\right)}{\sqrt{\xi}\sqrt{\xi^2+x^2}\sqrt{\nu^2-\xi^2}}\,d\xi\right\} \tag{B.12.78}$$

$$L_0(x) = K_0(x) + \ln x\, I_0(x) -$$

$$\frac{2}{\pi}\lim_{\nu\to\infty}\left\{\nu\int_0^{\infty}\frac{\sin\left[\nu\sin^{-1}\left(\frac{\xi}{\nu}\right)\right]\ln\left[\xi+\sqrt{\xi^2+x^2}\right]}{\sqrt{\xi^2+x^2}\sqrt{\nu^2-\xi^2}}\,d\xi\right\} \tag{B.12.79}$$

$$L_0(x) = I_0(x) - \frac{4}{\pi} \lim_{\nu \to \infty} \left\{ \nu \int_0^\infty \frac{\cos\left[\nu \sin^{-1}\left(\frac{\xi}{\nu}\right)\right] Y_0\left(\sqrt{2x\xi}\right) K_0\left(\sqrt{2x\xi}\right)}{\sqrt{\nu^2 - \xi^2}} d\xi \right\} \tag{B.12.80}$$

$$L_1(x) = -\frac{2}{\pi} + I_1(x) + \frac{2}{\pi x} \lim_{\nu \to \infty} \left\{ \nu^{\nu+1} \int_0^x \frac{\xi}{\sqrt{x^2 - \xi^2}\sqrt{\nu^2 + \xi^2}\left[\xi + \sqrt{\nu^2 + \xi^2}\right]^\nu} d\xi \right\} \tag{B.12.81}$$

$$\mathrm{L}_1(x) = \frac{x}{\pi} \lim_{\nu \to \infty} \left\{ \nu^{\nu+1} \int_0^x \frac{\sinh\left(x\sqrt{1 - e^{-\xi}}\right)}{\sqrt{e^\xi - 1}\sqrt{\nu^2 + \xi^2}\left[\xi + \sqrt{\nu^2 + \xi^2}\right]^\nu} d\xi \right\} \tag{B.12.82}$$

$$L_\mu(x) = I_\mu(x) - \frac{1}{2^{\mu-1}\sqrt{\pi}\,\Gamma\left(\mu + \frac{1}{2}\right) x^\mu} \lim_{\nu \to \infty} \left\{ \nu^{\nu+1} \int_0^x \frac{(x^2 - \xi^2)^{\mu - 1/2}}{\sqrt{\nu^2 + \xi^2}\left[\xi + \sqrt{\nu^2 + \xi^2}\right]^\nu} d\xi \right\} \tag{B.12.83}$$

$$\operatorname{Re}\mu > -\frac{1}{2}, \quad x > 0$$

$$J_\mu(x) = J_\mu(x) + \frac{\sin(\pi\mu)}{\pi} \lim_{\nu \to \infty} \left\{ \nu^{\nu+1} \int_0^\infty \frac{\left(\sqrt{\xi^2 + 1} - \xi\right)^\mu}{\sqrt{\xi^2 + 1}\sqrt{\nu^2 + \xi^2}\left[\xi + \sqrt{\nu^2 + \xi^2}\right]^\lambda} d\xi \right\} \tag{B.12.84}$$

$$J_\mu(x) = J_\mu(x) - \frac{x^\mu \sin(\pi\mu)}{\pi\mu} \lim_{\nu \to \infty} \left\{ \nu^{\nu+1} \int_0^\infty \frac{\left[\left(\xi + \sqrt{x^2 + \xi^2}\right)^{-\mu} - x^{-\mu}\right]}{\sqrt{\nu^2 + \xi^2}\left[\xi + \sqrt{\nu^2 + \xi^2}\right]^\lambda} d\xi \right\} \tag{B.12.85}$$

$$J_\mu(x) = J_\mu(x) + \frac{1}{\pi \csc(\pi\mu)\, x} \lim_{\nu \to \infty} \left\{ \nu^{\nu+1} \int_0^\infty \frac{e^{-\mu \sinh^{-1}\left(\frac{\xi}{x}\right)}}{\sqrt{\left(\frac{\xi}{x}\right)^2 + 1}\sqrt{\nu^2 + \xi^2}\left[\xi + \sqrt{\nu^2 + \xi^2}\right]^\lambda} d\xi \right\} \tag{B.12.86}$$

$$J_{-\mu}(x)=J_{-\mu}(x)-$$

$$\frac{1}{\pi\,\csc(\pi\mu)\,x}\lim_{\nu\to\infty}\left\{\nu^{\nu+1}\int_0^\infty\frac{\left[\left(\frac{\xi}{x}\right)+\sqrt{\left(\frac{\xi}{x}\right)^2+1}\right]^\mu}{\sqrt{\left(\frac{\xi}{x}\right)^2+1}\sqrt{\nu^2+\xi^2}\left[\xi+\sqrt{\nu^2+\xi^2}\right]^\lambda}d\xi\right\} \tag{B.12.87}$$

$$E_\mu(x)=-Y_\mu(x)+$$

$$x\lim_{\nu\to\infty}\left\{\nu^{\nu+1}\int_0^\infty\frac{\left[-\left(\frac{\xi}{x}+\sqrt{1+\frac{\xi^2}{x^2}}\right)^\mu+\cos(\pi\mu)\sqrt{1+\frac{\xi^2}{x^2}}\left(\sqrt{1+\frac{\xi^2}{x^2}}-\frac{\xi}{x}\right)^\mu\right]}{\sqrt{1+\frac{\xi^2}{x^2}}\sqrt{\nu^2+\xi^2}\left[\xi+\sqrt{\nu^2+\xi^2}\right]^\nu}d\xi\right\} \tag{B.12.88}$$

$$S_{-1,\mu}(x)=\frac{1}{\mu x}\lim_{\nu\to\infty}\left\{\nu^{\nu+1}\int_0^\infty\frac{\sinh\left[\mu\sinh^{-1}\left(\frac{\xi}{x}\right)\right]}{\sqrt{1+\left(\frac{\xi}{x}\right)^2}\sqrt{\nu^2+\xi^2}\left[\xi+\sqrt{\nu^2+\xi^2}\right]^\nu}d\xi\right\} \tag{B.12.89}$$

$$S_{-1,\mu}(x)=\frac{1}{\mu}S_{0,\mu}(x)-$$

$$\frac{1}{\mu x}\lim_{\nu\to\infty}\left\{\nu^{\nu+1}\int_0^\infty\frac{\left(\sqrt{\left(\frac{\xi}{x}\right)^2+1}-\left(\frac{\xi}{x}\right)\right)^\mu}{\sqrt{1+\left(\frac{\xi}{x}\right)^2}\sqrt{\nu^2+\xi^2}\left[\xi+\sqrt{\nu^2+\xi^2}\right]^\nu}d\xi\right\} \tag{B.12.90}$$

$$S_{0,\mu}(x)=\frac{1}{\mu x}\lim_{\nu\to\infty}\left\{\nu^{\nu+1}\int_0^\infty\frac{\sinh\left[\mu\sinh^{-1}\left(\frac{\xi}{x}\right)\right]}{\sqrt{\nu^2+\xi^2}\left[\xi+\sqrt{\nu^2+\xi^2}\right]^\nu}d\xi\right\} \tag{B.12.91}$$

$$S_{0,\mu}(x)=\frac{1}{x}\lim_{\nu\to\infty}\left\{\nu^{\nu+1}\int_0^\infty\frac{\cosh\left[\mu\sinh^{-1}\left(\frac{\xi}{x}\right)\right]}{\sqrt{1+\left(\frac{\xi}{x}\right)^2}\sqrt{\nu^2+\xi^2}\left[\xi+\sqrt{\nu^2+\xi^2}\right]^\nu}d\xi\right\} \tag{B.12.92}$$

$$S_{0,\mu}(x)=\frac{1}{2\mu}\lim_{\nu\to\infty}\left\{\nu^{\nu+1}\int_0^\infty\frac{\left(\sqrt{\left(\frac{\xi}{x}\right)^2+1}+\left(\frac{\xi}{x}\right)\right)^\mu-\left(\sqrt{\left(\frac{\xi}{x}\right)^2+1}-\left(\frac{\xi}{x}\right)\right)^\mu}{\sqrt{\nu^2+\xi^2}\left[\xi+\sqrt{\nu^2+\xi^2}\right]^\nu}d\xi\right\} \tag{B.12.93}$$

$$S_{1,\mu}(x)=\frac{1}{2}\lim_{\nu\to\infty}\left\{\nu^{\nu+1}\int_0^\infty \frac{\left(\sqrt{\left(\frac{\xi}{x}\right)^2+1}+\left(\frac{\xi}{x}\right)\right)^\mu+\left(\sqrt{\left(\frac{\xi}{x}\right)^2+1}-\left(\frac{\xi}{x}\right)\right)^\mu}{\sqrt{\nu^2+\xi^2}\left[\xi+\sqrt{\nu^2+\xi^2}\right]^\nu}d\xi\right\} \tag{B.12.94}$$

$$S_{1,\mu}(x)=\frac{1}{x}\lim_{\nu\to\infty}\left\{\nu^{\nu+1}\int_0^\infty \frac{\cosh\left[\mu\sinh^{-1}\left(\frac{\xi}{x}\right)\right]}{\sqrt{\nu^2+\xi^2}\left[\xi+\sqrt{\nu^2+\xi^2}\right]^\nu}d\xi\right\} \tag{B.12.95}$$

$$S_{2,\mu}(x)=x+\left(\mu-\frac{1}{\mu}\right)\lim_{\nu\to\infty}\left\{\nu^{\nu+1}\int_0^\infty \frac{\sinh\left[\mu\sinh^{-1}\left(\frac{\xi}{x}\right)\right]}{\sqrt{\nu^2+\xi^2}\left[\xi+\sqrt{\nu^2+\xi^2}\right]^\nu}d\xi\right\} \tag{B.12.96}$$

$$S_{\mu-1,\mu}(x)=\frac{x^\mu}{2^{\mu+1}}\lim_{\nu\to\infty}\left\{\nu\int_0^\infty \frac{e^{-\nu\xi+x^2/4q}}{q^{\mu+1}}d\xi\right\};\quad q=\nu\sinh\xi \tag{B.12.97}$$

$\mathrm{Re}\mu>-1$

$$s_{-1,\mu}(x)=\frac{1}{2\mu}\lim_{\nu\to\infty}\left\{\nu^{\nu+1}\int_0^\infty \frac{\left(\sqrt{\left(\frac{\xi}{x}\right)^2+1}+\left(\frac{\xi}{x}\right)\right)^\mu-\left(\sqrt{\left(\frac{\xi}{x}\right)^2+1}-\left(\frac{\xi}{x}\right)\right)^\mu}{\sqrt{\left(\frac{\xi}{x}\right)^2+1}\sqrt{\nu^2+\xi^2}\left[\xi+\sqrt{\nu^2+\xi^2}\right]^\nu}d\xi\right\} \tag{B.12.98}$$

$$s_{0,1/3}(x)=\frac{1}{x}\lim_{\nu\to\infty}\left\{\nu^{\nu+1}\int_0^\infty \frac{e^{-4\xi^2/27x}}{\sqrt{\nu^2+\xi^2}\left[\xi+\sqrt{\nu^2+\xi^2}\right]^\nu}d\xi\right\} \tag{B.12.99}$$

$$s_{2,0}(x)=\lim_{\nu\to\infty}\left\{\nu^{\nu+1}\int_0^\infty \frac{\left\{\frac{\sqrt{1+(\xi x)^2}}{x}-\xi\ln\left[\xi x+\sqrt{1+(\xi x)^2}\right]\right\}}{\sqrt{\left(\frac{\xi}{x}\right)^2+1}\sqrt{\nu^2+\xi^2}\left[\xi+\sqrt{\nu^2+\xi^2}\right]^\nu}d\xi\right\} \tag{B.12.100}$$

$$s_{0,\mu}(x)=\frac{1}{2\sqrt{\pi}}\lim_{\nu\to\infty}\left\{\nu^{\nu+1}\int_0^\infty \frac{e^{-x^2/8\xi}K_{\mu/2}\left(\frac{x^2}{8\xi}\right)}{\sqrt{\xi}\sqrt{\nu^2+\xi^2}\left[\xi+\sqrt{\nu^2+\xi^2}\right]^\nu}d\xi\right\} \tag{B.12.101}$$

$|\mathrm{Re}\mu|<\frac{1}{2}$

$s_{\mu,1/2}(x) =$

$$\frac{x^{\mu+1}}{\Gamma(\frac{1}{2}-\mu)} \lim_{\nu\to\infty}\left\{ \nu^{\nu+1} \int_0^\infty \frac{1}{\xi^{\mu+1/2}(x^2+\xi^2)\sqrt{\nu^2+\xi^2}\left[\xi+\sqrt{\nu^2+\xi^2}\right]^\nu}\, d\xi \right\} \tag{B.12.102}$$

$\operatorname{Re}\mu < \frac{1}{2}$

$$s_{\mu,\mu+1}(x) = -\frac{1}{x^{\mu+1}} \lim_{\nu\to\infty}\left\{ \nu \int_0^x \frac{\cos\left[\nu \sin^{-1}\left(\frac{\xi}{\nu}\right)\right] \xi\, (\xi^2-x^2)^{\mu-1/2}}{\sqrt{\nu^2-\xi^2}}\, d\xi \right\} \tag{B.12.103}$$

$\operatorname{Re}\mu > -\frac{1}{2}$

$$s_{\mu,1-\mu}(x) = \lim_{\nu\to\infty}\left\{ \nu^{\nu+1} \int_0^\infty \frac{J_\mu(x\sqrt{1-e^{-\xi}})}{(1-e^{-\xi})^{\mu/2}\sqrt{\nu^2+\xi^2}\left[\xi+\sqrt{\nu^2+\xi^2}\right]^\nu}\, d\xi \right\} \tag{B.12.104}$$

Appendix C Notation and Definitions of Special Functions

Mathematical Constants

$\pi = 3.14159265\ldots$
$\gamma = 0.577215664\ldots$ Euler's constant
$G = 0.915965594\ldots$ Catalan's constant

Exponential Integral and Related Functions

$$\mathrm{Ei}(x) = -\int_{-x}^{\infty} \frac{e^{-t}}{t}\,dt = li(e^{x}) \quad , \quad x > 0$$

$$\mathrm{Ei}(-x) = -\int_{x}^{\infty} \frac{e^{-t}}{t}\,dt = \gamma + \ln x + \sum_{n=1}^{\infty} (-1)^n \frac{x^n}{n!n} \quad , \quad x > 0$$

$$E_n(x) = \int_{1}^{\infty} \frac{e^{-xt}}{t^n}\,dt, \quad x > 0, \quad n = 0, 1, 2, \ldots$$

$$li(x) = \int_{0}^{x} \frac{1}{\ln t}\,dt, \quad x > 0$$

$$li(x) = \gamma + \ln(\ln x) + \sum_{n=1}^{\infty} \frac{(\ln x)^n}{n!n} = \mathrm{Ei}(\ln x), \quad x > 1$$

$$li(x) = \gamma + \ln(-\ln x) + \sum_{n=1}^{\infty} \frac{(\ln x)^n}{n!n}, \quad 0 < x < 1$$

Sine and Cosine Integrals

$$si(x) = -\int_{x}^{\infty} \frac{\sin t}{t}\,dt = -\frac{\pi}{2} + \sum_{n=1}^{\infty} \frac{(-1)^{n+1} x^{2n-1}}{(2n-1)!(2n-1)}$$

$$Si(x) = \int_{0}^{x} \frac{\sin t}{t}\,dt = si(x) + \frac{\pi}{2}$$

https://doi.org/10.1515/9783110681642-008

$$ci(x) = -\int_x^\infty \frac{\cos t}{t} dt = \gamma - \ln x + \sum_{n=1}^\infty \frac{(-1)^n x^{2n}}{(2n)!2n}$$

$$Ci(x) = \gamma + \ln x + \int_0^x \frac{\cos t - 1}{t} dt = ci(x)$$

Gamma Function and Related Functions

$$\Gamma(z) = \int_0^\infty t^{z-1} e^{-t} dt, \quad \operatorname{Re} z > 0$$

$$\Gamma(n+1) = n!$$

$$\Gamma(\alpha, x) = \int_x^\infty t^{\alpha-1} e^{-t} dt, \quad \operatorname{Re} \alpha > 0$$

$$\gamma(\alpha, x) = \int_0^x t^{\alpha-1} e^{-t} dt$$

$$B(x, y) = \int_0^1 t^{x-1}(1-t)^{y-1} dt = \frac{\Gamma(x)\Gamma(y)}{\Gamma(x+y)} = B(y, x)$$

$$\psi(z) = \frac{d \ln \Gamma(z)}{dz}$$

Error Functions and Fresnel Integrals

$$\operatorname{erf}(z) = \frac{2}{\sqrt{\pi}} \int_0^z e^{-t^2} dt$$

$$\operatorname{erfc}(z) = 1 - \operatorname{erf}(z)$$

$$i^n \operatorname{erfc}(z) = \int_z^\infty i^{n-1} \operatorname{erfc}(t) dt, \quad n = 0, 1, 2...$$

$$i^{-1} \operatorname{erfc}(z) = \frac{2}{\sqrt{\pi}} e^{-z^2}$$

$$i^0 \operatorname{erfc}(z) = \operatorname{erfc}(z)$$

$$C(z) = \int_0^z \cos\left(\frac{\pi}{2}t^2\right) dt$$

$$S(z) = \int_0^z \sin\left(\frac{\pi}{2}t^2\right) dt$$

Legendre Functions

$$P_\nu^\mu(x) = \frac{1}{\Gamma(1-\mu)} \left(\frac{x+1}{x-1}\right)^{\mu/2} {}_2F_1\left(-\nu, \nu+1; 1-\mu; \frac{1-x}{2}\right), \quad |x| < 1$$

$$Q_\nu^\mu(x) = \frac{\pi}{2\sin(\pi\mu)} \left[P_\nu^\mu(x)\cos(\pi\mu) - \frac{\Gamma(\nu+\mu+1)}{\Gamma(\nu-\mu+1)} P_\nu^{-\mu}(x)\right]$$

Complete Elliptic Integrals

$$E(k) = \int_0^{\pi/2} \sqrt{1-k^2(\sin\theta)^2}\, d\theta = \frac{\pi}{2}\, {}_2F_1\left(\frac{1}{2}, -\frac{1}{2}; 1; k^2\right)$$

$$K(k) = \int_0^{\pi/2} \frac{d\theta}{\sqrt{1-k^2(\sin\theta)^2}} = \frac{\pi}{2}\, {}_2F_1\left(\frac{1}{2}, \frac{1}{2}; 1; k^2\right)$$

Parabolic Cylinder and Whittaker Functions

$$D_\mu(z) = 2^{\mu/2+1/4} W_{\mu/2+1/4,\, -1/4}\left(\frac{z^2}{2}\right)$$

$$M_{\nu,\mu}(z) = z^{\mu+1/2} e^{-z/2}\, {}_1F_1(\mu-\nu+\frac{1}{2}; 2\mu+1; z)$$

$$W_{\nu,\mu}(z) = \frac{\Gamma(-2\mu)}{\Gamma(\frac{1}{2}-\mu-\nu)} M_{\nu,\mu}(z) + \frac{\Gamma(2\mu)}{\Gamma(\frac{1}{2}+\mu-\nu)} M_{\nu,-\mu}(z)$$

Jacobi, Gugenbauer, Chebyshev, Legendre, Laguerre and Hermite Orthogonal Polynomials

$$P_n^{\alpha,\beta}(x) = \frac{(-1)^n}{2^n n!(1-x)^\alpha(1+x)^\beta}\frac{d^n}{dx^n}\left\{(1-x)^\alpha(1+x)^\beta(1-x^2)^n\right\}$$

$$C_n^\alpha(x) = \frac{(-1)^n\Gamma(\alpha+\frac{1}{2})\Gamma(2\alpha+n)}{2^n n!\Gamma(2\alpha)\Gamma(\alpha+n+\frac{1}{2})(1-x^2)^{\alpha-1/2}}\frac{d^n}{dx^n}\left\{(1-x^2)^{\alpha-1/2}(1-x^2)^n\right\}$$

$$T_n(x) = \frac{(-1)^n\sqrt{\pi}(1-x^2)^{1/2}}{2^n\Gamma(n+\frac{1}{2})}\frac{d^n}{dx^n}\left\{(1-x^2)^{n-1/2}\right\}$$

$$U_n(x) = \frac{(-1)^n\sqrt{\pi}(n+1)}{2^{n+1}\Gamma(n+\frac{3}{2})(1-x^2)^{1/2}}\frac{d^n}{dx^n}\left\{(1-x^2)^{n+1/2}\right\}$$

$$P_n(x) = \frac{(-1)^n}{2^n n!}\frac{d^n}{dx^n}\left\{(1-x^2)^n\right\}$$

$$L_n^\alpha(x) = \frac{x^{-\alpha}e^x}{n!}\frac{d^n}{dx^n}\{x^{n+\alpha}e^{-x}\}$$

$$H_n(x) = (-1)^n e^{x^2}\frac{d^n}{dx^n}\left\{e^{-x^2}\right\}$$

$$He_n(x) = (-1)^n e^{x^2/2}\frac{d^n}{dx^n}\left\{e^{-x^2/2}\right\}$$

Riemann zeta, Lerch, Jonquière, Liouville, Mittag–Leffler and Möbius Functions

$$\zeta(z) = \sum_{n=1}^{\infty}\frac{1}{n^z}, \quad \operatorname{Re} z > 1$$

$$\zeta(z,\alpha) = \sum_{n=1}^{\infty}\frac{1}{(\alpha+n)^z}, \quad \operatorname{Re} z > 1$$

$$\Phi(z,s,\alpha) = \sum_{n=0}^{\infty}\frac{z^n}{(\alpha+n)^s}, \quad |z|<1, \quad \alpha \neq 0, -1$$

$$F(z,s) = \sum_{n=1}^{\infty}\frac{z^n}{n^s} = z\Phi(z,s,1)F(z,s) = \sum_{n=1}^{\infty}\frac{z^n}{n^s} = z\Phi(z,s,1)$$

$$\lambda(n) = \sum_{m=0}^{\infty}\frac{1}{(2m+1)^n} = \left(1-\frac{1}{2^n}\right)\zeta(n), \quad n = 2, 3, \ldots$$

$$E_\alpha(z) = \sum_{n=0}^{\infty} \frac{z^n}{\Gamma(\alpha n + 1)}$$

$$E_{\alpha,\beta}(z) = \sum_{n=0}^{\infty} \frac{z^n}{n!\Gamma(\alpha n + \beta)}$$

$$_mE_{\alpha,\beta}(z) = \sum_{n=0}^{\infty} \frac{z^n}{(n!)^m\Gamma(\alpha n + \beta)}, \qquad m = 1, 2, 3...$$

the Möbius function $\mu(n)$ is defined by

$\mu(n) = 1$ if $n = 1$.

$\mu(n) = (-1)^k$ if n is product of k distinct primes.

$\mu(n) = 0$ if n is divisible by a square bigger than unity.

Bernoulli and Euler Polynomials and Numbers

$$\frac{te^{zt}}{e^t - 1} = \sum_{n=0}^{\infty} B_n(x)\frac{t^n}{n!}, \qquad |t| < 2\pi$$

$$B_n = B_n(0)$$

$$\frac{2e^{zt}}{e^t + 1} = \sum_{n=0}^{\infty} E_n(x)\frac{t^n}{n!}$$

$$E_n = 2^n E_n\left(\frac{1}{2}\right)$$

Index

https://doi.org/10.1515/9783110681642-009

www.ingramcontent.com/pod-product-compliance
Lightning Source LLC
Chambersburg PA
CBHW082106220326
41598CB00066BA/5635
* 9 7 8 3 1 1 0 6 8 1 5 7 4 *